AA007249

WindEurope 2025

Journal of Physics: Conference Series Volume 3025

Copenhagen, Denmark
8-10 April 2025

ISBN: 979-8-3313-2260-1
ISSN: 1742-6588

Printed from e-media with permission by:

Curran Associates, Inc.
57 Morehouse Lane
Red Hook, NY 12571

Some format issues inherent in the e-media version may also appear in this print version.

This work is licensed under a Creative Commons Attribution 4.0 International Licence.
Licence details: http://creativecommons.org/licenses/by/4.0/.

No changes have been made to the content of these proceedings. There may be changes to pagination and minor adjustments for aesthetics.

Printed with permission by Curran Associates, Inc. (2025)

For permission requests, please contact the Institute of Physics
at the address below.

Institute of Physics
Dirac House, Temple Back
Bristol BS1 6BE UK

Phone: 44 1 17 929 7481
Fax: 44 1 17 920 0979

techtracking@iop.org

Additional copies of this publication are available from:

Curran Associates, Inc.
57 Morehouse Lane
Red Hook, NY 12571 USA
Phone: 845-758-0400
Fax: 845-758-2633
Email: curran@proceedings.com
Web: www.proceedings.com

TABLE OF CONTENTS

Preface

Peer Review Statement

Market-Driven Wind Resource Assessment 1
A. Bechmann, J. Quick

A Comparison Between Scanning LiDAR and Scada-Based Hyperparameter Tuning of Analytical Wake Models 11
Pieter-Jan Daems, Diederik Van Binsbergen, Jonathan Sterckx, Timothy Verstraeten, Jan Helsen

Techno-Economic Wind Energy Potential of Germany Considering State-Specific Regulations 20
Thorsten Reichartz, Georg Jacobs, Jil Hoffmann, Joscha Halber, Jannik Besseling, Lucas Blickwedel, Dustin Frings, Ralf Schelenz

The Effectiveness of a Vision-Based Condition Monitoring Framework for Determining Gear Health in Gearboxes 31
Djordy Van Maele, Jean C. Poletto, Roeland De Geest, Dieter Fauconnier, Patrick De Baets

Offshore Wakes Measured by an Adaptive Dual-Doppler Scanning Lidar System 40
Tabea Hildebrand, Florian Jäger, Nathalie Gloria, Luis Michaelis, Alexander Krasnov, Doron Callies, Jens Riechert, Lukas Pauscher

Numerical Evaluation of the Smearing Effect in a Spinner-Mounted Lidar for Wind Field Reconstruction 53
Esperanza S. Sagredo, Jennifer M. Rinker, Ásta Hannesdóttir, Michael Courtney

EDiGaPS: A Novel Machine Learning-Based Approach to Model the Influence of Turbulence Intensity on Wind Turbine Power Curves 64
Fernando S. Filho, Nathianne Andrade, Rodrigo Queiroz, Guido Santos, Samuel Lima, Gustavo Carvalho

Physics-Based, Data-Augmented Model for Wind Turbine Control Design 74
Jalal Heidari, Nezmin Kayedpour, Lieven Vandevelde, Guillaume Crevecoeur

A Parameter Study of Leading Edge Protection in a Wind Tunnel 85
Christian Bak, Anders S. Olsen, Jimmie Beckerlee, Sigurd Ildvedsen, Andreas Fischer, Oliver Lylloff, Michael D. Haag, Jonas R. Langebæk, Stefan Weber

Reduction of Crosstalk Effects in Multicomponent Force and Moment Transducers for Wind Turbine Test Benches via Geometrical Design Optimisation 94
Lukas Schmitz, Georg Jacobs, Christian Hollas, Maximilian Zweiffel, Julian Röder, Fabian Renger, Ernst Manner

Virtual Load Sensors Based on Calibrated Wind Turbine Strain Sensors for Damage Accumulation Estimation: A Gap-Filling Technique 105
Bruno R. Faria, Nikolay Dimitrov, Victor Perez, Athanasios Kolios, Asger B. Abrahamsen

Sensitivity of Drivetrain Condition Monitoring Signals to Wind Turbine Blade Damage 116
Simon J. Watson, Sumit K. Pal, Donatella Zappalá, Amir R. Nejad

Dynamic Reliability Estimation of Wind Turbine Gearboxes using Real-Time Operational Data.................. 127
Georgo Angelis, Laurens Krudde

Data Imputation for SCADA Data using Graph Neural Networks.. 136
Florian Hammer, Sarah Barber

Push Power Where You Can and Prevent Damage Where You Need... 147
J. Pacheco, M. Berezyak, J. Martins, N. Teixeira, C. Pereira, R. Figueiredo, C. Oliveira

Performance Analysis of Offshore Wind O&M Activities Based on Vessel and Helicopter Movement Data.. 156
David Baumgärtner, Jonas Kaczenski, Jürgen Mackeprang, Thomas Pracejus

A Bin-Wise Approach to Wake Steering Optimization Under Fatigue Load Constraints................................ 165
Ibtihel Ben Gharbia, Nina Khvoenkova, Paolino Tona, Fabrice Guillemin, Marine Dupoiron, Emilie Quenedey, Alexis Cousin

Investigation of the Effects of Individual Pitch Control of a Wind Turbine on Loads, Power and Acoustic Emission.. 176
Anna Wegner, Ashutosh Sharma, Sebastian Mechler, Paula Ludwig, Corinna Kraasch

Impact of Innovative Maintenance Technologies on Offshore Wind Farm Performance................................ 184
Christian Schläger, Torge Lorenz, David Baumgärtner

Examining the Potential for Floating Offshore Wind Farms in Irish Waters to Achieve the 37 GW Offshore Wind Energy Goal by 2050.. 198
Krzysztof Szczepaniec, Fergal O'Rourke, Peter Ryan

Field Investigation into Mooring Load Reduction Device Performance in Hurricane Conditions at an Offshore Test Site.. 210
Tom Doyle, Danny Golden, Syed Hasan

Optimal Day-Ahead Trading and Power Control for a Hybrid Wind-Hydrogen Plant with Multi-Agent Reinforcement Learning.. 220
Stijn Ally, Timothy Verstraeten, Ann Nowé, Jan Helsen

Reliable Operation of Wind-Storage Systems for Baseload Power Production.. 231
Jenna Iori, Michiel Zaaijer, Jasper Kreeft, Dominic Von Terzi, Simon Watson

First Experience in Operation and Data Acquisition at the Hydrogen Lab Bremerhaven............................... 241
Jannes Vervoort, Anna Heuschmann, Kevin Schalk, Marieke Stein, Karsten Korsch, Nora Denecke

Author Index

PREFACE

This Special Issue presents selected contributions from the **WindEurope Annual Event 2025 Conference** that took place 08-10 April in Copenhagen. A total of 24 papers are included, addressing one (or more) of the conference themes:

- Resource assessment
- Turbine technology
- Wind farm: installation and operations
- Floating offshore wind
- Digitalisation
- Environmental sustainability and social acceptance
- Grid and system integration
- Renewable hydrogen and Power-to-X.

Conference Committee for the Scientific Program

- Amir R. Nejad, Professor, Norwegian University of Science and Technology, Norway
- Stephan Barth, Managing Director, ForWind - Center for Wind Energy Research, Germany
- Simon Watson, Professor, Delft University of Technology, The Netherlands
- Julia Gottschall, Chief Scientist, Fraunhofer IWES, Germany
- Jan Helsen, Professor, Vrije Universiteit Brussel, Belgium

We thank all 15 reviewers from European Academy of Wind Energy (EAWE) for their engagement and effort, without which these proceedings would not have been possible. As with previous WindEurope events, the European Academy of Wind Energy (EAWE) has been responsible for organizing the review process for scientific contributions and has contributed to the development of the conference sessions. WindEurope has been responsible for the

Content from this work may be used under the terms of the Creative Commons Attribution 4.0 licence. Any further distribution of this work must maintain attribution to the author(s) and the title of the work, journal citation and DOI.
Published under licence by IOP Publishing Ltd

organization and logistics of the conference, and we thank their highly professional staff and their associates for the excellent collaboration.

Proceedings editor

Amir R. Nejad, Professor, Norwegian University of Science and Technology (NTNU), Norway

Journal of Physics: Conference Series **3025** (2025) 011002 doi:10.1088/1742-6596/3025/1/011002

Peer Review Statement

All papers published in this volume have been reviewed through processes administered by the Editors. Reviews were conducted by expert referees to the professional and scientific standards expected of a proceedings journal published by IOP Publishing.

- **Type of peer review:** Single Anonymous
- **Conference submission management system:** Morressier
- **Number of submissions received:** 27
- **Number of submissions sent for review:** 27
- **Number of submissions accepted:** 25
- **Acceptance Rate (Submissions Accepted / Submissions Received × 100):** 92.6
- **Average number of reviews per paper:** 3.28
- **Total number of reviewers involved:** 15
- **Contact person for queries:**
  Name: Stergiani Nikou
  Email: stergiani.nikou@windeurope.org
  Affiliation: WindEurope asbl vzw

Content from this work may be used under the terms of the Creative Commons Attribution 4.0 licence. Any further distribution of this work must maintain attribution to the author(s) and the title of the work, journal citation and DOI.
Published under licence by IOP Publishing Ltd

Journal of Physics: Conference Series **3025** (2025) 012001 doi:10.1088/1742-6596/3025/1/012001

Market-Driven Wind Resource Assessment

A Bechmann[1], and J Quick[1]

[1] Department of Wind and Energy Systems, Technical University of Denmark (DTU)

E-mail: andh@dtu.dk

Abstract. In open electricity markets, mechanisms of supply and demand set the hourly spot price. In markets with high penetration of wind energy, prices drop during high wind speed periods due to an oversupply of wind-generated electricity. The fluctuating nature of wind energy leads to price volatility. Due to increasing wind energy penetration and price volatility, wind farm operators in Denmark have experienced a 15% drop in revenue during the last 15 years, and the profitability of new projects is uncertain. But how can we accurately assess the electricity value of future wind farms?

This paper presents a new wind resource assessment method for estimating the market value of wind farm projects—considering the site-specific correlation between wind speed and electricity price. The study demonstrates the process by assessing the "value curve" at a site in Denmark over the past 24 years, and it proposes that traditional wind resource assessment methods adopt a similar approach.

1. Introduction

Wind resource assessment (WRA) aims to estimate the conditions needed to evaluate wind turbines' lifetime electricity production— often measured as the annual energy production (AEP). As wind turbine power is mainly determined by wind speed, standard WRA estimates the probability density function (PDF) of wind speed at the wind turbine's location [1]. When wind turbines are placed in large modern farms, energy production no longer depends only on the ambient wind conditions but also on neighbouring turbines via the wakes they generate. Therefore, the WRA process is evolving: today, it must also estimate how turbines affect each other to derive the PDF of the wake-reduced wind speed [2, 3].

The thesis of this paper is that the increasing share of wind in many energy markets is now changing the requirements for WRA again. In open electricity markets, mechanisms of supply and demand set the hourly spot price, which tends to drop during high wind speed periods [4, 5]. Today, wind farm operators are, therefore, not only interested in the AEP but also the average selling price, often measured as the value factor (VF)— the ratio of the mean selling price to the mean yearly electricity price.

Estimating the VF requires knowledge of the production-price correlation from concurrent time series. The historical wind energy production and electricity price in a bidding zone are often available from the Transmission System Operator (TSO) [6]. However, the derived VF represents a market average, not a specific wind farm, and does not necessarily represent future market dynamics.

It is common to estimate wind farm power production by applying a wind farm power curve to a time series of measured or forecasted wind speeds in the location of interest [7]. Combining the production with the measured or forecasted electricity prices for the same period allows for

Content from this work may be used under the terms of the Creative Commons Attribution 4.0 licence. Any further distribution of this work must maintain attribution to the author(s) and the title of the work, journal citation and DOI.
Published under licence by IOP Publishing Ltd

estimating the wind farm-specific VF. However, since the generation of time series for each wind turbine position with the correct correlations is challenging and computationally expensive, it is hard to use for traditional wind farm optimisation, where the turbine type and position are iteratively updated. Due to the high dimensionality and associated computational burdens of the time series approach, we seek a simple method for market-driven WRA based on the PDF of wind speed and electricity price. This paper, therefore, seeks to answer the research question:

How can we include price volatility in traditional PDF-based wind resource assessment?

This paper suggests a method for evaluating the correlation between wind speed and electricity price as part of WRA. The speed-price correlation, named the "energy value curve", is independent of the turbine type and can be simplified to a linear function with only a single parameter— the value curve slope. The steeper the slope, the greater the price volatility caused by wind speed variations. As a demonstration, the slope's historical evolution at a Danish site is estimated from 24 years of measurements. The study shows how price volatility causes large revenue drops for wind farms. Therefore, evaluating price volatility during the WRA phase is essential, and the paper presents a method applicable to current PDF-based WRA practices.

2. Market-driven Wind Resource Assessment

This section presents the mathematical foundation for including price volatility in the traditional PDF-based WRA process. Our proposed WRA method requires estimating the energy value curve in addition to the PDF of wind speed and the wind turbine power curve. Section 2.1 introduces the concept of energy value, and Section 2.2 defines the energy value curve.

For the sake of simplicity, this work only considers a single wind turbine to avoid wakes and other effects that depend on wind direction. Therefore, the method presented is written as a function of wind speed only but can be expanded with wind direction for real-life cases. Furthermore, this analysis is limited to the day ahead spot market price, although there are opportunities for participation in other electricity markets [8].

2.1. The annual energy value

First, we will briefly introduce the annual energy value (AEV) and the value factor (VF), which Bechmann et al. 2024 [4] also describe. The AEV is similar to the well-known AEP— they measure the annual energy produced and have the same units (Wh). However, contrary to the AEP, AEV evaluates energy produced at different wind speeds differently. The energy produced at high wind speeds is typically assigned a reduced value since large amounts of wind energy tend to reduce its market value. AEV, therefore, tends to be smaller than AEP.

In a traditional fixed-price market, the gross annual energy revenue (AER) is estimated as follows:

$$\begin{aligned} AER &= \overline{p} \cdot 8766 \int_0^\infty P(u)\ f_u(u)\ du \\ &= \overline{p} \cdot AEP \end{aligned} \tag{1}$$

where $f_u(u)$ is the PDF of wind speed, and $P(u)$ is the density-corrected power curve. The 8766 factor is the number of hours per year, $\overline{p}$ is the annual average electricity sales price, and u is the hub height wind speed determined during the WRA flow modelling step [9]. As the electricity price is assumed to be constant, traditional WRA is primarily concerned with estimating the AEP.

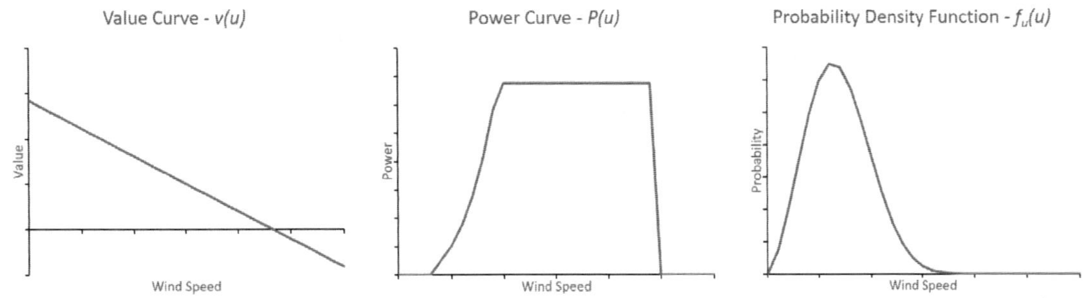

Figure 1: Illustration of the three models needed to estimate the annual energy value: the energy value curve (left), the power curve (middle), and the PDF of wind speed (right). The value curve is the new addition to the "market-driven" WRA process; it describes the correlation between wind speed and the mean electricity price.

In volatile markets, like day-ahead spot markets, the electricity price is not constant— there is an inverse correlation between wind speed and electricity price [4]. We can update our AER equation to reflect the price-speed correlation:

$$\begin{aligned} AER &= 8766 \int_0^{\infty} p\,(u)\; P\,(u)\; f_u\,(u)\; du \\ &= \overline{p} \cdot 8766 \int_0^{\infty} v\,(u)\; P\,(u)\; f_u\,(u)\; du \\ &= \overline{p} \cdot AEV \end{aligned} \tag{2}$$

p is the spot price for the produced electricity, and $v = p\,/\,\overline{p}$ is the energy value with an annual mean of $\overline{v} = 1$.

Comparing Equations 1 and 2, we see that the only difference between the AEP and the AEV is the addition of the energy value curve, $v(u)$, which evaluates production at different wind speeds. For fixed-price power purchase agreements or contracts for difference, there is no correlation between the electricity price and the wind speed. If the agreed price is set to the average spot price, $p(u) = \overline{p}$, then the value curve is $v\,(u) = 1$.

Equation 2 shows that it is possible to include price volatility caused by wind speed variations in a traditional PDF-based WRA process by including the energy value curve; see also Figure 1. To compare AEP and AEV, we define the value factor,

$$VF = \frac{AEV}{AEP}, \tag{3}$$

that expresses the ratio of the mean achieved selling price to the mean market price. If the value factor is below 1, the captured value is lower than the average wholesale price.

2.2. The wind energy value curve

To include price volatility in PDF-based WRA, we need a way to estimate the value curve at every wind turbine position. This section describes how we can estimate it from measurements and how the value curve can be modelled using a single parameter— the slope of the value curve.

The wind energy value curve is defined by:

$$v\,(u) = \frac{p\,(u)}{\overline{p}} = \frac{\mathbb{E}\,(p\,|\,u)}{\overline{p}} \tag{4}$$

where $\mathbb{E}(p \mid u)$ denotes the expected price given the wind speed at a wind turbine position.

We can represent the value curve as a histogram if wind speed and price measurements are available. The wind speed range is divided into 1-m/s bins, and each bin's mean energy value is estimated (see Figure 2). The value curve is a natural addition to the traditional WRA method as it is established from wind speed data, which is already part of WRA.

To simplify the analysis further, we suggest representing the energy value curve as a linear function, $v(u) \approx \alpha\, u + \beta$, where α and β are model parameters. We first apply $\overline{v} = 1$:

$$\begin{aligned}\int_0^\infty v(u)\, f_u(u)\, du &= 1 \\ \int_0^\infty (\alpha u + \beta)\, f_u(u)\, du &= 1 \\ \alpha \int_0^\infty u f_u(u)\, du + \beta \int_0^\infty f_u(u)\, du &= 1 \\ \alpha \overline{u} + \beta &= 1\end{aligned}$$

where $\overline{u} = \int_0^\infty u f_u(u)\, du$ is the average wind speed at the wind turbine position. Then, inserting $\beta = 1 - \alpha\overline{u}$ into the linear function gives us the final energy value curve model:

$$v(u) \approx 1 + \alpha\,(u - \overline{u}) \tag{5}$$

The wind energy value slope, α, can be found by fitting to the measured histogram (see Figure 2).

Suppose the value slope is constant for all wind turbine positions in a wind farm. In that case, we have successfully included price volatility in PDF-based WRA using one parameter that can be estimated as a natural part of the WRA process. The method makes it fast for wind farm developers to perform sensitivity studies of changing market conditions (or power purchase agreements) and investigate the revenue potential of different wind turbine types or farm layouts. As an example of the method's utility, we will demonstrate how the value curve has developed at a site in Denmark during the last 24 years.

3. The Risø Case Study

This section describes a case study that demonstrates the new WRA method and illustrates the importance of including market volatility. The data is openly available, allowing readers to reproduce all presented results (Section 3.1).

The case study is based on an onshore meteorology mast located on the Risø campus of the Technical University of Denmark (DTU) in eastern Denmark. The Risø-site has been chosen due to 24 years (2000-2023) of available data and because the Danish wind power share has risen from around 20% to 60% in this period [6]. This makes it a unique dataset to investigate the long-term effects of increasing wind energy penetration. The sections below describe the dataset and how the value curve and the value factor for a single wind turbine are estimated.

3.1. The dataset

In Denmark, the electricity is traded on the Nord Pool power exchange. There are two bidding areas with separate hourly prices: one for Eastern (DK1) and one for Western Denmark (DK2). The spot price data is available from the website "Energi Data Service", managed by Danish transmission system operator (TSO) "Energinet" [10]. We use data on wind speed from a cup-anemometer at 76.6-meter height located at the Risø peninsular in eastern Denmark. The data is available from the "Online Meteorological Data Website" by DTU [11].

Journal of Physics: Conference Series **3025** (2025) 012001 doi:10.1088/1742-6596/3025/1/012001

Based on visual inspections of the wind speed time series for 2000-2023, periods of missing data were found. However, instead of trying to restore the missing data, the years 2000, 2005, 2009, 2015, 2020, and 2023 with near-full data coverage were selected and analysed instead. Wind speeds larger than 30 m/s were removed, and the time series were finally resampled from 10-minute to 1-hour averages to "merge" with the hourly spot price data. The wind speed was not extrapolated to the wind turbine hub height ($H = 110\,\mathrm{m}$); all results presented below are based directly on these six 1-year-long time series of wind speed and electricity prices.

3.2. Estimating the value curve
Based on the wind speed and electricity price time series for the six selected years, we want to estimate the energy value curve (histogram) and the slope of the linear value curve model. This process is similar to estimating the PDF of wind speed for traditional WRA: First, the wind speed histogram is determined, followed by a step where the Weibull distribution is fitted to the histogram.

To estimate the value curve for this case study, we first group the price data into 1-m/s wind speed bins and find the mean energy value for each bin. Secondly, using linear regression for wind speeds between $u = 0 - 19\,\mathrm{m/s}$, we determine the value slope. This wind speed range was chosen since the value curve deviates from the linear assumption at higher wind speeds (see section 4.1). However, as wind speeds above 19 m/s are seldom, this approximation has little effect on the AEV estimate. Finally, having estimated the value slope, the value curve is given by Equation 5. Figure 2 shows the estimated value curves, and Table 1 gives the estimated mean slope and mean wind speed for each of the six years.

3.3. The numerical setup
To estimate how price volatility, as expressed by the energy value curve, affects the value factor, we must define a wind turbine power curve and a PDF of wind speed (Figure 1).

This case study uses the IEA 3.7-MW on-shore wind turbine [12]. It is a low-wind-speed turbine with a rated power of 3.37 MW and a rotor diameter of $D = 130\,\mathrm{m}$ giving it a specific power of $SP = 254\,\mathrm{W/m^2}$. The relatively large rotor allows it to produce more power during low-wind periods when electricity is expensive, increasing its value factor [4]. The wind turbine power in the full load region is set to the rated power, while we use the power equation in the partial load region, assuming the optimal power coefficient ($C_p = 0.49$):

$$P(u) \approx \frac{1}{2}\rho u^3 C_p \pi (D/2)^2 \tag{6}$$

where $\rho = 1.225\,\mathrm{kg/m^3}$ is the air density is .

The PDF of wind speed is represented in compact form by the Weibull distribution:

$$f_u(u) \approx \frac{k}{A}\left(\frac{u}{A}\right)^{k-1} \exp\left(-\left(\frac{u}{A}\right)^k\right) \tag{7}$$

where A and k are the all-sector Weibull scale and shape parameters. We apply a fixed shape parameter of $k = 2.5$ and set the scale parameter such that the mean wind speeds correspond to the mean measured wind speed at a particular year as given in Table 1 ($A = 7.7 - 8.4\,\mathrm{m/s}$).

4. Results
The results section will first analyse how price volatility has changed during the last 24 years by examining the wind energy value curves and the PDF of energy value. Then, the historical estimates of the capacity and value factors for the case study will be presented, and their effect on WRA will be discussed.

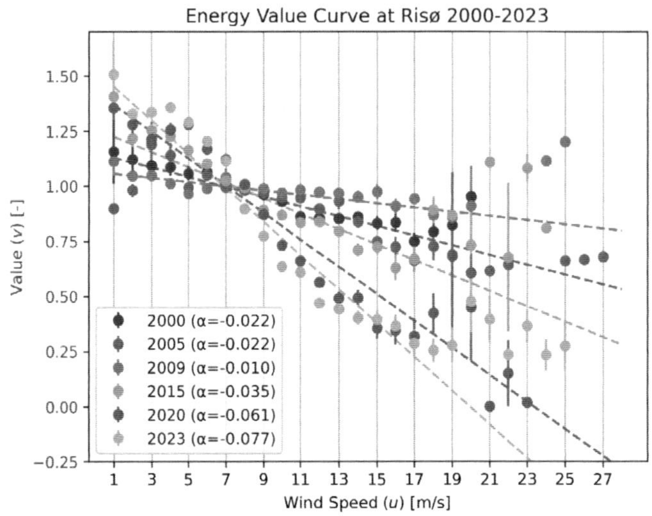

Figure 2: The measured value curves for different years at the Risø site. The symbols represent the mean energy value as a function of wind speed and year (indicated with colour). The error bars are the standard error, and the dashed lines are fitted value curve models—the slopes steepen with time.

4.1. The value curve at Risø

To understand the historical development of price volatility in Eastern Denmark and validate the linear value curve model, Figure 2 shows the measured energy value as a function of wind speed for the six selected years (the value curve). The symbols represent the mean values, the error bars are the standard error of the means, and the dashed lines are the linear model fits—the estimated value slopes are given in the legend and Table 1.

Figure 2 shows that from 2009 to 2023, there is a clear tendency for the slope of the value curve to steepen. In 2009, the slope was $\alpha = -0.01$, dropping to $\alpha = -0.08$ in 2023. During these 15 years, Denmark experienced a large expansion of wind energy— the wind energy share increased from about 25% to 60% [6]. Between 2000 and 2009, the slope increased slightly despite a steady increase in wind power. This trend illustrates that not only does a market's wind energy penetration (its energy supply) determine the value curve slope, but also factors, like the market's ability to absorb the produced wind energy (its energy demand).

At wind speeds below 19 m/s, the linear energy value model is a good fit to the data. The energy value increases again at speeds above 19 m/s (Figure 2). This trend was also observed by [4] and is contributed to other Danish wind farms at offshore sites reaching their cut-out wind speed. While this may seem like an opportunity for onshore wind to produce high-value energy, these wind speeds happen so seldom that the actual revenue potential is small. Bechmann et al. [4] found that only 3% of the potential revenue comes from the 20-25 m/s wind speed range. The error bars that grow for high wind speeds show the lack of high wind cases.

4.2. The PDF of value

The energy value curve expresses how the mean energy value depends on the wind speed. To understand how the uncertainty of energy value has developed historically, Figure 3 shows the PDF of the energy value for three different wind speeds (2, 7, and 12 m/s). The top and bottom graphs show the PDFs for 2009 and 2023, respectively, and the legends provide the means and standard deviations.

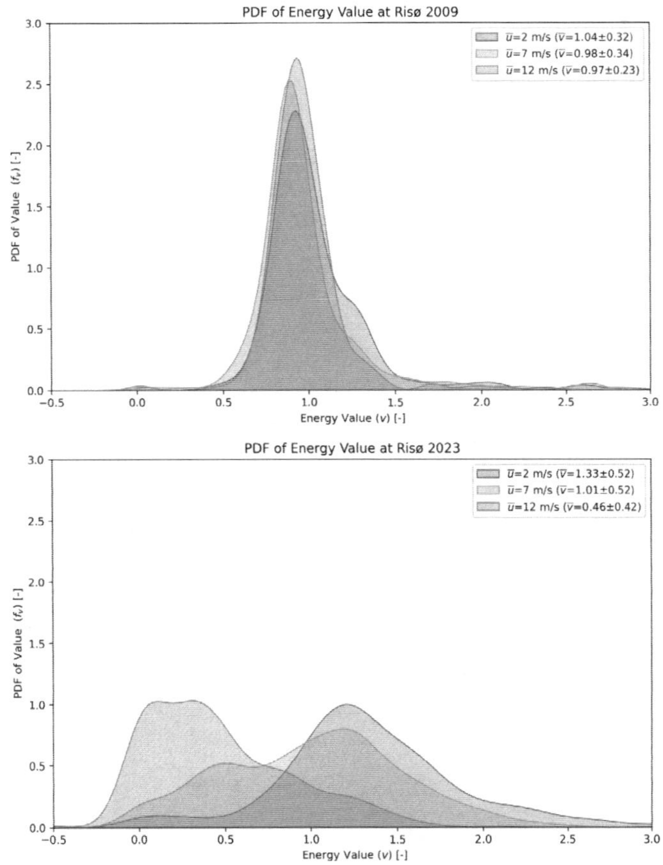

Figure 3: The measured PDF of the energy value at the Risø site for three wind speeds in 2009 (top) and 2023 (bottom). In 2009, the price volatility (the PDF width) was much smaller than in 2023. In 2023, the energy value depends on wind speed and becomes negative during high wind speed periods.

Figure 3 clearly illustrates how the uncertainty of energy value (the standard deviation) increased from 2009 to 2023. For 12 m/s, the uncertainty increased from 24% (0.23/0.97) in 2009 to 91% (0.42/0.46) in 2023. Therefore, not only is the mean energy value reduced at high wind speed, as expressed by the value curve, but in 2023, we also see frequent incidents of negative electricity prices. While the mean energy value is needed for wind farm design and WRA, the PDF of energy value may be valuable in day-to-day wind farm operation.

4.3. The value factor

To investigate the error made by not including market volatility in WRA and its historical effect on the profitability of wind energy in Denmark, we calculate the VF for the IEA 3.7-MW onshore wind turbine. Since many wind farm developers are familiar with the capacity factor, we also define the value-corrected capacity factor:

$$\hat{CF} = CF \cdot VF = \frac{AEP}{8766\, P_r} \cdot \frac{AEV}{AEP} = \frac{AEV}{8766\, P_r} \tag{8}$$

where $P_r = 3.7MW$ is the turbine's rated power.

Journal of Physics: Conference Series **3025** (2025) 012001 doi:10.1088/1742-6596/3025/1/012001

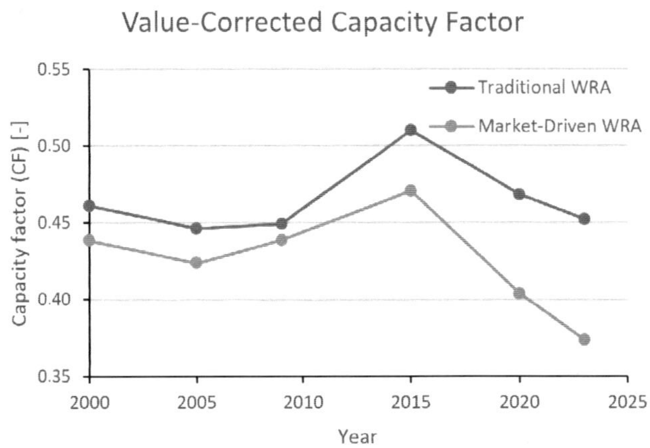

Figure 4: Modelled historical development of the capacity factor (value-corrected) for a wind turbine located at the Risø site. The blue curve is for the traditional WRA approach that does not account for price volatility, while the orange curve is the suggested "market-driven" approach. Due to Denmark's growing wind energy penetration, the two methods deviate increasingly with time.

Table 1: The historical development of the value- and capacity factors for at Risø location in eastern Denmark.

Year	$\overline{u}$ [m/s]	$\overline{\alpha}$ [s/m]	CF [-]	VF [-]	$\hat{CF}$ [-]
2000	6.99	-0.022	0.461	0.951	0.438
2005	6.85	-0.022	0.446	0.950	0.424
2009	6.88	-0.010	0.449	0.977	0.439
2015	7.44	-0.035	0.510	0.923	0.471
2020	7.05	-0.061	0.468	0.863	0.404
2023	6.91	-0.077	0.452	0.826	0.373

Figure 4 compares the capacity factor estimated using traditional WRA and the value-corrected capacity factor that includes price volatility. The conventional capacity factor has a natural year-to-year variation around a mean of about $CF = 0.46$. The value-corrected capacity factor, however, has a clear decreasing trend. Therefore, the error made by not including market volatility in WRA has increased over the years. For 2023, for example, the capacity factors for the traditional and new WRA methods were 0.452 and 0.373, respectively— representing a 17% error in annual energy revenue using the traditional WRA approach (Equation 1).

Table 1 gives the VF for the six selected years. Even though the case study uses a low-wind-speed turbine that increases the VF, it still drops from 0.95 to 0.83. The drop in VF and, consequently, wind farm revenue is due to Denmark's increasing share of wind energy (from 20% to 60%). While no other market has a 60% share of its electricity consumption covered by wind energy, it is still essential to include price volatility in the WRA; even a VF of 0.95 represents a substantial loss that should be accounted for by the WRA method.

Journal of Physics: Conference Series **3025** (2025) 012001
doi:10.1088/1742-6596/3025/1/012001

5. Discussion

As the wind energy sector matures into a major energy source in the modern grid, it is becoming increasingly important to consider the energy value. Wind energy requires huge investments, which take a long time to pay off. As demonstrated, electricity prices strongly depend on wind speed in markets with high wind energy penetration. This presents new difficulties, as electricity price forecasting is challenging and not standardised.

In practice, energy systems models can estimate region-specific spot market price time series, given assumed policy scenarios and technology costs [13]. However, the proposed method presents a practical tool for representing complex market dynamics in WRA. The alpha parameter represents a sort of latent space where complex grid dynamics are summarised with this single parameter. In practice, trajectories of the mean electricity price and energy value curve could be estimated to find a value curve representative of evolving market dynamics. The impacts of these market assumptions can be easily examined by changing the single-parameter value curve representation in Equation 5.

The present paper investigated how the alpha parameter has changed historically in Denmark. However, the model has not been applied to other markets. Future research could investigate if the suggested WRA approach is also valid for other markets— could the energy value curve concept even apply to solar energy?

6. Conclusions

This study presented a novel approach for incorporating electricity price volatility into PDF-based wind resource assessment (WRA). The method only adds one new parameter to the traditional WRA— the slope of the energy value curve. The slope can be estimated from existing wind speed measurements, and the market-driven WRA framework can, therefore, immediately help wind farm developers maximise the wind energy value.

A demonstration case showed that price volatility has profoundly impacted the market value of wind turbines in Denmark over the last two decades. The example wind turbine had a 15% reduction in the value factor— a loss not included in traditional WRA. The reduction is due to negative production-price correlations that increase with the share of wind energy. These significant discrepancies illustrate the importance of the proposed WRA framework.

References

[1] Troen I and Petersen E L 1989 *European wind atlas* (Roskilde, Denmark: Risø National Laboratory) ISBN 87-550-1482-8
[2] Clifton A, Smith A and Fields M 2016 Wind plant preconstruction energy estimates. current practice and opportunities Tech. rep. National Renewable Energy Lab.(NREL), Golden, CO (United States)
[3] Göçmen T, Van der Laan P, Réthoré P E, Diaz A P, Larsen G C and Ott S 2016 *Renewable and Sustainable Energy Reviews* **60** 752–769
[4] Bechmann A, Barlas A and Madsen H A 2024 *Journal of Physics: Conference Series* **2745**
[5] Loth E, Qin C, Simpson J G and Dykes K 2022 *Advances in Applied Energy* **8** 100112 ISSN 26667924
[6] Grohnheit P E and Sneum D M 2023 *Energy Policy* **179** 113631 ISSN 03014215
[7] Leon J P M, Habbou H, Friis-Møller M, Gupta M, Zhu R and Das K 2024 *Wind Energ. Sci.* **9** 759–776
[8] Nguyen T H, Quick J, Réthoré P E, Toubeau J F, De Jaeger E and Vallée F 2024 *Wind Energy Science Discussions* **2024** 1–26
[9] Bechmann A, Leon J P M, Olsen B T and Hristov Y V 2020 *Wind Energy Science* **5** 1679–1688
[10] Energinet 2024 Energi Data Service URL `https://www.energidataservice.dk/tso-electricity/Elspotprices`
[11] DTU Wind and Energy Systems 2024 DTU Online Meteorological Data URL `http://rodeo.dtu.dk/rodeo/ProjectOverview.aspx?&Project=5&Rnd=684499`
[12] Bortolotti P, Tarres H C, Dykes K, Merz K, Sethuraman L, Verelst D and Zahle F 2019 IEA wind task 37 on systems engineering in wind energy – WP2.1 reference wind turbines Tech. rep. International Energy Agency tex.howpublished: NREL/TP-73492

[13] Quick J, Leon J P M, Kanellas P, Yamujala S, Das K and Koivisto M J 2024 *Electric Power Systems Research* **234** 110675

Journal of Physics: Conference Series **3025** (2025) 012002 doi:10.1088/1742-6596/3025/1/012002

A comparison between scanning LiDAR and scada-based hyperparameter tuning of analytical wake models

Pieter-Jan Daems[1], Diederik Van Binsbergen[1,2], Jonathan Sterckx[1], Timothy Verstraeten[1], Jan Helsen[1]

[1] Vrije Universiteit Brussel, Pleinlaan 2, 1020 Brussels, Belgium
[2] Norwegian University of Science and Technology (NTNU), Otto Nielsens veg 10, 7052 Trondheim, Norway

E-mail: pieter-jan.daems@vub.be

Abstract. Proper calibration of engineering wake models is essential to accurately estimate the energy yield of wind farms. This study has as main goal to calibrate analytical wake models using two different data sources, i.e. scanning LiDAR and scada data. This calibration is applied on a Gauss-Curl Hybrid model, and uses different methodologies depending on the used data source(s). In case LiDAR measurements are used, reconstruction is first applied to estimate the horizontal wind speed based on the line-of-sight wind speeds measurements. These are then combined with the local wind direction and farm lay-out in order to identify wake velocity deficits, which serve as the basis for the hyperparameter tuning. In case scada data is used, the wind speed and active power measurements are used for performing a time-series-based calibration procedure. These methodologies are compared with one another. It is shown that both methods estimate slower wake recovery in the far-wake region compared to reference parameters. In the near-wake region, more variability is present between the different methodologies, which can in part be explained by the unobservability of the near-wake region using scada data. This study aims to highlight the use-case of LiDAR data in measurement campaigns to perform wake validation, compares the outcomes with readily available scada data.

1 Introduction

The growth rate of wind energy installations has shown a steady increase in Europe. This is especially true for offshore wind, for which an additional 2.5 GW has been installed in 2022. These increasing growth rates are mainly attributable to the achieved reductions in levelized cost of electricity (LCOE) throughout the last decade [1]. The capacity of offshore farms is typically larger than that of their onshore counterparts, reaching up to 1.4 GW. Therefore, it is critical to characterize how the different turbines affect each other, for example, by quantifying wake losses [2]. Wind turbine wakes result in velocity deficits and increased turbulence behind the wind turbine [3]. These aspects are important to consider during the design, as both result in power losses and increased fatigue loading [4]. Therefore, these effects are typically modeled using models with different levels of fidelity [5]. Although these capture similar trends, the exact magnitude of the effects is prone to uncertainty for several reasons [6]. First, the magnitude of wake effects is inherently dependent on a multitude of environmental parameters, such as atmospheric stability and ambient turbulence intensity, among other factors [7]. These are not always trivial to incorporate in low-fidelity models, whereas they drastically increase the number of required simulation cases for high-fidelity models. Second, low-fidelity models typically require the definition of several

Content from this work may be used under the terms of the Creative Commons Attribution 4.0 licence. Any further distribution of this work must maintain attribution to the author(s) and the title of the work, journal citation and DOI.
Published under licence by IOP Publishing Ltd

parameters that influence wake recovery, which are typically tuned using computational fluid dynamics models and field data [8]. However, recent research has shown that this approach can underestimate effects such as far wakes in large wind farm clusters [9]. This introduces uncertainty in the yield estimation of large wind farms.

Therefore, the calibration of engineering wake models using field data is becoming increasingly important, as it allows to validate the design models. In this way, adjustments can be made to the business case of existing wind farms and changes can be made to the available models, increasing their accuracy for future designs. In practice, this calibration can be done using different data sources. Scada data is readily available to wind farm operators, allowing to compare the measured power and/or wind speed at each turbine within the farm with the values predicted by the model. Based on discrepancies, the parameters of this model are then tuned to match, on average, the behavior captured by the scada data [6]. A secondary method is to deploy dedicated measurement equipment to gain information on the wake effects. For example, long-range scanning LiDARs are capable of measuring line-of-sight wind speeds up to 10 kilometers [10]. These are then reconstructed into horizontal wind speeds, based on which the wake velocity deficit is identified.

1.1 Objectives

The main objective of this study is to perform model calibration of engineering wake models using scada and/or scanning LiDAR data. Within the scada framework, wind speed, wind direction, and active power measurements are used to tune the wake parameters [6]. The advantage of this approach is that it allows continuous and accurate assessment of the power losses due to wakes, ensuring that these match with the model on average. However, this is an indirect way to observe wake effects, allowing little physical insight into the correctness of the shape and recovery of the wakes. To this end, a scanning LiDAR is used. These two different approaches are compared with one another, as indicated in Figure 1a. This is illustrated for a wind farm within the Belgian-Dutch offshore zone, where the research team has conducted a scanning LiDAR measurement campaign. The correct estimation of wake effects in this concession is especially important due to its high power density ($\geq 8\,\mathrm{MW/km^2}$ on Belgian side). This paper uses the following structure to achieve the aforementioned objectives: first, the scanning LiDAR measurement campaign is discussed in Section 2. In addition, this section discusses the extraction of wake velocity deficits from this data and the methodology used to tune the wake model parameters. Second, Section 3 discusses the methodology to tune the same wake model using scada data using a time-series-based optimization method. Finally, Section 4 compares the results obtained from these two methodologies, focusing both in the change in near-wake and far-wake behavior as a result of the performed tuning.

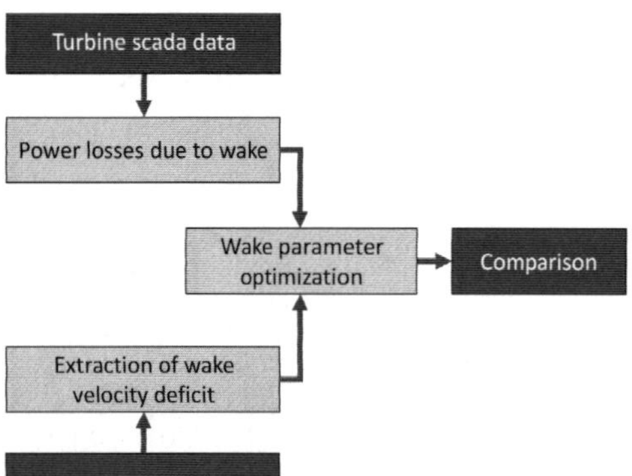

(a)

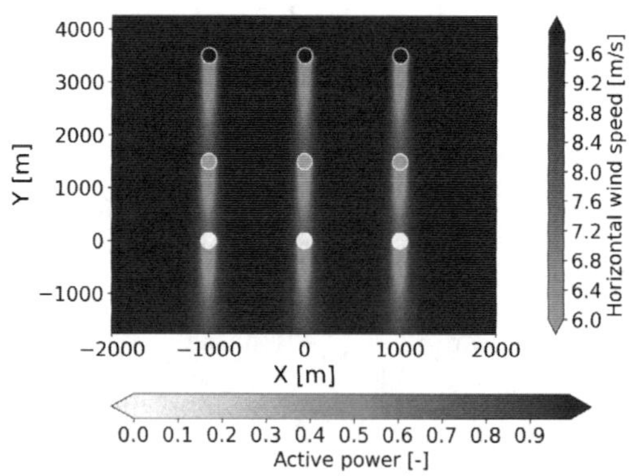

(b)

Figure 1: (a) Schematic overview of the paper outline. (b) Overview of the data used within the paper. With the scanning LiDAR data, the velocity deficits are measured, whereas the scada-based approach identifies wake effects by the decrease in power production.

2 Extraction of wake profiles using scanning LiDAR measurements

This section discusses the extraction of wake profiles using scanning LiDAR measurements. First, the measurement setup with the scanning LiDAR is discussed. Second, filtering is performed at the level of individual scans to ensure proper data quality. Third, the line-of-sight wind speeds are reconstructed into horizontal wind speeds. Finally, the wake velocity deficit is extracted around several wind turbines, based on which calibration is performed. These results are then compared with the ones obtained from the scada-based calibration procedure in Section 4.

2.1 Measurement set-up

Figure 2 shows the position of the scanning LiDAR with respect to neighbouring wind turbines. It can be observed that the LiDAR is oriented to measure the wind turbine outflow of four neighbouring wind turbines along the dominant wind direction (SWW direction). These turbines are all of the same type, and have a rated power of more than 8 MW. Throughout the measurement campaign, Plane Position Indicator (PPI) scans are performed. These scans maintain a constant elevation angle while varying the azimuth angle of the scanning head. In total, more than four months of data is processed.

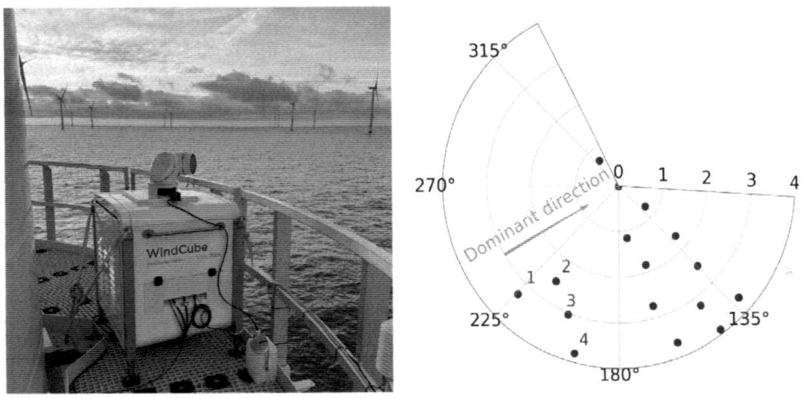

(a) (b)

Figure 2: (a) Installed LiDAR on the transition piece of the offshore wind turbine. (b) Orientation of the LiDAR with respect to neighbouring wind turbines. The radial distances are expressed in km.

2.2 Filtering

The scanning LiDAR is capable of measuring line-of-sight wind speeds up to 10 km. However, this range can be drastically reduced in the event of heavy rain or fog. Therefore, a filtering procedure is first performed to remove poor quality measurements. This is done in the following manner [11]:

1. A threshold is imposed to the carrier-to-noise ratio (CNR), which represents the strength of the backscattered signal compared to background noise. All scan values with a CNR value lower than -27 dB are discarded for further processing.
2. Line-of-sight velocity data outside the interval $[\mu - 2.5 \cdot IQR, \mu + 2.5 \cdot IQR]$ is filtered. Here, μ is the mean of the data, and IQR its interquantile range. In this way, outliers are discarded for further analysis.

After applying these filtering criteria, the data availability can be assessed within a specific range of interest. For the chosen analysis, it is imposed that the data availability should be at least 90% within the scan section containing the outflow of the four turbines indicated in Figure 2b. If this is not the case, the scan is discarded for further analysis. Otherwise, wind field reconstruction is applied, which is the subject of the next section.

2.3 Wind field reconstruction

LiDARs measure line-of-sight wind speeds (U_{LOS}), which typically are not directly used for wake validation. To translate measurements into a local wind field u, reconstruction is needed, solving the so-called cyclops dilemma [12]. The measured line-of-sight speed can be expressed using Equation 1 [13]. Here, θ and ψ are respectively the elevation and azimuth angle of the scan, and ϕ the local wind direction. U and W are, respectively, the horizontal and vertical velocity components of the wind field vector u.

$$U_{LOS} = U\cos(\theta - \phi)\cos(\psi) + W\sin(\psi) \tag{1}$$

Using a single measurement, this cannot be resolved in a general manner. However, since the used PPI scans are acquired at low elevation angles ($\psi \leq 3°$), it can be assumed that the contribution of the vertical velocity component is negligible compared to the horizontal one, resulting in Equation 2.

$$U \approx \frac{U_{LOS}}{\cos(\theta - \phi)\cos(\psi)} \tag{2}$$

The validity of this assumption depends on the misalignment angle between the azimuth angle of the scanner and the local wind direction. In practice, this assumption therefore requires to limit the domain of the reconstruction e.g. by imposing $|\theta - \phi| < 60°$. An example of a reconstructed scan can be seen in Figure 3a, where the black dots indicate the positions of neighboring wind turbines.

In order to remove the effects of variability in inflow conditions, a coordinate transformation and normalization is applied after the wind field reconstruction. To identify the wake profile of a wind turbine $T(x_0, y_0)$, Equation 3 is applied, where α is the rotation angle chosen as the local wind direction. In this way, a situation is mimicked in which the target wind turbine is the origin of the transformed coordinate system with the wind coming from the North, as illustrated in Figure 3b.

$$\begin{bmatrix} x' \\ y' \end{bmatrix} = \begin{bmatrix} \cos(\alpha) & -\sin(\alpha) \\ \sin(\alpha) & \cos(\alpha) \end{bmatrix} \begin{bmatrix} x - x_0 \\ y - y_0 \end{bmatrix} \tag{3}$$

Finally, the wind speeds are normalized by the estimated freeflow wind speed (U_∞) of the turbine (Equation 4).

$$U^* = \frac{U}{U_\infty} \tag{4}$$

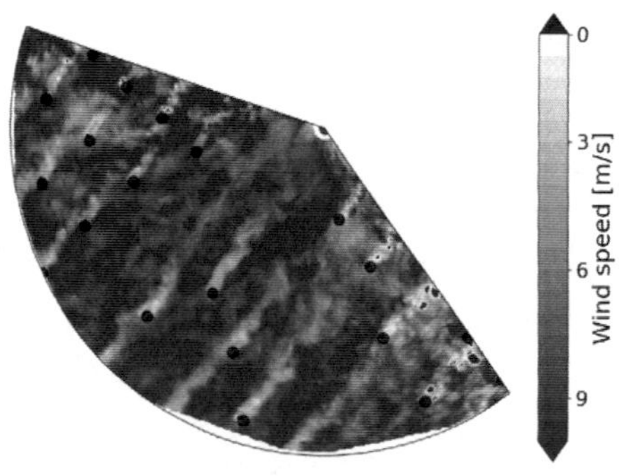

(a)

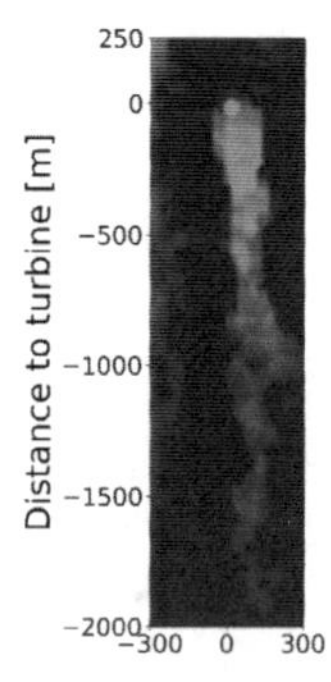

(b)

Figure 3: Example of a scanning LiDAR measurement after wind field reconstruction.

2.3.1 Comparison between reconstruction and scada data In order to assess the validity of the made assumptions during the reconstruction process, the obtained wind speed and direction estimates are compared with the available scada measurements. This is illustrated in Figure 4, where the blue dotted line depicts the median trend in the data, and the shaded area the associated uncertainty bounds. It can be noted that the general trend in terms of wind speed and direction are accurately followed, but that a significant variability is present. This cannot be attributed solely to the errors in the reconstruction, but also to the fact that the anemometer measures behind the rotor. These measurements are compensated by means of a transfer function by the manufacturer. Nevertheless, this process also introduces inherent uncertainty [14].

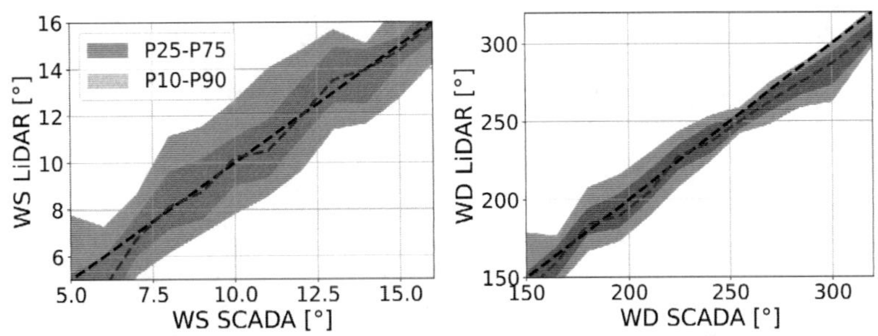

(a) (b)

Figure 4: Comparison between the reconstructed values and the scada measurements for the (a) wind speed and (b) wind direction.

2.3.2 Extraction of wake velocity deficits and hyperparameter tuning Figure 3b shows the outflow measurements of one of the neighboring wind turbines. Due to meandering effects, the largest velocity deficit will not always be directly aligned with the centerline of the rotor. These nonstationary effects are commonly not considered in engineering wake models. To further reduce the dimensionality of the data, the minimum velocity deficit is therefore tracked along the turbine centerline. This profile is classified per wind direction and wind speed, allowing to estimate the median behavior together with its uncertainty bounds. This is shown in Figure 7 for a particular inflow condition. This data is used to minimize the error $e(x, U)$ between the observations $o(x, U)$ and the engineering wake model $f(x, U)$ (Equation 5) . In this work, a Gauss-Curl Hybrid model (GCH) is optimized within the FLORIS framework [15], which requires the definition of four parameters: α, β, k_a, and k_b. The former two parameters alter the near-wake behavior, whereas the latter two alter the far wake. A window $w(x)$ is used to weigh the fit quality around specific zones (e.g. near wake, far wake). The different investigated cases are further discussed and compared in Section 4.

$$\text{minimize} \qquad |e(x,U)| = \Big| w(x) \cdot \big[o(x,U) - f_{GCH}(x,U,\alpha,\beta,k_a,k_b) \big] \Big| \tag{5}$$

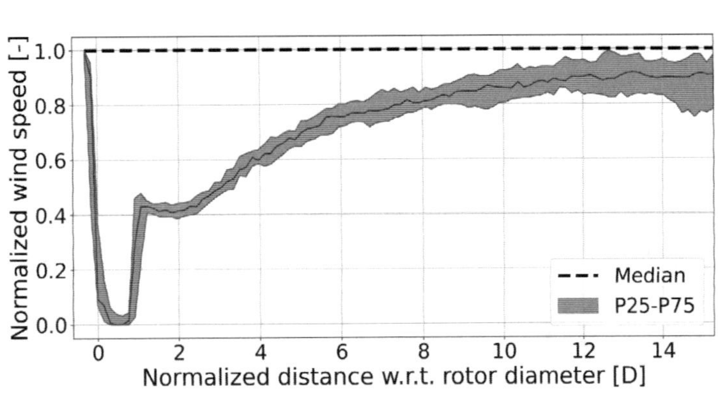

Figure 5: Example of the wake velocity deficit identified using the scanning LiDAR by aggregating different measurements.

3 Scada-based wake model calibration

3.1 Calibration procedure

To facilitate a comparison between analytical wake models and LiDAR observations, it is started from the calibration procedure discussed in [6]. This is a time-series-based calibration procedure that minimizes the error between the wake model power production ($\bar{P}$) and the observations from scada ($\hat{P}$) at the level of the individual turbines (cost function $L^2_{turbine}$) and the farm-wide level power production (cost function L^2_{farm}) by tuning the inflow wind field (speed U_∞ and direction ϕ_∞) and wake parameters ($\mathbf{\Omega}$), which consist of α, β, k_a, and k_b for the GCH model. These errors are assessed over the total number of wind turbines within the wind farm, N_T, and are relatively weighed with one another using the parameters a and b (Equation 6). A more in-depth discussion regarding this procedure can be found in [6, 16, 17].

$$\text{minimize} \quad a \cdot L^2_{turbine} + b \cdot L^2_{farm} \tag{6}$$

$$\text{with} \quad L^2_{turbine} = \frac{1}{N_T} \sum_{i=1}^{N_T} \left(\bar{P}_i - \widehat{P}_i \left(U_\infty, \phi_\infty, \mathbf{\Omega} \right) \right)^2 \tag{7}$$

$$\text{and} \quad L^2_{farm} = \left(\sum_{i=1}^{N_T} \bar{P}_i - \sum_{i=1}^{N_T} \widehat{P}_i \left(U_\infty, \phi_\infty, \mathbf{\Omega} \right) \right)^2 \tag{8}$$

3.2 Reference data and filtering

The calibration is conducted for the wind farm where the scanning LiDAR is deployed, using one year of scada data. It is specifically performed along the dominant wind direction, matching the wind direction of the PPI scans conducted by the scanning LiDAR for all below-rated wind speeds. Filtering of the data is performed by removing power-curve mismatches of individual turbines and is further detailed in [6, 16, 17]. The freeflow turbulence intensity, TI, is calculated using Equation 9, with I_{ref} and b fitted to match the freeflow turbulence from the dominant wind direction at the site.

$$TI = \frac{I_{ref} \cdot (0.75 U_\infty + b)}{U_\infty} \tag{9}$$

4 Results and discussion

This section discusses the results obtained from different wake parameter optimization procedures using data from the scanning LiDAR and/or the scada system. The following are compared with each other:

1. Optimization of far-wake using scanning LiDAR data. As it has been observed that the parameters β and k_b did not change the quality of fit with a meaningful amount, they are opted to be kept constant. A rectangular window is chosen for $w(x)$ in Equation 5, which is 0 for distances up until 7D, and 1 outside this range.

2. Optimization of the near wake (α) using scanning LiDAR data, followed by optimization of the far wake (k_a) using scada data. For the former procedure, a rectangular window is chosen for $w(x)$, which is 1 between 0 and 7D, and 0 outside this range. This is done because the goal of this step is to optimize the near-wake behavior.

3. Optimization of far-wake behavior (k_a) using scada data.

4. Optimization of far-wake behavior (k_a, k_b) using scada data.

Figure 7 shows the results of these different procedures. It can be observed that different trends are obtained for tuning the parameter α depending on the choice of weighing function $w(x)$. In case optimization is performed solely between 0 and 7D, the values of α decrease below the reference values, resulting in slower wake recovery in the near-wake region. The opposite trend is seen when optimization is performed for the far wake. To assess the impact of these differences, the fitting error in near-wake is assessed in Section 4.2.

Furthermore, it can be observed that all investigated procedures decrease the far-wake parameter k_a with respect to the reference values. This parameter is inversely correlated with the wake recovery in the far wake region. This change will therefore result in increased wake losses with respect to the baseline parameters. When both far wake parameters (k_a and k_b) are tuned using scada data, the values

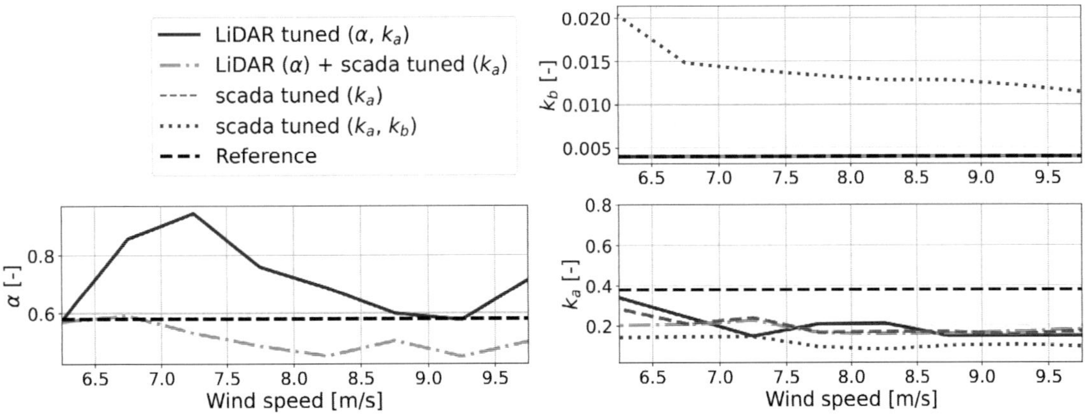

Figure 6: Overview of the fitted parameters for the different procedures .

of k_a decrease compared to when only the parameter k_a is varied. This can be explained by k_b, which is positively correlated with k_a, increasing with respect to its reference value. Due to this correlation between the different parameters (α, k_a, k_b) in the far-wake region, it is difficult to objectively assess the impact of these changes on the resulting wake profile and the far-wake velocity deficit. This is therefore further discussed in Section 4.1.

4.1 Comparison between wake profiles from different tuning procedures

The impact of the change in wake parameters on the simulated wake profile at a particular wind speed can be observed in Figure 7. It can be noted that the main variability occurs near the transition point (around 3.5D), reaching 14.6% . The variability nevertheless decreases further downstream. At 20D, the maximum relative error between the four models amounts to 2.1%. This is significant compared to the downward shift that takes place with respect to the reference values available in literature, which is on average 5.9%. While Figure 7 illustrates a single wind speed, the above mentioned observations remain valid across the investigated wind speed range, especially when looking at the far wake region. These observations are similar to the ones made in [9], i.e. that far wake effects are commonly underestimated in dense wind farms. These insights can have an important effect on the energy yield estimation of wind farms, and highlight the need of proper calibration of these models for new generation, energy-dense wind farms.

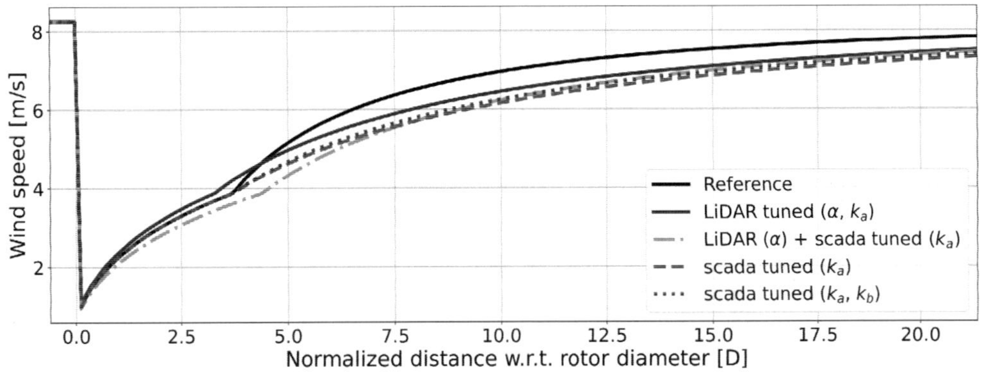

Figure 7: Comparison between the different obtained wake profiles at 8.25 m/s.

4.2 Error reduction in near-wake

The previous section has highlighted the consistency of the different tuning procedures when looking at far-wake conditions. Nevertheless, significant variability is present in Figure 7 up to a distance of 7D. Therefore, this section explores the differences between the different calibration methodologies by comparing them with the scanning LiDAR measurements, which are considered as the ground truth in this section. First, the cumulative fitting error between the scanning LiDAR observations and the model using baseline parameters is calculated at a given wind speed using Equation 10, where the error $e(x, U)$ is defined according to Equation 5.

$$\epsilon_{baseline}(U) = \sum_{i=1}^{N} \left| e_{baseline}(x_i, U) \right| \qquad D \leq x_i \leq 7D \tag{10}$$

This procedure is then compared for the different calibration procedures, based on which the relative change in fitting error (E) can be calculated according to Equation 11.

$$E(U) = \frac{\epsilon_{baseline}(U) - \epsilon_{tuned}(U)}{\epsilon_{baseline}(U)} \tag{11}$$

The relative change in fitting error E is shown in Figure 8. It can be observed that the most consistent reduction is achieved by performing the optimization in two different steps, i.e. by decoupling the optimization problem into first optimizing the near-wake, followed by optimizing the far wake. Furthermore, it can be seen that the reduction mainly takes place for wind speeds above 7.5 m/s, at which a change in wake parameters takes place for both cases (reduction in α and/or in k_a). Below this, the reduction in fitting error is relatively limited, showcasing that the baseline parameters are properly able to capture the near-wake behavior. These findings highlight the importance of properly choosing the weighing function $w(x)$ in Equation 5 based on the envisioned use case, as optimization of the far wake can come at the cost worse fit in near wake. Using only scada data, this near-wake optimization is moreover difficult, as the minimum spacing is around 7D within this concession zone.

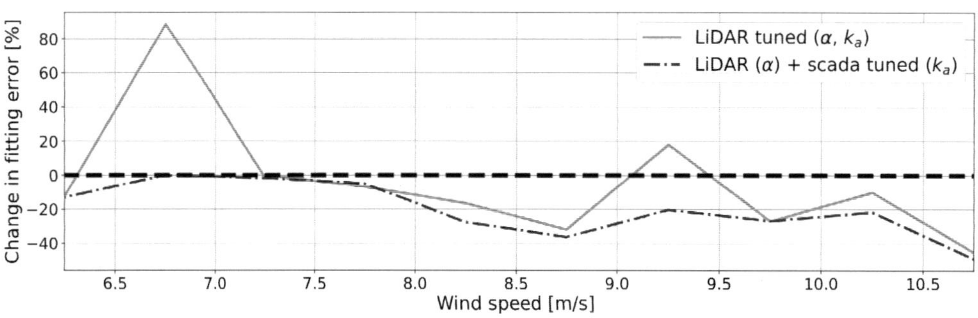

Figure 8: Change in fitting error $E(U)$ for two different calibration procedures.

5 Conclusions

This paper has compared the results from different calibration procedures to tune the Gauss-Curl Hybrid model using scada and/or scanning LiDAR data. The tuning based on scanning LiDAR data has been done by first reconstructing the line-of-sight wind speed measurements into vertical horizontal wind speeds. Based on the wind direction and wind farm lay-out, the wake velocity deficits have been identified, which have been used to fit the wake model parameters. Scada data has been used to perform a time-series-based hyperparameter optimization. These different procedures have then been compared with one another, showing decreased wake recovery with respect to baseline parameters. This is an important finding, as it directly influences the expected energy yield of the wind farm during energy yield assessments. The shift in far-wake recovery between baseline models and the obtained tuned models therefore warrants further research to investigate the validity and/or required changes for calibrating these models in the context of current generation wind turbines.

In the near-wake region, less consistency has been found between the results of the different fitting techniques. This can, among other, be attributed by the fact that this behavior is in large part unobservable for the neighbouring wake turbines using scada data. It is therefore not expected that this region can be properly tuned using solely this data, highlighting the benefit of incorporating scanning LiDAR data for validating the behavior in this region.

Acknowledgements
The authors acknowledge the support of De Blauwe Cluster through the support of projects Cloud4Wake and Supersized 5.0.

References

[1] International Renewable Energy Agency 2020 Irena renewable cost database [Online; accessed May 31, 2022;] URL `https://www.irena.org/statistics`

[2] Zhan L, Letizia S and Iungo G V 2020 *Wind Energy Science* **5** 1601–1622

[3] Lissaman P B 1979 *Journal of energy* **3** 323–328

[4] van Binsbergen D W, Wang S and Nejad A R 2020 Effects of induction and wake steering control on power and drivetrain responses for 10 mw floating wind turbines in a wind farm *Journal of Physics: Conference Series* vol 1618 (IOP Publishing) p 022044

[5] Zhou L, Wen J, Wang Z, Deng P and Zhang H 2023 *Energy* **275** 127525

[6] van Binsbergen D, Daems P J, Verstraeten T, Nejad A R and Helsen J 2024 *Wind Energy Science* **9** 1507–1526

[7] Abkar M and Porté-Agel F 2015 *Physics of fluids* **27**

[8] Niayifar A and Porté-Agel F 2016 *Energies* **9** 741 ISSN 1996-1073

[9] Nygaard N G, Steen S T, Poulsen L and Pedersen J G 2020 Modelling cluster wakes and wind farm blockage *Journal of Physics: Conference Series* vol 1618 (IOP Publishing) p 062072

[10] Hung L Y, Santos P and Gottschall J 2022 A comprehensive procedure to process scanning lidar data for engineering wake model validation *Journal of Physics: Conference Series* vol 2265 (IOP Publishing) p 022091

[11] Bodini N, Zardi D and Lundquist J K 2017 *Atmospheric Measurement Techniques* **10** 2881–2896

[12] Borraccino A, Schlipf D, Haizmann F and Wagner R 2017 *Wind Energy Science* **2** 269–283

[13] Zhan L, Letizia S and Valerio Iungo G 2020 *Wind Energy* **23** 501–527

[14] Shin D and Ko K 2019 *Energies* **12** 1087

[15] King J, Fleming P, King R, Martínez-Tossas L A, Bay C J, Mudafort R and Simley E 2021 *Wind Energy Science* **6** 701–714

[16] van Binsbergen D, Daems P J, Verstraeten T, Nejad A and Helsen J 2024 *Journal of Physics: Conference Series* **2745** 012014 ISSN 1742-6596

[17] van Binsbergen D, Daems P J, Verstraeten T, Nejad A and Helsen J 2024 *Journal of Physics: Conference Series* **2767** 092059 ISSN 1742-6596

Techno-economic wind energy potential of Germany considering state-specific regulations

Thorsten Reichartz[1*], Georg Jacobs[1], Jil Hoffmann[1], Joscha Halber[1], Jannik Besseling[1], Lucas Blickwedel[1], Dustin Frings[1], and Ralf Schelenz[1]

[1] Chair for Wind Power Drives, RWTH Aachen University, Aachen, Germany

*thorsten.reichartz@cwd.rwth-aachen.de

Abstract. Further expansion of wind energy is essential to achieve climate neutrality in the German electricity sector. However, in a densely populated country like Germany, providing sufficient land for wind energy remains challenging. Rapid technology advancements in wind turbine technology, as well as changing setback regulations, necessitate continuous reassessment of wind energy potential, as knowledge of potential sites is key for policymakers and project developers. Germany's federal structure requires a state-specific consideration of setback distances, a factor that has not been adequately addressed in previous studies. This study therefore assesses the current techno-economic wind energy potential, taking these into account. The results indicate that between 1.13 % and 12.52 % of Germany's land area is available for wind energy, resulting in a techno-economic installable capacity between 227 GW and 1,456 GW. Areas already occupied by wind turbines are not excluded from potential sites in this analysis. The wide range of results is due to the lack of clarity in some of the setback distances regulations. The results therefore highlight the significant challenges that project developers face due to the need for case-by-case assessments of the majority of available sites.

1. Introduction

To achieve the 2 °C goal of the Paris Agreement [1], a rapid transition towards a 100 % renewable energy system is essential [2]. Germany has set ambitious targets for installed renewable energy source capacities, aiming to increase onshore wind energy capacity to 160 GW by 2040 [3], from the current 63 GW [4]. Achieving this target requires allocating an estimate of 2 % of Germany's land area for wind energy [5]. Providing sufficient space is complex in a country as densely populated as Germany, due to land-use conflicts and the need to consider social acceptance, as well as for example preserving nature reserves [6]. Detailed knowledge of legally and technically suitable areas for wind energy is crucial for energy system stakeholders, especially wind farm developers, and policymakers.

Wind potential analysis is a well-established field of research, with several studies conducted on a global, European or a national scale [7–9]. In Germany, however, regulations and laws regarding wind energy are changing rapidly. In addition, technical progress in wind turbine (WT) technology has been immense over the last decade. For instance, while turbines with rotor diameters well below 125 m were installed in Germany in 2015, by 2024 rotor diameters are well exceeding 170 m [10]. As a result, potential studies conducted in the past are quickly outdated.

Content from this work may be used under the terms of the Creative Commons Attribution 4.0 licence. Any further distribution of this work must maintain attribution to the author(s) and the title of the work, journal citation and DOI.
Published under licence by IOP Publishing Ltd

Existing potential analyses, e.g. [11], often apply uniform setback distances for WTs across Germany despite significant state-specific differences [12]. Additionally, complete results in form of geospatial datasets are typically not available [13].

In this study, a model is developed to assess the techno-economic wind energy potential of Germany at a municipal level using up-to-date setback distances (as of 2023/2024) and comprehensive geodata. The model accounts for deviating setback distances in each of the 16 German federal states. Wind energy potential sites are categorized into those suitable under current regulations, those requiring case-by-case assessment, and sites within forested areas. No differentiation is made between Bremen and Lower Saxony, as Bremen accounts for less than 0.1 % of Germany's total area and less than 0.7 % of Lower Saxony's area. This study focuses on greenfield potential. Therefore, existing WTs do not lead to the exclusion of sites.

The results of this analysis provide citizens, municipalities, wind project developers and other stakeholders in the energy sector with comprehensive insights into the effects of setback distances. Project developers entering new markets, such as different federal states, can utilize this information to identify potential viable areas for new wind farms. Politicians are equipped with the ability to make informed decisions regarding land-use planning and regulatory adjustments to support wind energy deployment, which are necessary in light of the Wind Energy Land Requirement Act of 2022, which requires each federal state to designate sufficient land for WTs [14].

2. Methodology

This section describes the model developed to calculate the geographically viable area, the installable capacity, and the annual energy product (AEP) for WTs in each municipality in Germany. The model is implemented in Python and leverages the GeoPandas package for efficient geospatial data processing. The datasets used are explained in Sections 2.2, 2.3 and Table 2.

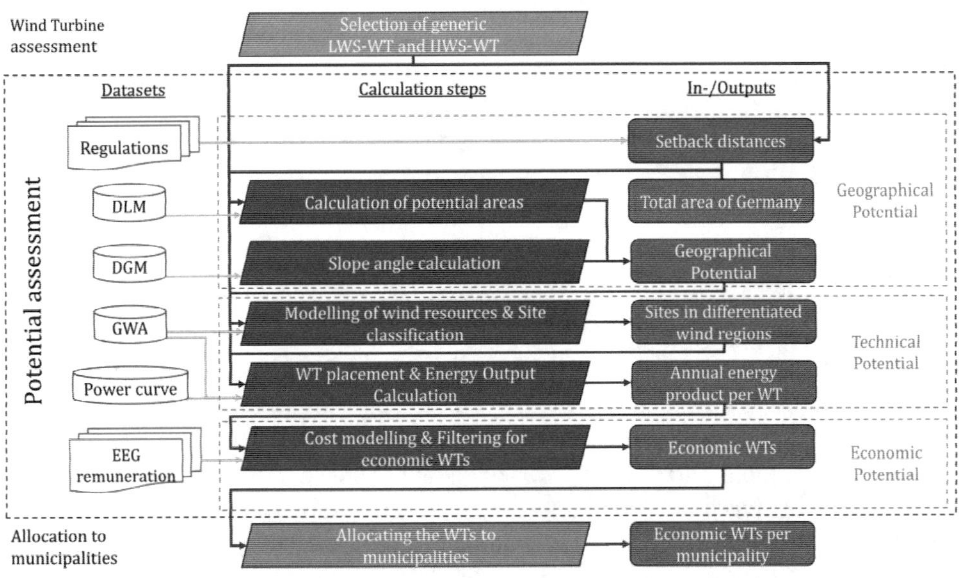

Figure 1. Overview of the steps applied to assess the techno-economic wind potential on a municipal level, based on federal regulations (LWS-WT: low wind-speed turbine, HWS-WT: high-wind-speed turbine, DLM: digital basic landscape model, DGM: digital terrain model, GWA: global wind atlas)

Figure 1 gives an overview of the calculation steps taken to obtain the techno-economic wind energy potential. Initially, the geographical potential is calculated, followed by the technical potential, and subsequently, the economic potential, which is further referred to as the techno-economic potential. In accordance with existing literature, the geographical potential encompasses areas suitable for WTs, taking into account technical, ecological and regulatory constraints, such as setback distances from residential buildings [7, 15, 16]. The technical potential is the amount of energy that can be generated within the geographically suitable areas, considering factors such as local wind conditions, electrical losses or WT power curves and wake effects. The techno-economic potential is defined as the subset of this amount of energy that can be generated economically.

2.1 Wind Turbine assessment

Figure 1 illustrates the initial step of the potential analysis, which involves identifying a generic low-wind speed turbine (LWS-WT) and high-wind speed turbine (HWS-WT) based on forecasts for WT size and rated power development by 2030 [17–19]. The configurations of these turbines are given in Table 1. Forecasting the average turbine configuration for 2030 causes uncertainties in model results, largely due to the rapid technological advancements in wind energy [20]. While onshore turbines with a rated power of more than 7 MW are already available, in 2023, the average turbine installed in Germany had a total height of 206 m, a rated power P_{el} of 4.7 MW, a rotor diameter RD of 141 m, and a hub height H of 136 m [21].

Table 1. Wind turbine configurations used for the potential analysis

	LWS-WT[a]	HWS-WT[b]
Rated Power P_{el} [kW]	6,000	6,500
Rotor Diameter RD [m]	170	160
Hub Height H [m]	175	130
Total Height TH [m]	260	210

[a] low-wind speed turbine
[b] high-wind speed turbine

2.2 Geographical potential

In the second step, the geographical potential is evaluated using geodata from the Digital Basic Landscape Model (DLM, [22]) alongside several supplementary data sources. These include infrastructure objects or areas, such as weather radars [23] or wetlands as defined by the Ramsar Convention [24], for which setback distances for WTs must be met. The complete list of geodata sources is provided in Table 2 ([22–30]). Setback distances specific to each federal state, outlined in [31] and valid as of March 2023, are applied. However, it should be noted that some setback distances are missing from this list, e.g. due to their lack of clarity in legal texts for certain federal states. For instance, no setback distances for residential buildings are specified for Saxony-Anhalt in [31]. However, it is implausible that distances of 0 m are possible. Consequently, in such cases, the setback distances are supplemented by determining upper and lower boundaries from other federal states with known regulations. Potential sites falling between these boundaries are assigned to sites requiring case-by-case assessment. Sites with a greater distance than the upper boundary are added to sites that are suitable under current regulations, while those below the lower boundary are excluded. A comprehensive list of the applied setback distances is provided in the Supplement. An additional differentiation to the distinction between sites requiring case-by-case assessment and those suitable under current regulations is made between potential sites in forested and non-forested areas. Despite the fact that some states do exclude forests from wind potential sites [32], forested areas are not categorically excluded in this analysis in any federal state, in order to be able to evaluate the effect of the exclusion of forest areas.

Table 2. Sources for geodata

Data	Source
Digital Basic Landscape Model (DLM)	[22]
Digital Terrain Model (DGM) Grid Width 50 m	[25]
DWD Weather radars	[23]
Wetlands as per Ramsar convention	[24]
Water protection areas[a]	[26]
German Regional Seismic Network	[27]
Military Radars[b]	[28]
Natura2000	[25]
Very High Frequency Omnidirectional Range Stations (VOR)	[30]

[a] without data for North Rhine-Westphalia, Rhineland-Palatinate and Baden-Wuerttemberg
[b] Own Assessment by Fachagentur Wind, as no official data is available

The remaining viable sites for WTs are further reduced by considering their slope angle. The Digital Terrain Model (DGM, [25]) is utilized to evaluate the slope angles. Typically, a slope angle of 20-30 ° is considered to be too high for the construction of WTs [8, 33]. In this study, any area with an angle of 30 ° or higher is excluded from the geographical potential.

2.3 Techno-economic potential

The remaining potential areas are further evaluated to determine the techno-economic potential, as described below. Therefore, geographically suitable sites are categorized into high-wind-speed and low-wind-speed sites using weather data from the Global Wind Atlas (GWA) [34]. No further differentiation is made between sites. Classification is based on the average wind speed at a height of 130 m above ground level at the center of gravity of each potential area. Sites with an average wind speed exceeding 7.5 m/s at this height are classified as high-wind speed regions, in accordance with existing literature [35].

In the following step, individual potential sites are grouped into clusters, where neighboring sites within 4 RD are combined (see Figure 2). Clustering enables efficient WT placement by optimizing land use within each cluster while ensuring a minimum spacing of 4 RD between turbines. This distance is chosen in all directions since wind direction is not considered in the model, instead of a 5 RD in the main wind direction, 3 RD perpendicular approach often used. Clustering enhances the efficiency of the computation, as computation time increases quadratically with each additional site considered in a cluster, and therefore reduces the overall computation time without negatively affecting the quality of the results. As high- and low-wind speed sites can exist within a cluster, the algorithm utilizes the RD of the LWS-WT at 170 m, to ensure a conservative estimate of the number of turbines that can be placed in total in a cluster.

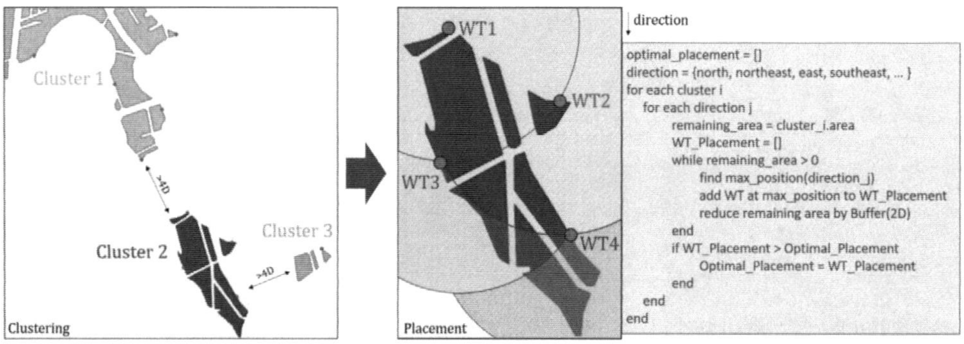

Figure 2. Wind turbine placement algorithm and exemplary visualization of clustering and WT placement

In the next step, the AEP for each WT and its position is calculated following Equation 1. The total efficiency η_{total} is determined using Equation 2. Assumptions for individual efficiency factors are: aerodynamic efficiency at η_{ae} 90 %, electrical efficiency η_{el} at 97 %, efficiency due to availability η_{av} at 97 %, and an additional factor η_{add} at 98 %, accounting for blade contamination and other losses, in line with current literature [33, 36]. No variation in the efficiency factors is considered depending on different sites. In Equation 1 $P(v)$ represents the power curve of the LWS-WT or the HWS-WT, while $f(v)$ is the wind speed frequency distribution on hub height for the respective turbine type, where v is the wind speed (see Table 1). Turbine types for each site are picked based on the classification of the site into a low- or high-wind speed site. The power curves for the generic turbines are derived from their specific power density following [9].

$$AEP_{WT} = 8760\,h \cdot \eta_{total} \cdot \int \big(f(v) \cdot P(v)\big)\, dv \tag{1}$$

$$\eta_{total} = \eta_{ae} \cdot \eta_{el} \cdot \eta_{un} \cdot \eta_{add} \tag{2}$$

To assess the economic potential, the Levelized Cost of Electricity (LCOE) for each WT is calculated using Equation 3. According to [37], the specific capital expenditures (CAPEX) for the LWS-WT is set at 1,676 €/kW, covering project planning, turbine costs, grid connection and all other expenses. For the HWS-WT, the specific CAPEX is 1,566 €/kW, due to the lower hub height and rotor diameter. Operational expenditures (OPEX) are assumed to be 44 €/kW for the first 10 years and 53 €/kW for the next 10 years [37] for both turbine types. The interest rate i is set at 8 %, consistent with other literature [7], and the turbine lifetime n is assumed to be 20 years, aligning with the subsidy period under the German Renewable Energy Act (EEG) [3].

$$LCOE = \frac{CAPEX + \sum_{t=1}^{n} \frac{OPEX}{(1+i)^t}}{\sum_{t=1}^{n} \frac{AEP_{WT}}{(1+i)^t}} \tag{3}$$

To evaluate the economic viability of the WTs, a threshold value $LCOE_{th}$ of 0.0735 €/kWh is defined. This value is based on the maximum allowable bid in onshore wind energy auctions for 2024 and 2025, as specified in EEG §85a [3, 38]. Additionally, it aligns with awarded auction values from previous years [39]. Since site-specific correction factors apply in the auction process, adjustments are incorporated according to EEG §36h [3]. WTs with an LCOE exceeding $LCOE_{th}$, even after correction, are considered economically infeasible. Finally, the techno-economically feasible WTs and the potential sites are allocated to the municipalities according to their location.

3. Results

3.1 Nationwide scenario analysis

This Section presents the results of the wind energy potential analysis for Germany, as calculated based on the methodology described in Section 2. Four different scenarios are considered:

- *STR_NFOR* (Strict rules, No Forest): Strict setback distances are applied, all sites requiring case-by-case assessment are excluded, and forest areas are not considered (setback distances to forests apply).
- *STR_FOR* (Strict Rules, With Forest): Strict setback distances are applied, all sites requiring case-by-case assessment are excluded, but forest areas are included.
- *CBC_NFOR* (Case-by-Case, No Forest): Sites requiring case-by-case assessment are included, while forest areas are excluded (setback distances to forests apply).
- CBC_FOR (Case-by-Case, With Forest): Sites requiring case-by-case assessment and forest areas are included.

These scenarios provide a comprehensive overview of the impact of different land-use restrictions on Germany's overall techno-economic wind energy potential. Given the approximately 11,000 municipalities in Germany, individual potentials are not presented in the paper but a comprehensive list is provided in the Supplement. The techno-economic installable capacity for the STR_NFOR scenario for each municipality is shown in Figure 3 (a) while a nationwide overview for each scenario is given in Table 3. Figure 3 (b) shows exemplary results for potential sites and WT placement based on the methodology described in Section 2.

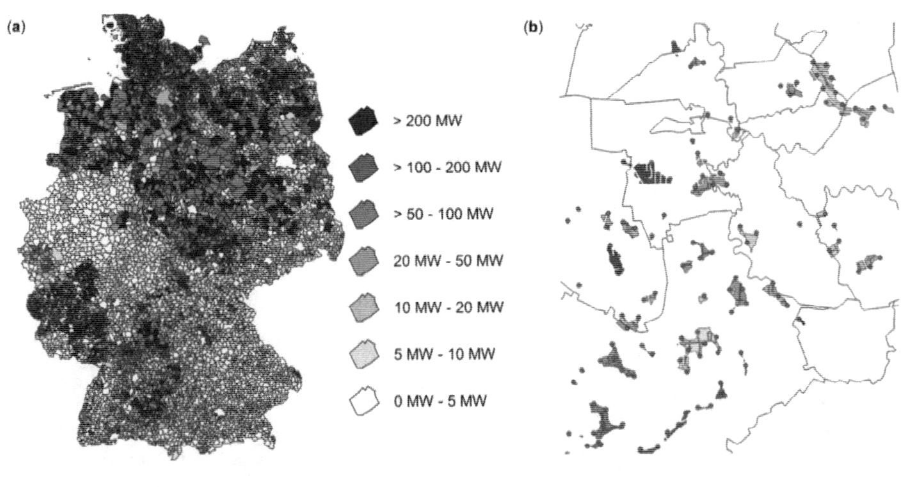

Figure 3. (a) Techno-economically installable wind capacity by municipality in Germany (STR_NFOR), (b) Exemplary result for arbitrary municipalities in Lower Saxony. Each red dot corresponds to a WT. Clusters are indicated by color. (Black lines are municipal borders)

Table 3. Installable wind capacity, AEP and percentage of available land area for Germany in each scenario

	STR_NFOR	STR_FOR	CBC_NFOR	CBC_FOR
Capacity [GW]	227	566	869	1456
AEP [TWh]	898	2143	3208	5216
% of total land area	1.13	4.11	5.09	12.52

In Figure 4, the techno-economic wind energy potential is presented for each federal state and scenario. Among the scenarios considered STR_NFOR imposes the greatest restrictions and therefore yields the most conservative results. For the STR_NFOR scenario, a total installable wind capacity of 227 GW is estimated, corresponding to an AEP of 898 TWh. The highest capacities are found in LSA (71 GW), SHO (39 GW), and MWP (29 GW). City-states BER, HAM, and BRE have negligible potential. Next to those, SAA and BAV have the lowest potential, with BAV reaching only 0.29 GW, increasing to 11 GW in the STR_FOR scenario. Regarding the geographical potential in STR_NFOR, SHO ranks the highest with 3.71 % of available land, followed by SAN (3.15 %) and MWP (2.49 %), while BAV has nearly 0 %, increasing to 0.49 % in STR_FOR, as a result of strict setback distances. A list of federal states that currently permit, restrict or prohibit wind energy in forests is available here: [32]. Including forests in potential sites has a particularly strong impact

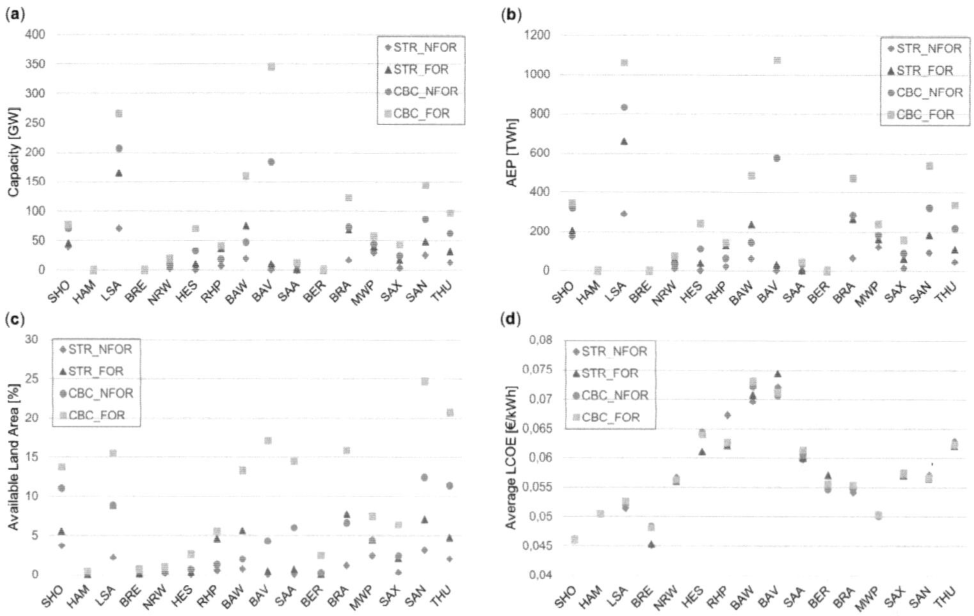

Figure 4. (a) installable capacity, (b) AEP, (c) percentage of available land area and (d) average LCOE, for each federal state and scenario [SHO: Schleswig-Holstein, HAM: Hamburg, LSA: Lower Saxony, BRE: Bremen, NRW: North Rhine-Westphalia, HES: Hesse, RHP: Rhineland-Palatinate, BAW: Baden-Württemberg, BAV: Bavaria, SAA: Saarland, BER: Berlin, BRA: Brandenburg, MWP: Mecklenburg-Western-Pomerania, SAX: Saxony, SAN: Saxony-Anhalt, THU: Thuringia]

in BAW, where the installable capacity in the STR_FOR scenario is approximately four times higher than in STR_NFOR, even exceeding the capacity for BAW in CBC_NFOR. Nationwide the STR_FOR scenario leads to an increase in the total installable wind capacity to about 565 GW, the CBC_NFOR scenario to about 869 GW and the CBC_FOR scenario to 1455 GW.
In the least restrictive CBC_FOR scenario, BAV achieves the highest installable capacity (345 GW), surpassing LSA (266 GW), as it results in the largest total available land area of all states, with 17.1 % of its landmass being geographically available. However, despite BAVs installable capacity being 1.3 times greater than LSAs, its total AEP is within 1 % of LSAs due to significantly weaker wind conditions. This is also reflected in the average LCOE, which is the highest in the southern states (BAW and BAV), exceeding 0.007 €/kWh in all scenarios. As a result, BAV and BAW are the only federal states where WTs are deemed economically infeasible. More details on the number and municipalities of economically infeasible WTs are given in the full list of results in the Supplement.

In the least restrictive CBC_FOR scenario, a total of about 733,000 individual potential sites are considered, resulting in about 12,500 clusters.

3.2 Sensitivity analysis for NRW

This section presents the results of a sensitivity analysis examplary conducted for NRW based on the STR_NFOR scenario, to asses the effect of setback distances. Therefore, several setback distances are varied. Detailed information is given in Table A1 in the Supplement. In scenario STR_NFOR_A, minimum distance values are reduced to a maximum of 800 m and the minimum distance for industrial areas is reduced to 300 m. In scenario STR_NFOR_B, the minimum distances are reduced based on the night time imission guideline values of the Technical Instructions on Noise (TA Lärm) [40]. The associated sound pressure level are also indicated in Table A1. The sound pressure levels of the generic WTs are estimated based on [41]. The corresponding potential results are shown in Figure 5. All calculations in Figure 5 refer to the STR_NFOR scenario.

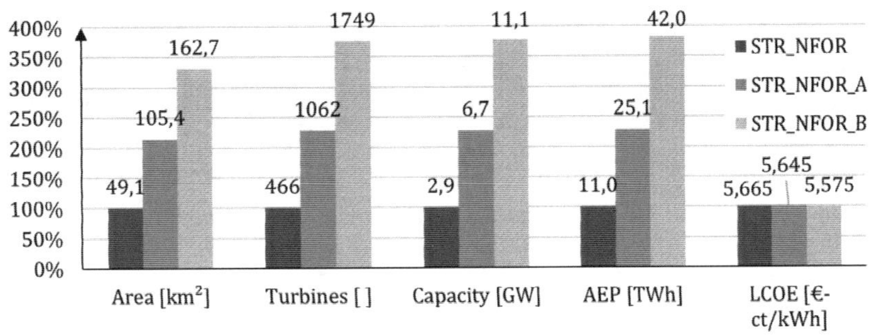

Figure 5. Results for NRW based on varied setback distances

The reduction in the minimum setback distances required lead to an increase in the available area of 115 % in the scenario A and 231 % in the scenario B. At the same time, installable capacity increases disproportionally in this case by 127 % for STR_NFOR_A and by 277 % for STR_NFOR_B. Due to the yield calculation methodology, installable capacity and energy yield increase almost linearly in relation to the number of turbines. Only small variations occur due to different turbine types and local wind conditions. Therefore, the LCOE decreases marginally between scenarios.

4. Discussion

In this section the results of this study are compared with findings from other research, highlighting differences and model limitations. However, variations in assumptions in setback distances, WT specifications, as well as geospatial and wind data make direct comparison particularly challenging.

Pape et al. (2022) [35] estimate an installable capacity of up to 2,086 GW with 26 % of Germanys land being geographically available. In our study, however, in the least restrictive scenario CBC_FOR an installable capacity of 1,456 GW and an available land area of 12.52 % is yield in contrast. One reason for the reduced land availability is the exclusion of all roads, including access roads of existing WTs, and agricultural roads. This likely leads to an underestimation of available land. On the other hand, the impact on turbine placement, and therefore installable capacity, is limited. The dominant factor in turbine spacing is the required distance of 4 RD between WTs. Small available potential sites remain suitable for wind turbines in our study, as the rotor-out principle is applied in all states. This means rotor overhang from potential site is allowed. This explains the comparatively high installable capacity with significantly lower available land mass. Additionally, Pape et al. (2022) [35] report average full load hours (FLH) of 2,654 h, whereas our study finds 3,951 h (for the STR_FOR scenario). This difference is partly due to the assumption of larger WTs, though potential overestimations in our study need to be further examined. Possible reasons are wake losses that are assumed at 90 % at all sites and the absence of correction factors for turbine shutdowns due to environmental protection regulations, such as species conservation. However, the potential overestimation of the AEP does not affect the results of the available area or installable capacity.

McKenna et al. (2014) [33] estimate a techno-economic potential of 855 TWh/a based on 41,623 km^2 of geographical potential, while the most restrictive scenario of our study (STR_FOR) identifies 898 TWh/a with only 4,036 km^2 of suitable land. However, they consider significantly smaller turbines (max. hub height 100 m, rotor diameter 117 m and rated power 3 MW) and larger turbine spacing (8 RD in the main wind direction, 5 RD perpendicular) at the same time, resulting in an estimated installed capacity of 8.8 MW/km^2, with regards to the available area, significantly lower than the 56 MW/km^2 (STR_FOR) found in our study. Combined with the reasons discussed above, this partially explains the significant differences in results.

In our study, rotor-out is currently allowed for all states. However, some regional regulations mandate full placement within potential sites, not allowing rotor overhang (often referred to as rotor-in) [42]. This affects the turbine placement and needs to be refined in future research, to improve accuracy of results regarding the techno-economic potential. Additionally, the assumed LCOE may be overestimated, as a turbine lifetime of 20 years is considered, despite 25-30 years being more realistic. However, this aligns with the EEG's feed-in tariff period [3], ensuring a conservative estimate. Furthermore, CAPEX values are not site-specific, which leads to site specific cost variations being neglected.

The model does not account for existing WTs, meaning identified sites may already be occupied. Therefore, results do not allow conclusions regarding whether Germany can meet its expansion targets as of the EEG [3] for onshore wind at all times.

The developed model provides a valuable and Up-to-date assessment of the geographical and techno-economic wind energy potential across Germanys ~11,000 municipalities. The highly variable results from state to state highlight the importance of considering state-specific setback distances when accurately assessing Germanys onshore wind energy potential. By analyzing different scenarios, this study offers a comprehensive overview of the potential, particularly with respect to variations in setback distances. As a result, the model and its findings serve as a valuable resource for policymakers, project developers, and researchers, enabling them to identify suitable locations while accounting for different setback distances. Depending on the applied setback scenario, between 1.13 % and 12.52 % of Germanys land area is deemed suitable for wind energy. Utilizing this area with state-of-the-art WTs could enable the installation of between 227 GW and 1,456 GW of wind capacity nationwide.

Future model improvements should focus on enhancing the calculation of the AEP for each WT. This requires the implementation of a method that accounts for site-specific wake effects, which in turn necessitates considering the wind direction at each turbine location. Additionally, a future version of the model should incorporate an analysis of existing WTs to assess the remaining potential for further wind energy expansion.

Acknowledgements

This work was funded by the German Federal Ministry for Economic Affairs and Climate Action (grant number: 03EI5253).

Supplement

The supplement is available at https://doi.org/10.5281/zenodo.14831880.

References

[1] UNFCCC 2015 *The Paris Agreement | UNFCCC* https://unfccc.int/process-and-meetings/the-paris-agreement (accessed 6 Sep 2024)
[2] IPCC 2023 *Climate Change 2023 Synthesis Report: Summary for Policymakers*
[3] EEG 2023 *Erneuerbare-Energien-Gesetz): Bundesministerium der Justiz sowie des Bundesamts für Justiz (*vol 21.07.2014*)*
[4] BWE 2025 *Windenergie in Deutschland - Zahlen und Fakten* https://www.wind-energie.de/themen/zahlen-und-fakten/deutschland/ (accessed 21 Jan 2025)
[5] UBA 2023 *Windenergie an Land* https://www.umweltbundesamt.de/themen/klima-energie/erneuerbare-energien/windenergie-an-land#flaeche (accessed 11 Dec 2023)
[6] Deloitte 2023 *Baurecht: Neuer Regelungsmechanismus für Windenergie an Land: Änderung im Baugesetzbuch (BauGB) und das neue Windenergieflächenbedarfsgesetz* https://www2.deloitte.com/dl/de/pages/legal/articles/baurecht-neuer-regelungsmechanismus-fuer-windenergie.html?id=de:2ps:3gl:4cbi-sus-baurecht-windenergie:5:6legal:20230920::&gad_source=1&gclid=EAIaIQobChMI5tqsvabDgwMVDdF3Ch2r-wSFEAAYASAAEgJg6fD_BwE (accessed 4 Jan 2024)
[7] McKenna R *et al* 2022 High-resolution large-scale onshore wind energy assessments: A review of potential definitions, methodologies and future research needs *Renewable Energy* **182** 659–84
[8] Lütkehaus I, Salecker H and Adlunger K 2013 Potenzial der Windenergie an Land: Studie zur Ermittlung des bundesweiten Flächen- und Leistungspotenzials der Windenergienutzung an Land
[9] Ryberg D S, Caglayan D G, Schmitt S, Linßen J, Stolten D and Robinius M 2019 The future of European onshore wind energy potential: Detailed distribution and simulation of advanced turbine designs *Energy* **182** 1222–38
[10] MaStR 2024 *Aktuelle Einheitenübersicht | MaStR* https://www.marktstammdatenregister.de/MaStR/Einheit/Einheiten/OeffentlicheEinheitenuebersicht (accessed 6 Jul 2024)
[11] Agora 2021 *Photovoltaik- und Windflächenrechner* https://www.agora-energiewende.de/daten-tools/photovoltaik-und-windflaechenrechner (accessed 21 Jun 2024)

[12] Fachagentur Windenergie an Land e.V. 2024 *Überblick – Abstandsvorgaben und -empfehlungen zur Ausweisung von Windenergiegebieten in den Ländern*
[13] Pelser T, Weinand J M, Kuckertz P, McKenna R, Linssen J and Stolten D 2024 Reviewing accuracy & reproducibility of large-scale wind resource assessments *Advances in Applied Energy* **13** 100158
[14] Bundesministerium der Justiz 2022 *Gesetz zur Festlegung von Flächenbedarfen für Windenergieanlagen an Land Gesetz zur Festlegung von Flächenbedarfen für Windenergieanlagen an Land (Windenergieflächenbedarfsgesetz - WindBG): WindBG*
[15] Jäger T, McKenna R and Fichtner W 2016 The feasible onshore wind energy potential in Baden-Württemberg: A bottom-up methodology considering socio-economic constraints *Renewable Energy* **96** 662–75
[16] Hoogwijk M, Vries B de and Turkenburg W 2004 Assessment of the global and regional geographical, technical and economic potential of onshore wind energy *Energy Economics* **26** 889–919
[17] Beiter P, Rand J T, Seel J, Lantz E, Gilman P and Wiser R 2022 Expert perspectives on the wind plant of the future *Wind Energy* **25** 1363–78
[18] Bošnjaković M, Katinić M, Santa R and Marić D 2022 Wind Turbine Technology Trends *Applied Sciences* **12** 8653
[19] Pape C and Geiger D 2023 Regionalisierung des Ausbaus der Erneuerbaren Energien
[20] Enevoldsen P and Xydis G 2019 Examining the trends of 35 years growth of key wind turbine components *Energy for Sustainable Development* **50** 18–26
[21] Lüers S 2023 Status des Windenergieausbaus an Land in Deutschland – Jahr 2023
[22] GeoBasis-DE 2020 Dokumentation Digitales Basis-Landschaftsmodell: Basis-DLM
[23] Deutscher Wetterdienst 2018 Metadaten zu den Radaren des Radarverbunds des DWD
[24] Bundesamt für Naturschutz 2024 *Feuchtgebiete internationaler Bedeutung in Deutschland (Ramsar-Gebiete)* https://www.bfn.de/daten-und-fakten/feuchtgebiete-internationaler-bedeutung-deutschland-ramsar-gebiete (accessed 9 Jun 2024)
[25] GeoBasis-DE 2016 Digitales Geländemodell Gitterweite 50 m: DGM50
[26] Bundesanstalt für Gewässerkunde 2024 *Geoportal der BfG* https://geoportal.bafg.de/inspire/download/AM/waterProtectionArea/datasetfeed.xml (accessed 9 Jun 2024)
[27] Personal Communication: Mitarbeiter der Bundesanstalt für Geowissenschaften und Rohstoffe 2024 Datensatz seismologische Messstationen (accessed 8 Apr 2024)
[28] Personal Communication: Mitarbeiter der Fachagentur Windenergie an Land 2024 Datensatz Militärradare (accessed 2 Apr 2024)
[29] Bundesamt für Naturschutz 2019 *FFH-Gebiete Deutschland* https://www.govdata.de/suche/daten/ffh-gebiete-deutschland (accessed 27 Jan 2025)
[30] Personal Communication: Mitarbeiter der Fachagentur Windenergie an Land 2024 Datensatz Drehfunkfeuer (VOR) 02.04.2024
[31] Fachagentur Windenergie an Land e.V. 2023 *Überblick – Abstandsempfehlungen und Vorgaben zur Ausweisung von Windenergiegebieten in den Bundesländern*
[32] Fachagentur Windenergie an Land e.V. 2024 *Entwicklung der Windenergie im Wald* (Berlin)
[33] McKenna R, Hollnaicher S and Fichtner W 2014 Cost-potential curves for onshore wind energy: A high-resolution analysis for Germany *Applied Energy* **115** 103–15
[34] Davis N N *et al* 2023 The Global Wind Atlas: A High-Resolution Dataset of Climatologies and Associated Web-Based Application *Bulletin of the American Meteorological Society* **104** E1507-E1525
[35] Pape C 2022 20220920_BWE_Flaechenpotentiale_Windenergie_an_Land
[36] Staffell I and Green R 2014 How does wind farm performance decline with age? *Renewable Energy* **66** 775–86
[37] Lüers S and Wallasch A-K 2023 *Kostensituation der Windenergie an Land - Stand 2023*
[38] Bundesnetzagentur 2024 *Festlegung des Höchstwert 2025* https://www.bundesnetzagentur.de/SharedDocs/Downloads/DE/Sachgebiete/Energie/Unternehmen_Institutionen/Ausschreibungen/Onshore/FestlegungOnshore2025.pdf?__blob=publicationFile&v=4 (accessed 29 Jan 2025)
[39] Bundesnetzagentur 2025 *Bundesnetzagentur - Beendete Ausschreibungen / Statistiken: Ergebnisse und Hintergrundinformationen zu Ausschreibungen für Windenergieanlagen an Land* https://www.bundesnetzagentur.de/DE/Fachthemen/ElektrizitaetundGas/Ausschreibungen/Wind_Onshore/BeendeteAusschreibungen/start.html (accessed 29 Jan 2025)
[40] TA Lärm 2017 *Technische Anleitung zum Schutz gegen Lärm): Die Bundesregierung* (vol 01.06.2017)
[41] Hau E 2016 *Windkraftanlagen: Grundlagen, Technik, Einsatz, Wirtschaftlichkeit* 6th edn (Berlin: Springer Vieweg)
[42] Bons M, Jakob M, Sach T, Pape C, Zink C, Geiger D, Wegner N, Boinski O, Benz S and Kahles M 2023 *Abschlussbericht – Flächenverfügbarkeit und Flächenbedarfe für den Ausbau der Windenergie an Land* (Umweltbundesamt)

Journal of Physics: Conference Series **3025** (2025) 012004 doi:10.1088/1742-6596/3025/1/012004

The effectiveness of a vision-based condition monitoring framework for determining gear health in gearboxes

Djordy Van Maele[1,2], Jean Carlos Poletto[1,2], Roeland De Geest[3], Dieter Fauconnier[1,2] and Patrick De Baets[1,2]

[1]Soete Laboratory, Department of Electromechanical, Systems and Metal Engineering, Ghent University, Technologiepark 46, 9052, Zwijnaarde, Belgium
[2]Core Lab Miro, Flanders Make @Ugent, Ghent, Belgium
[3]MotionS, Flanders Make, Diesterbaan 133, 3920, Lommel, Belgium

E-mail: Djordy.vanmaele@ugent.be

Abstract. Gearbox failures in wind turbines significantly impact operational reliability and maintenance costs, particularly in offshore environments. Traditional condition monitoring (CM) methods, such as vibration and oil analysis, often fail to detect defects early due to limitations caused by the indirect measurement approach. This is further amplified by external influences, such as changing weather conditions, on system dynamics. This study presents a vision-based condition monitoring (VCM) framework for gear health assessment. The proposed system integrates a camera system within a back-to-back gear tester equipped and utilizes an automated auxiliary system to mitigate oil splash interference on image quality. The test rig was also equipped with accelerometers to compare visual and vibrational monitoring. The effectiveness of the VCM is evaluated through eight accelerated lifetime tests, inducing gear pitting, scuffing, and root cracks. Results demonstrate that visual monitoring enables early detection, straightforward qualitative analysis of gear defects, and quantification of damage progression by leveraging deep learning. Compared to vibrational monitoring, visual monitoring offers an alternative with improved diagnostic capabilities and reduced susceptibility to external factors, showing its potential for application in condition monitoring. These findings suggest that vision-based monitoring can enhance predictive maintenance strategies, reducing downtime and optimizing gearbox longevity in wind energy applications.

1 Introduction

Gearboxes, crucial for power transmission in machines like wind turbines, sometimes fail earlier than expected due to high power demands from this industry, leading to excessive loading or caused by inadequate lubrication, oil contamination, improper maintenance, or manufacturing and material defects. In offshore settings, this is aggravated by high maintenance and intervention costs. Wind turbines use condition monitoring (CM) systems for remote gearbox health checks to address this. Current CM techniques rely on indirect measurements like vibrations [1], oil [2] and temperature monitoring [3] but often miss defect detection and are affected by external factors (e.g. weather conditions). This paper introduces a vision-based CM system (VCM) and assesses its effectiveness for monitoring gears inside a gearbox. This is done using eight accelerated lifetime datasets in which gear pitting, scuffing, and tooth cracks occur. These datasets will be the basis for evaluating the VCM's effectiveness qualitatively and quantitatively. Additionally, it allows for comparing its performance with vibrational measurements.

Content from this work may be used under the terms of the Creative Commons Attribution 4.0 licence. Any further distribution of this work must maintain attribution to the author(s) and the title of the work, journal citation and DOI.
Published under licence by IOP Publishing Ltd

2 Methodology

The effectiveness of a vision-based approach was investigated by integrating and automating a camera system into a back-to-back gear tester for helical gears and by synchronising the images captured with accelerometer measurements. This allows for the assessment of the effectiveness of VCM and comparison with vibrational measurements. The installed VCM is shown in Figure 2. It consists of a vision system (i.e., camera sensor, lens and lighting) and an auxiliary system to deal with the lubricant splash inside the gearbox, which would otherwise hinder good-quality images. The key elements of this auxiliary system are:

- An automated slider iris to create a temporary viewport for the camera to see the gear flanks.
- An air knife to momentarily clean the gear surface before image acquisition.
- An air curtain to protect the camera lens from oil vapour escaping through the temporary opening.
- A vacuum generator to extract as much oil vapour before creating the temporary opening.

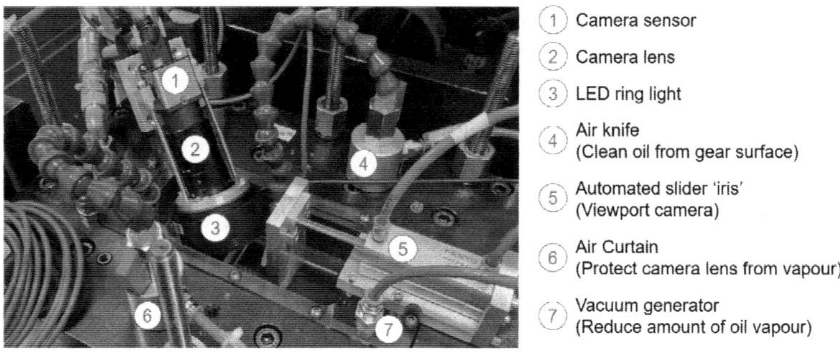

Figure 1: The VCM installed on top of a gearbox, including the auxiliary system to allow for image capture during operation

The vision system was fully automated, and images were captured at regular intervals during the whole test (every 1 to 30 minutes, depending on the test). Additionally, the entire system is cloud-connected, allowing the generated data to be accessed remotely. An overview of the image acquisition process is shown in Figure 2. Three stages can be defined during image acquisition. Firstly, the camera captures a burst of images during one full gear rotation (+/- 1500 images). The test rig rotates at 1 rpm to avoid motion blur while the auxiliary system deals with the lubricant splash inside the gearbox. Secondly, the image burst is filtered by calculating a sharpness score for each image [4]. As the gear is rotating and the camera has a fixed working distance, the gear flanks keep moving in and out of focus. This allows for an algorithm to extract one image per tooth per burst. Finally, each captured image can be assigned to a specific tooth using image correlation, allowing the images to be grouped by tooth. This is possible as each gear flank has a unique visual fingerprint resulting from distinct visual identifiers such as machining marks. Further details on the integration of the VCM and image acquisition can be found in [5]. To analyse the effectiveness of the vision-based monitoring system and compare it to vibration monitoring, eight accelerated lifetime tests were performed in which pitting, scuffing, and root cracks were induced. Damage was constrained to only the visually monitored gear due to its lower hardness of 250 HV compared with the other hardened gears of 550 HV.

3 Effectivness analysis of the VCM

The effectiveness of the VCM framework was analyzed at three levels:

- qualitative analysis, i.e., capability of detecting gear damage;
- quantitative analysis, i.e., capability of extracting gear damage metrics;
- comparative analysis, i.e., benefits compared to vibrational monitoring.

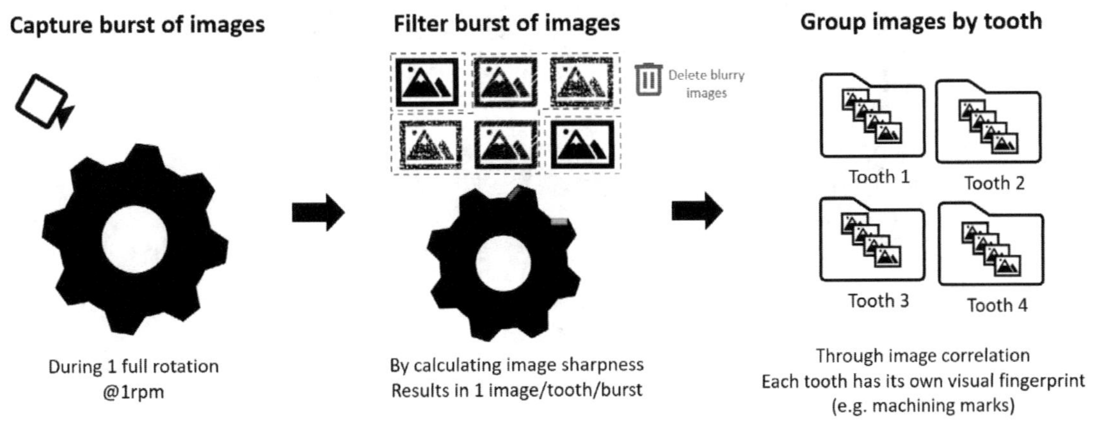

Figure 2: Schematic overview of the image acquisition during the tests

3.1 Qualitative analysis of gear's health

In contrast to other condition monitoring techniques, images enable straightforward qualitative surface analysis, allowing differentiation between the damage type and its severity. However, the caveat is that the damage needs to be visible at the gear surface. Each type of damage has its characteristic look, making it possible to distinguish between various types. This is illustrated in Figure 3, where three different types of damage induced during the performed accelerated lifetime tests can be differentiated: scuffing, pitting, and root cracks.

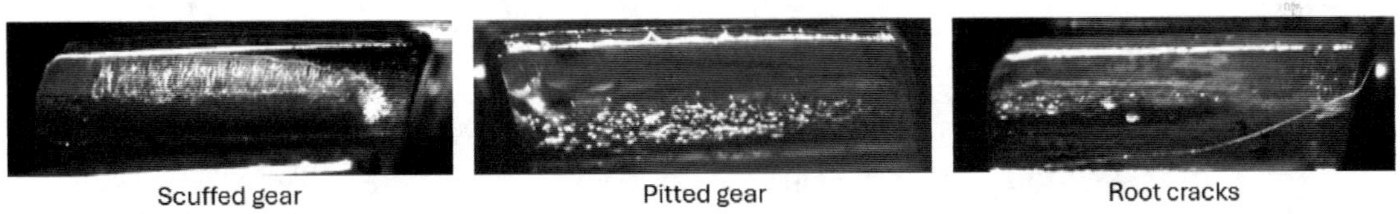

Figure 3: Different gear damage type capture during accelerated lifetime tests using the vision-based CM system.

The analysis of gears subjected to pitting confirmed that different severity levels can be distinguished. However, this is not limited to only gear pitting. This can be noted for other types of damage. For example, the growth of a root crack. In Figure 4, four different stages of increasing pitting damage are observed:

a) No pitting: the gear surface is undamaged, or the damage is too small and falls below the camera's resolution.

b) Pitting initiation: tiny white specks can be noted on the gear surface, corresponding to the first initial pits forming on the surface of the gear.

c) Progressive pitting: the pits start growing, causing more distinct craters to appear.

d) Destructive pitting: different pits begin to coalesce, forming larger pits (often referred to as spalls), combined with new pits forming over the entirety of the gear surface

Journal of Physics: Conference Series **3025** (2025) 012004 doi:10.1088/1742-6596/3025/1/012004

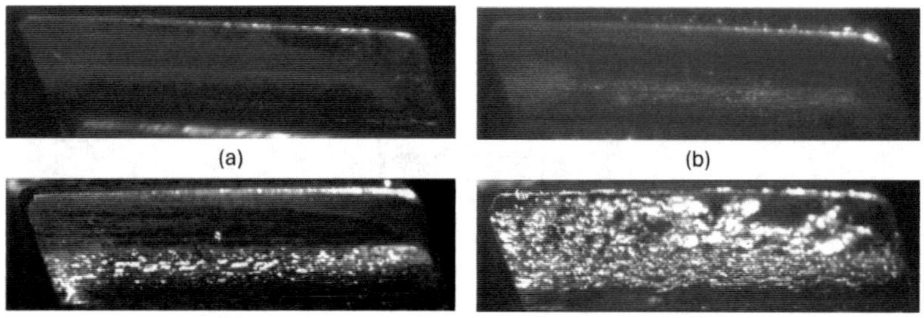

Figure 4: Different stages of pitting of a gear surface.

3.2 Qualitative analysis of gear's health

Annotating the images enables quantitative analysis, facilitating the extraction of relevant indicators related to gear damage. However, the annotation process has two limitations:

1. The minimal defect size that can be quantitatively analysed is directly related to the vision system's pixel size. According to our experience, defects should be larger than 10 pixels to be annotated, which equals 150 µm for this system. This limitation does not apply to the qualitative analysis of the gear surface.

2. Image annotation is labour-intensive, slow, and requires a trained individual. However, it can be expedited by leveraging deep learning models, where a manually annotated dataset is used to train a deep learning neural network. An active learning approach establishes the manually annotated dataset for model training. A schematic overview of this approach is shown in Figure 5, and further details can be found in [6, 7]. The deep learning model architecture used is the Feature Pyramid Network [8], starting from the pre-trained weights after being trained on the ImageNet database [9].

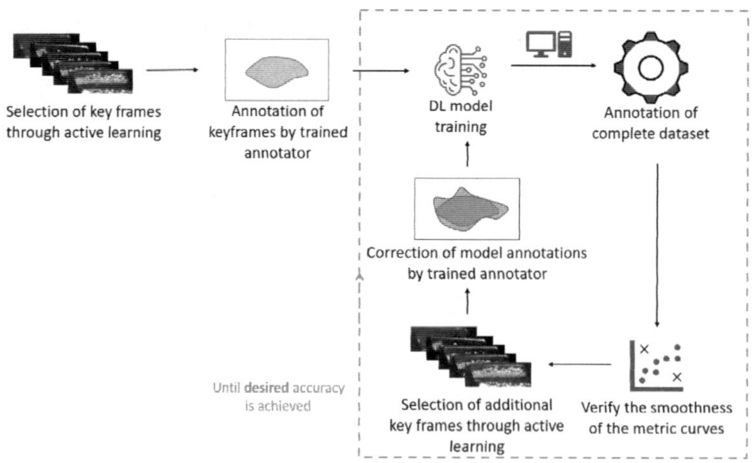

Figure 5: Active learning model training approach for image segmentation

Once the trained deep learning model fully annotates a dataset, various indicators can be extracted. Figure 6 showcases an example: horizontal and vertical position, length and width of the pits, pitted area.The evolution of these indicators can be calculated throughout the test, as shown in Figure 7. This figure shows the total pitted area relative to the flank surface area for individual gear teeth. This provides insights into the evolution of gear damage. Additionally, it can be noted that the pitted area grows in stages during the test, reaching a temporary equilibrium before transitioning to the next stage. Four stages can be discerned, as previously indicated from the qualitative analysis shown in Figure 4. The terminology for the types of pitting (progressive, non-progressive, and spall pitting) follow the definitions from the ISO 10825:2021 standard [10].

1. Stage 1 (S-1): Undamaged gear surface or the damage is too small to be annotated (<10 px);
2. Stage 2 (S-2): Progressive macropitting as a result of the gears meshing;
3. Stage 3 (S-3): Non-progressive macropitting as a result of wear debris;
4. Stage 4 (S-4): Destructive spall pitting of a select number of teeth.

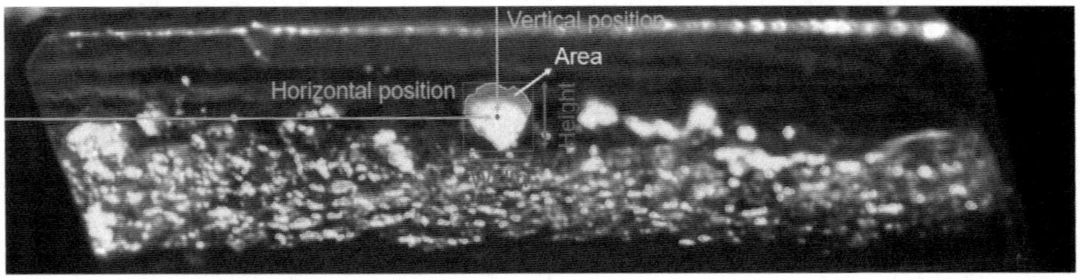

Figure 6: Extracted indicators of a pitted gear surface based on an annotated image

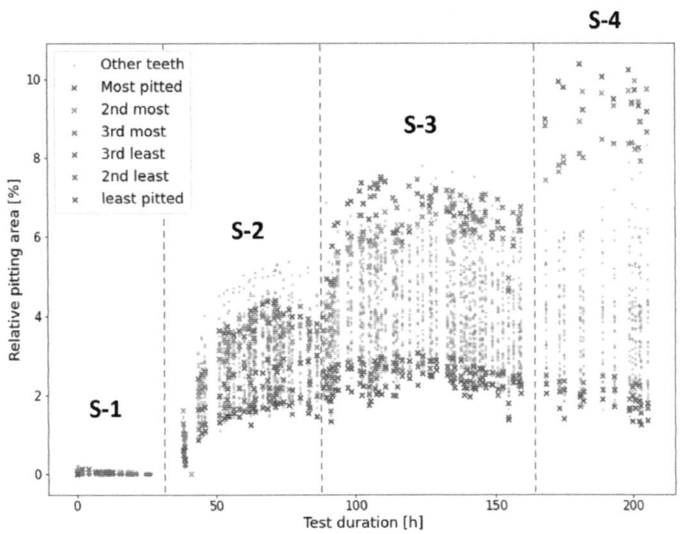

Figure 7: Evolution of the relative pitted area over the full duration of an accelerated lifetime test

3.3 Comparison of effectiveness VCM against vibration monitoring
Vibrations were captured throughout the test and synchronised with the images captured to enable comparison with the VCM. This allows for a direct comparison of the two methods. Four triaxial accelerometers were installed on the test rig, as shown in Figure 8. Three accelerometers aligned each with a shaft in the test gearbox, and the fourth was used to monitor the secondary gearbox, which serves as a healthy reference.

The primary benefit of the VCM is its ability to detect damage earlier than traditional vibrational monitoring. Vibrational monitoring typically only detects damage once it has progressed to a sufficiently

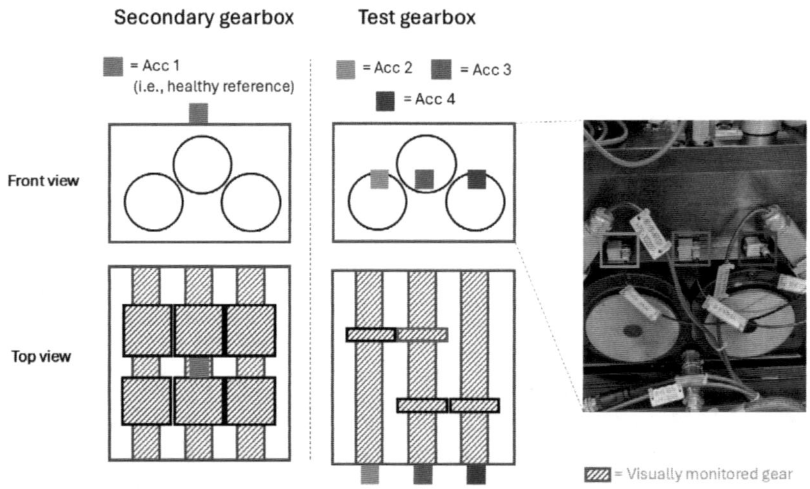

Figure 8: Configuration of the accelerometers on the test rig

severe level where the damage impacts the system's dynamics. For instance, in one of the tests, the VCM detected a tooth root crack 96 minutes before that tooth snapped off, while the root-mean-square (RMS) of the vibration signal detected it 48 minutes after the tooth snapped off, as shown in Figure 9. This delay in the detection using vibrations is likely due to multiple teeth being engaged simultaneously, which undermines the impact of a missing tooth in the overall RMS signal. Early detection of such cracks by VCM can prevent catastrophic damage to the entire system by the dislodged tooth. The root crack in this test was detected by manually inspecting the dataset post-mortem. However, in the future, it will be possible to detect this crack automatically utilizing a deep learning model.

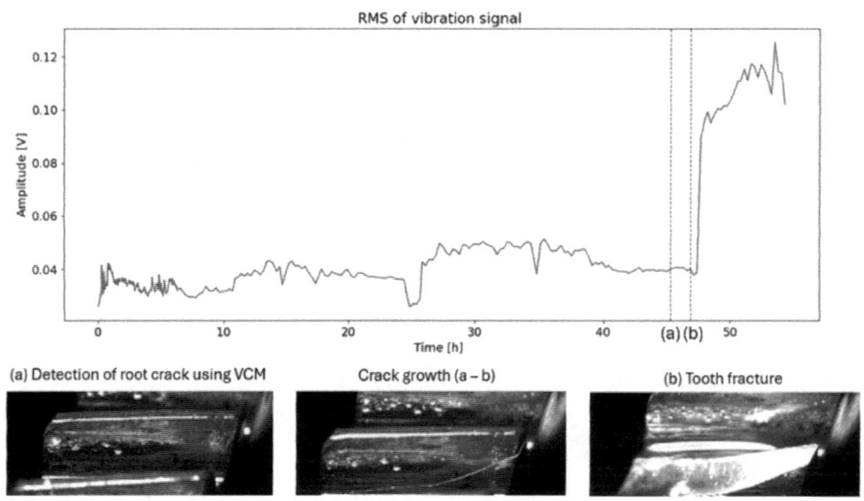

Figure 9: Comparison of root crack detection between VCM and RMS of vibration signal

Table 1 highlights a second benefit of the VCM: it provides confirmation of damage, detailed insights into the damage, and quantitative damage measurements for each tooth compared to vibration analysis. While visual monitoring is still conducted in-situ and needs more research, it can become a real-time alternative to vibration monitoring.

Journal of Physics: Conference Series **3025** (2025) 012004 doi:10.1088/1742-6596/3025/1/012004

Table 1: Comparison of the information provided by vibration analysis and visual monitoring

Information provided by measurement type	Vibration analysis	Visual monitoring
Detection capability	Indication of possible damage	Confirmation of damage/no damage
Information on damage	Fault severity estimation for the full gear	Type, severity, area and location for each individual tooth
Absolute/Relative measurement	Relative (Requires healthy baseline)	Absolute
Real-time monitoring capability	Real-time	In-situ

Thirdly, the VCM is less affected by external factors, such as changes in temperature, wind speeds and wind direction, which influence system dynamics and, consequently, the vibration signal. This is illustrated in Figure 10, where the RMS (Root Mean Square) values (in volts) of the secondary and test gearbox are plotted normalised by torque. This normalization was performed to reduce the effect of torque changes in the system on the vibration signal. After normalization, the secondary gearbox, assumed to be healthy, shows changes in its vibration signal that are attributed to unknown external factors. These changes, highlighted in Figure 10 with red dashed ellipses, align with shifts in the RMS of the test gearbox, demonstrating the impact of external factors on vibrations.

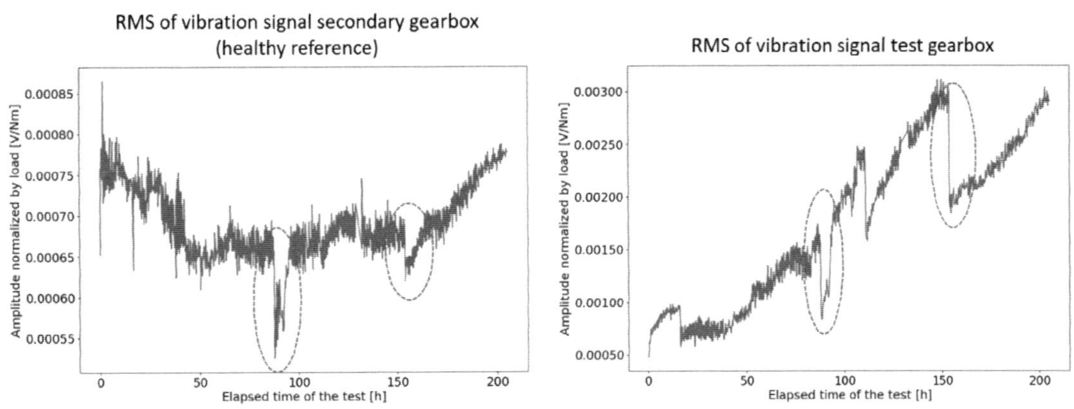

Figure 10: Influence of unknown external influences on the vibration signal

Finally, the VCM can sometimes reveal the root cause of surface damage. For example, Figure 11, shows larger, circular pits higher up on the gear's surface caused by wear debris trapped between meshing gears. This non-progressive pitting is also visible in Figure 6, where the pitted area gradually increased between 25h and 90h, followed by a sharp spike in the pitted area due to non-progressive pits appearing. This sudden spike results from wear debris caught between the meshing gear, creating small indentations, which create high-stress concentration zones for these pits to appear rapidly.

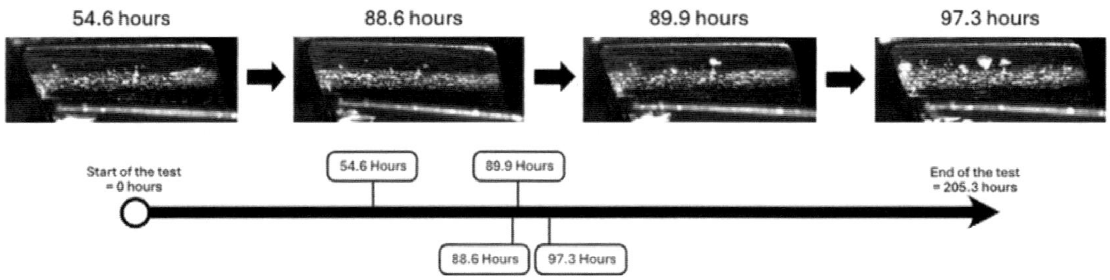

Figure 11: Appearance of non-progressive pits due to wear debris

4 Conclusion

This paper showed that visual monitoring of gears inside a gearbox is an effective CM tool for gear defect monitoring. It allows for the straightforward qualitative analysis of gear damage, where different types of damage and severities can be differentiated. Additionally, it provides a method to quantitatively analyse gear damage by leveraging deep learning, allowing for further insight into the initiation and evolution of gear pitting. Furthermore, when compared with vibration analysis, visual monitoring offers benefits such as earlier detection, more detailed insight into damage through qualitative and quantitative analysis, a monitoring method that is less influenced by external factors and the ability to, at times, allow for root cause analysis. These various aspects show that a vision-based condition monitoring system (VCM) effectively assesses gear health and could enable data-driven and tailored operation and maintenance actions (O&M). Additionally, it can potentially enhance Research and development (R&D) with better, more insightful data. In the future, it could offer a faster, more detailed, and more robust real-time monitoring alternative to the currently used techniques, such as vibration monitoring.

Acknowledgment

This research was supported by Flanders Make, the strategic research centre for the manufacturing industry, and explored during the SBO (Strategic Basic Research) project for QED (Quantified Evolution of Degradation of Gears, NO.: 2020- 1138). Additionally, this author would like to thank the other partners in this project, as this work results from collaboration between all partners involved.

References

[1] Amin A, Bibo A, Panyam M, Tallapragada P. Vibration-based Condition Monitoring in Wind Turbine Gearbox Using Convolutional Neural Network. In: 2022 American Control Conference (ACC); 2022. p. 3777-82.

[2] Coronado D, Kupferschmidt C. Assessment and Validation of Oil Sensor Systems for On-line Oil Condition Monitoring of Wind Turbine Gearboxes. Procedia Technology. 2014;15:747-54. 2nd International Conference on System-Integrated Intelligence: Challenges for Product and Production Engineering. Available from: `https://www.sciencedirect.com/science/article/pii/S2212017314001625`.

[3] Fu Z, Zhou Z, Zhu J, Yuan Y. Condition Monitoring Method for the Gearboxes of Offshore Wind Turbines Based on Oil Temperature Prediction. Energies. 2023;16. Available from: `https://www.mdpi.com/1996-1073/16/17/6275`.

[4] De K, Masilamani V. Image Sharpness Measure for Blurred Images in Frequency Domain. Procedia Engineering. 2013;64:149-58. Available from: `https://www.sciencedirect.com/science/article/pii/S1877705813016007`.

[5] Maele DV, Poletto JC, Neis PD, Ferreira NF, Fauconnier D, Baets PD. Online vision-assisted condition monitoring of gearboxes. In: 8th European Conference and Exhibition on Lubrication, Maintenance and Tribotechnology (LUBMAT 2023), Proceedings; 2023. .

[6] Maele DV, Poletto JC, Geest RD, Ferreira NF, Neis PD, Fauconnier D, et al. Strategic Annotation and Quantification Approach for Gear Pitting Analysis with a Vision Condition Monitoring System. In: 9th European Conference and Exhibition on Lubrication, Maintenance and Tribotechnology (LUBMAT 2024); 2024. .

[7] Liao W, Geest RD. Active learning for gear defect detection in gearboxes. PHM Society European Conference. 2024 6;8:10.

[8] Lin TY, Dollár P, Girshick RB, He K, Hariharan B, Belongie SJ. Feature Pyramid Networks for Object Detection. CoRR. 2016;abs/1612.03144. Available from: `http://arxiv.org/abs/1612.03144`.

[9] Deng J, Dong W, Socher R, Li LJ, Li K, Fei-Fei L. ImageNet: A large-scale hierarchical image database. In: 2009 IEEE Conference on Computer Vision and Pattern Recognition; 2009. p. 248-55.

[10] ISO. ISO/DIS 10825-1: Gears - Wear and damage to gear teeth - Part 1: Terminology; 2021.

Offshore wakes measured by an adaptive dual-Doppler scanning lidar system

Tabea Hildebrand[1,2,3], Florian Jäger[1,2,3], Nathalie Gloria[1], Luis Michaelis[1], Alexander Krasnov[1], Doron Callies[1,2,3], Jens Riechert[4] and Lukas Pauscher[1,3,5]

[1]Fraunhofer Institute for Energy Economics and Energy System Technology (IEE), Kassel, Germany
[2]Department of Integrated Energy Systems, University of Kassel, Kassel, Germany
[3]Department of Sustainable Electrical Energy Systems, University of Kassel, Kassel, Germany
[4]EnBW Energie Baden-Württemberg AG, Stuttgart, Germany
[5]Vrije Universiteit Brussel, Brussels, Belgium

E-mail: tabea.hildebrand@iee.fraunhofer.de

Abstract. Dual-Doppler scanning lidars offer the possibility to measure wakes at scales ranging from a few hundred meters to several kilometers. They can therefore facilitate comprehensive coverage of atmospheric dynamics in and around wind farms at several highly relevant scales ranging from the individual turbine to cluster wakes. High fidelity wake measurements with scanning lidars can significantly support the efforts towards high fidelity wake modelling by using the data as validation.

This work demonstrates offshore dual-Doppler wake measurements at different scales, performed with a control software that allows to synchronise and adapt the scan pattern to the observed wind conditions. Two scan strategies are evaluated: a dual PPI (plan position indicator) scan to capture fine scale dynamics within the wind farm and a step-stare campaign to measure the spatial evolution of the wake at larger scales behind the wind farm.

An analysis of the deviations in measurement height due to the inclined PPI planes and a thrust-induced turbine tilt was carried out and showed that caution is advised when interpreting dual PPI data. Nevertheless, the dual PPI scans showed to be an effective tool in capturing the spatial characteristics of wakes within the wind farm on the scale of a single turbine or a few neighbouring turbines. Step-stare scans on the other hand have the advantage of lower uncertainties due to less height deviations, but the spatial resolution is limited. The results of the step-stare measurement were compared to widely used engineering wake models (TurbOPark and Jensen) and showed a good agreement regarding the trends, but the models overestimated the deficits by around five to eight percentage points. The control software to dynamically adapt the scanning patterns to the observed wind conditions ensured a highly efficient and tailored measurement design that optimised measurement time in the area of interest while increasing the sampling rate.

Content from this work may be used under the terms of the Creative Commons Attribution 4.0 licence. Any further distribution of this work must maintain attribution to the author(s) and the title of the work, journal citation and DOI.
Published under licence by IOP Publishing Ltd

Journal of Physics: Conference Series **3025** (2025) 012005 doi:10.1088/1742-6596/3025/1/012005

1 Introduction

In several offshore areas of the world, arrays of wind farms are developed. Especially in the North Sea, wind farms are particularly densely spaced. This makes a good understanding of wake dynamics of different scales imperative for an efficient planning and development process (Schneemann et al., 2020; Cañadillas et al., 2022). Relatively small differences in the estimate of the wind speed deficit caused by wake effects can have significant influence on the design of the layout and calculated AEP of large wind farms. One of the challenges for wake modelling is that physical effects at different scales are important and interact with each other. This makes high quality wind measurements at different scales ranging from individual turbine's wakes to wake interactions in and between wind farms important to improve wake models.

Generally, there is a challenge to measure spatially resolved in an offshore environment. Often SCADA data are used including nacelle-based cup anemometry to evaluate the performance of wake models, but these data can be inaccurate due to the positioning of the sensors on top of the nacelle behind the rotor (Barthelmie et al., 2009). The wind speed and direction measurements on the nacelle are affected by the rotor and misalignment of wind vanes can lead to inaccuracies. Using energy production data introduces uncertainties related to the power curve.

Independent high quality wind measurements are thus a highly valuable additional basis for wake model development. Particularly, scanning Doppler lidar measurements offer the possibility to retrieve high precision estimates of wind speeds over scales ranging approximately from a hundred meters up to several kilometers. If used as a single device, retrievals of the horizontal wind speed often rely on the assumption of homogeneous wind conditions over a larger area. In waked conditions this assumption is always violated to some degree. As a result, expected errors/uncertainties are generally difficult to quantify. Nevertheless, single scanning lidar have been used successfully to yield valuable insights into wind turbine wakes. Several measurement campaigns aimed at capturing cluster wakes with scanning lidar (Schneemann et al., 2020; Cañadillas et al., 2022). In these studies single scanning lidar operated in a PPI (plan position indicator) mode. In Schneemann et al. (2020), measurements were taken at the transition piece height (where the lidar was installed) with 0° elevation angle rather than nacelle height. The purpose of the measurement was to assess the overall induction effect, known as the global blockage of the wind farm, in connection with atmospheric stability and the operational states of the wind farm. Cañadillas et al. (2022) retrieved data with five different elevation angles. Single lidar PPI scans are also used to analyze the wake of a single turbine. Angelou et al. (2023) mounted a scanning lidar on top of the nacelle to be able to measure at nacelle height while reducing the effect of varying height with distance. In Goit et al. (2020), PPI measurements from the transition piece were combined with additional wind direction data from a measurement mast in order to retrieve horizontal wind speeds while at the same time avoiding to average lidar measurements over a larger area. Other studies analysed radial velocities instead of horizontal wind speeds (Smalikho et al., 2013; Aitken et al., 2014; Banta et al., 2015).

In contrast to single scanning lidar measurements, dual-Doppler scanning lidar measurement techniques can avoid assumptions about the horizontal homogeneity of the wind speed, and yield highly accurate measurements even in complex flows (Pauscher et al., 2016; Mann et al., 2017; Coimbra et al., 2025). Moreover, their high accuracy has been validated in offshore conditions (e.g. Shimada et al., 2022). Wildmann et al. (2018) used a multi-lidar measurement in a step-stare mode to measure the wake of a single turbine while adapting automatically to the wind direction and showed good agreement of their experimental results with the Jensen wake model (Jensen, 1983). Other studies have relied on dual-PPI setups to measure wakes (Wu et al., 2016; Van Dooren et al., 2016).

While providing high quality information of wind speed and wind directions at the intersection points, dual-Doppler (or more general multi-Doppler) measurements also have several limitations. If the lidars are operated in step-stare mode, the devices can be programmed to intersect their beams in the points of interest with high accuracy. However, the number of measurement points is limited due to the time the devices need for moving and measuring wind speeds when steering to multiple measurement points (MP). Moreover, in an offshore setting, they are often mounted on the transition piece which exhibits some thrust-induced (and wave-induced) movement. This means that beam alignment accuracy is affected and induces errors in height. If operated in PPI mode, the spatial coverage of the dual-Doppler measurements can be significantly increased. However, in addition to the motion-induced beam misalignment, there is a significant height difference between the two lidar measurements for most of the measurement plane.

Finally, the uncertainty of a dual-Doppler measurement also depends on the angle between the lidar beams and on their angle to the wind direction (Stawiarski et al., 2013). In consequence, careful evaluation is necessary to ensure the potential for high accuracy measurements with high spatial coverage by dual-Doppler measurements.

This paper addresses the constraints and evaluates and demonstrates how dual-Doppler wake mea-

Journal of Physics: Conference Series **3025** (2025) 012005 doi:10.1088/1742-6596/3025/1/012005

surements at different scales in an offshore wind farm can be performed using different scan strategies. In this context, it combines the classical uncertainty estimation approach (Stawiarski et al., 2013) from error propagation which has been used by previous studies (e.g. Vasiljević et al., 2020; Hung et al., 2023) with considerations of height differences between the lidar beams. This allows to derive a more complete picture of the capabilities and limitations of the approach. The paper also introduces a control software that makes the dual-Doppler measurements even more versatile by allowing to adapt the scan pattern to the observed wind conditions. This control is used to automatically switch between scan patterns and maximise the time during which measurements are performed in the wake. Based on a dual-Doppler scanning lidar measurement in an offshore wind farm using two different dual-Doppler scanning techniques, this work addresses in a practical example how wake dynamics at different scales can be captured and achieve accurate spatially resolved measurements. Finally, the observed wind speed deficits of one of the scan techniques are compared to two commonly used engineering wake models.

2 Methods

2.1 Experimental setup

The data presented in this paper were collected during an offshore measurement campaign conducted from March until November 2023. Two scanning lidars of the type WLS 400S by Vaisala were positioned on the transition pieces of monopile-founded offshore turbines. Scanner 1 was deployed in the southeast corner of the wind farm whereas scanner 2 was deployed around 4800 m north of scanner 1. An overview of the experimental setup together with the two scanning strategies which were consecutively performed and are analysed in this paper are given in figure 1 and figure 2.

The beam alignment of the devices was carefully calibrated to reduce uncertainties in the wind speed measurements by performing regular CNR (carrier-to-noise ratio) mappings (hard target tests as well as sea surface levellings (Rott et al., 2022; Gramitzky et al., 2024)). To estimate the dynamic parts of the beam pointing variation, rigorous testing regarding thrust-induced turbine tilt was performed using co-located high frequency inclinometers (Hildebrand et al., in preparation).

2.2 Scan strategies

The two following scan strategies were used to put the focus of the measurements on two different aspects and scales of wind turbine wakes:

1. A dual-Doppler plan-position-indicator (PPI) scan to capture the fine scale dynamics within the wind farm,
2. A dual-Doppler step-stare campaign to measure the spatial evolution of the wake at larger scales behind the wind farm.

The two measurement setups and the device configurations regarding the two different strategies are explained in the following two subsections. To be able to interpret the reconstructed wind speeds, the uncertainty of the measurement is an important aspect to consider. For both measurement approaches a standard error propagation was conducted according to the JCGM guide to express uncertainty in measurement (JCGM, 2008). Similar analysis were made by Stawiarski et al. (2013), Hill et al. (2010) and Van Dooren et al. (2016).

2.2.1 Dual-Doppler PPI scans

During a PPI scan, the scan head moves with a constant speed under a fixed elevation angle, just varying the azimuth angle, where measurement data are aggregated according to a given accumulation time (here 1 s). For a dual-Doppler PPI scan, two scanning lidars simultaneously perform PPI scans that cover an overlapping area. Note that the measurement planes of both devices for the most part do not have the same measurement height, if they measure with elevation angles other than 0° (here 1.35° and 1.96°).

The dual-Doppler PPI measurements analysed in this paper were conducted in the time period March 22 to March 31, 2023. Horizontal wind directions and wind speeds were reconstructed from the dual-PPI scans in the (horizontally) overlapping area based on weighted averages from the nearest measurement points using a simplified version of the approach of Van Dooren et al. (2016). Since it is not possible to measure the shear inside the wind farm with the measurement setup, a value for α has to be assumed. Here $\alpha = 0.17$ was chosen. This value was derived in the upstream flow of the wind farm by a nacelle-mounted lidar for the time period for which the PPI measurements are shown in section 3.2.

The extrapolation of the radial wind speed to hub height before the wind vector is reconstructed can also have a significant impact on the uncertainty of the reconstructed horizontal wind speed. Uncertainties are therefore higher the larger the extrapolation distance is, e.g. in areas around the devices and far away.

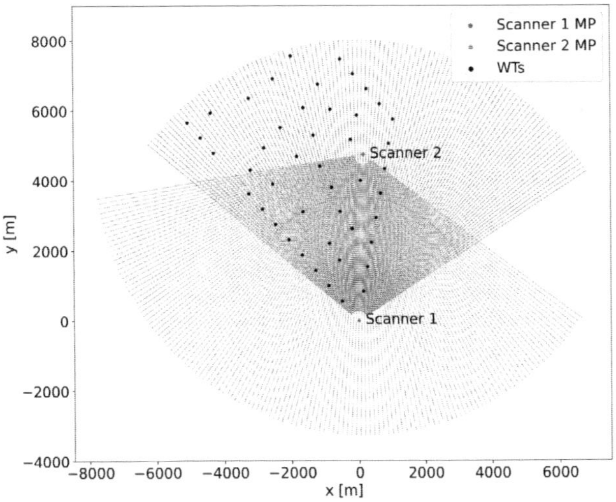

Figure 1: Measurement layout of PPI scans, with measurement points of both scanning lidar PPI planes

Parameter	Setting
movement speed	1°/s
accumulation time	1 s
range gate length	150 m
Sc1: scan sector	[309°, 58°]
Sc1: elevation angle	1.35°
Sc2: scan sector	[126°, 261°]
Sc2: elevation angle	1.96°

Table 1: Scanner configuration during the PPI scans

2.2.2 Dual-Doppler step-stare scans

The second part of the measurement campaign focused on capturing wake dynamics in a larger scale behind the wind farm. This measurement phase relied on a step-stare configuration combined with an adaptive scan strategy, which automatically adapts the measurement layout depending on the observed wind direction within the past 10-minute interval. This assured that the measurement grid was always located downstream of the wind farm for either westerly or easterly winds. The campaign was conducted in the time period August 8 to November 1, 2023 (with some planned interruptions for another measurement phase). The beams of the two scanning lidar were intersected at 13 measurement points. Two of these points were located upstream of the wind farm at different heights. The upstream measurement at hub height was used to determine the inflow wind speed and direction for the calculation of wake deficits and wake modelling. All other measurement points on hub height were positioned downstream of the wind farm. The 10 wake measurement points were arranged in three lines downstream of the wind farm.

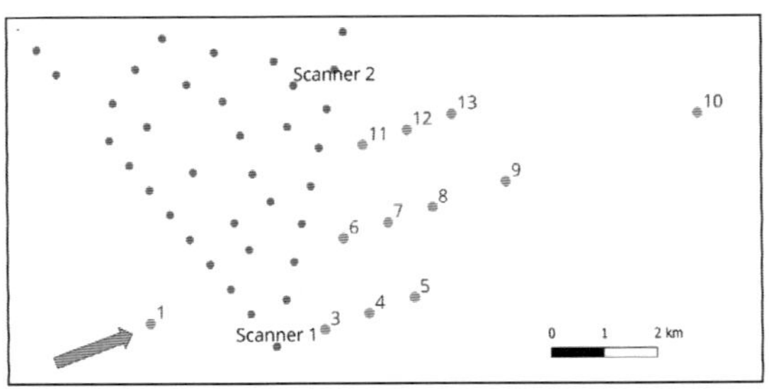

(a) for west wind [180°, 360°) marked in orange

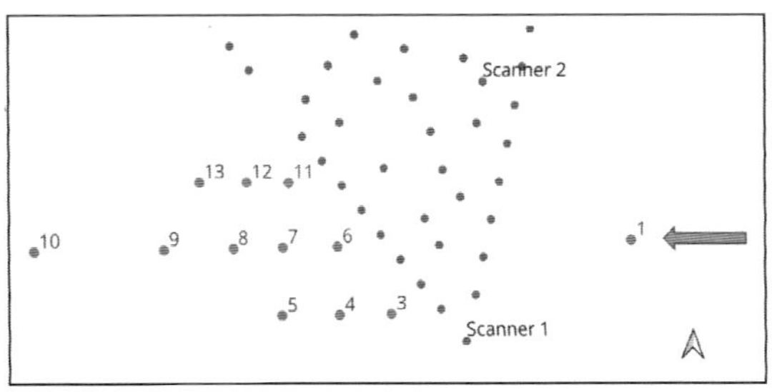

(b) for east wind [0°, 180°) marked in green

Figure 2: Measurement layout of adaptive dual-Doppler campaign, turbine positions are marked in blue, a reference measurement upstream of the wind farm is taken at two heights

2.3 Adaptive control software

The step-stare measurement was performed using ArgusScan - a comprehensive control software for temporally synchronised dual-Doppler measurements, which also facilitates the automatic adaptation of the scan pattern to the current wind conditions.

Using the adaptive capabilities of the software, automated measurements in the wake of the wind farm could be performed for different wind directions, while maximising the points available in the measurement grid. For this purpose, two scan patterns were pre-configured. A reference point was defined in each scan pattern upstream of the wind farm. During the measurement campaign, the wind vector at this point was instantaneously and automatically derived for each 10-minute interval by the control software. Based on this information, the control software then chose between the two pre-programmed scan patterns. For westerly wind directions $[180°, 360°)$, the measurement grid was placed downstream of the wind farm in its east. For easterly wind direction $[0°, 180°)$, the measurement grid was placed in the west of the wind farm (figure 2).

The control software allows for much more complex adaptive strategies including more scan scenarios and more advanced conditions for selecting a scan pattern based on other measurement variables like wind speed or CNR. However, this simple setup was chosen to ensure sufficient samples at each measurement location in the grid.

The time synchronisation (both scanners measure at the same measurement point) is performed based on a scheduler functionality. Synchronisation is achieved by calculating schedules based on the estimated times that the lidar devices require for measuring, moving the scanner heads between the locations and a "thinking" time that is required if the scanner heads move position. At the beginning of the measurement campaign schedules are transferred to the individual lidar devices. After that the individual lidar devices perform these schedules independently. As long as the clocks on the devices are synchronised, time synchronisation is ensured even if the network connection to the devices is lost. This is in contrast to e.g. the WindScanner software (Vasiljević et al., 2016), which requires network connection between the devices and the control unit to facilitate time synchronisation.

2.4 Data preparation and quality control

In order to get high quality data and to avoid implausibilities, filters were applied to the step-stare data. Before aggregating the data to 10-minute mean values, upper (e.g. when hitting an obstacle such as a wind turbine) and lower (too low backscatter signal) thresholds for the CNR signal were defined. In addition, the radial wind speed status as given in the raw lidar data of the Vaisala WLS 400S was checked for the step-stare campaign. Within each 10-minute interval, the standard deviation σ_{u_r} of the radial velocity u_r was calculated and outliers with deviations of more than $3\sigma_{u_r}$ from the mean were removed. A lower threshold of 30 % of data points relative to the maximum possible amount (recovery rate) was applied to each 10-minute period of the step-stare campaign. The analysis was limited to wind speeds at which the turbines were operating ($\geq$4 ms^{-1}) and wind directions for which wake effects were expected. An overview over all filters and thresholds for the step-stare setup is given in table 2.

Table 2: Overview over the post processing of the data for quality control

Filter	Prerequisites
CNR	[-25 dB, -5 dB]
Radial wind speed status	1
Standard deviation of radial wind speeds	$[\overline{u}_r - 3\sigma_{u_r}, \overline{u}_r + 3\sigma_{u_r}]$
Min. recovery rate in 10-minute interval	30 %
Wind speed (cut-in wind speed)	≥ 4 ms^{-1}
Wind direction - west	$[180°, 320°]$
Wind direction - east	$[20°, 180°]$

For the PPI scans no automated quality control was done except for the CNR filter. Since the focus of the analysis is on individual 10-minute intervals rather than on the whole measurement period, reliability of the reconstructed data was ensured by considering the freestream wind speed of a single timestamp and filtering out unreasonable high wind speed values especially close to turbines, where erroneous data is likely due to the measurement volume being partly within the areas of the rotor. To avoid high measurement errors, no grid points with a beam intersection angle of more than 150° are considered in the analyses of the PPI campaign.

2.5 Wake modelling

For this analysis wake modelling was conducted using PyWake (Pedersen et al., 2023). Simulations were performed using two different models - TurbOPark (Nygaard et al., 2020, 2022; Pedersen et al., 2022) and the Jensen (Jensen, 1983; Katic et al., 1987) model. As inflow conditions for wind speed and wind direction, the reference measurement point at hub height upstream of the wind farm was used. For TurbOPark, the turbulence intensity (TI) is set to 6 % and the simulation was performed based on standard parameters (Nygaard et al., 2022). During the measurement campaign itself, no nacelle-mounted lidar was available to derive an estimate of the turbulence intensity. Also, the sampling rate of the dual-Doppler step-stare measurement was relatively low, which is likely to affect the reliability of TI measurements with this approach. Therefore, the value of 6 % was derived from the median TI upstream (west) of the wind farm during a different measurement phase using a nacelle-mounted lidar. We acknowledge that TI measurements from lidars tend to underestimate the measurements from cup anemometers. Which introduces some uncertainty in the parameterisation. For the device used in this study this effect has been reported to be a factor of 0.9 (Liang et al., 2022). The effect on the wake model results is, thus, estimated to be small. For the Jensen model, the wake expansion parameter was set to $k = 0.02$, because Nygaard et al. (2022) detected a reduction in the overall wake model bias in comparison to the previously recommended values of 0.04 or 0.05 for offshore cases (Porté-Agel et al., 2020; Barthelmie et al., 2009). The simulation results at measurement point locations are queried and the speed deficits of the mean wind speeds from measurement and simulation are then compared.

3 Results and discussion

This section presents wind speed measurements from two different setups - dual-Doppler PPI and dual-Doppler measurements in step-stare mode. The first section discusses the advantages and challenges of both measurement strategies from the perspective of uncertainties resulting from the measurement setup. In the second section measurement results from both dual-Doppler setups are presented and the measurements are compared to two engineering wake models.

3.1 Beam geometry induced uncertainties

Dual-Doppler measurements offer the great advantage that wind speeds can be retrieved with a high spatial coverage and resolution. However, the measurement quality varies depending on measurement location and the measurement geometry associated with it. The uncertainty is mainly influenced by two factors: the difference of the height of the probe-volumes of the lidars to each other and the height of interest (here hub height) as well as the intersection angles between the beams and between beams and the wind vector.

Height differences of the lidar probe volumes are mainly caused by two effects. Firstly, for PPI scans with an elevation angle not equal to 0° the two planes which are scanned will only intersect in a single line. Secondly, the thrust-induced movement of the platform will cause errors in beam alignment, which in turn results in height deviations from the targeted height. These depend on the tilt angle of the mounting platform and thus the wind direction and speed during the measurement. They are present in both PPI and step-stare mode. For the analysed measurement campaign, in which the lidars are mounted on the transition piece of a bottom fixed monopile wind turbine, the observed wind-induced tilt showed variations of up to approximately 0.3° (peak-to-peak inclination) (Gramitzky et al., 2024).

The impact of the thrust-induced height variations for a wind direction of 240° and a tilt of 0.15° (since only the amplitude is of importance) is shown in figure 3a. The magnitude of the effect strongly depends on the measurement location, i.e. the distance to the lidar and the angle between the laser beam and the wind direction. The height deviation reaches a maximum of approximately 10 m for the presented experimental PPI setup. This value will vary with thrust and thus with wind speed. It should be noted that, for northerly or southerly winds, the height differences of scanner 1 and scanner 2 are in opposite directions and the resulting height difference between the beams has to be added up. For westerly and easterly wind directions, height differences are in the same directions and both beams measure either higher or lower than the targeted height. To some degree, the effect of thrust induced tilt can be mitigated, if it is accounted for in the programmed elevation angles. However, this is only applicable if the application is focused on specific wind conditions (speed and direction) and a relatively small range of azimuth angles.

Figure 3b shows the height differences between the two PPI planes without a thrust-induced tilt. A comparison again shows that the relative importance of both effects strongly depends on the measurement location. For most of the area the height difference inherent in the dual-PPI setup causes greater height differences.

The differences in height from the targeted measurement height make assumptions about the vertical profile of wind speed and wind direction necessary and thus increase the uncertainties in the retrieved wind speeds. An exact derivation of uncertainties would require prior knowledge about the vertical wind profile itself. While for undisturbed flows well-established profile laws like the power law can be applied

Journal of Physics: Conference Series **3025** (2025) 012005 doi:10.1088/1742-6596/3025/1/012005

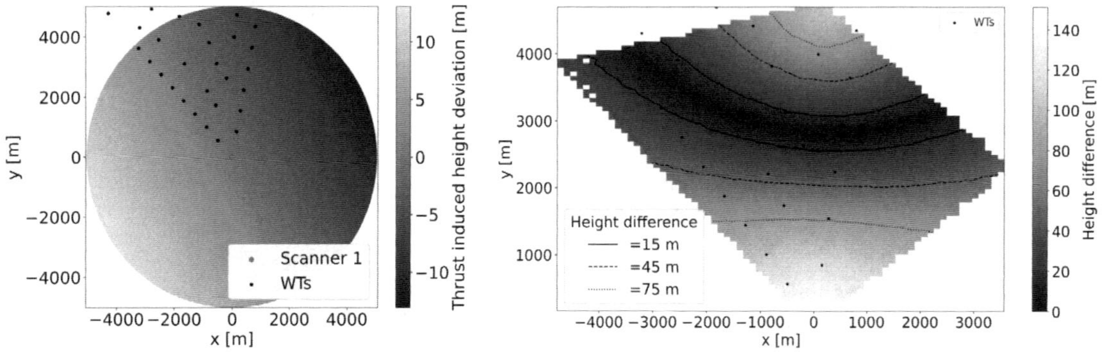

(a) (b)

Figure 3: Measurement height deviations for dual-Doppler measurements: **(a)** deviation in height caused by the thrust-induced tilt of the mounting platform (transition piece) for scanner 1; numbers change with angle between beam and wind direction and thrust/wind speed; 240° wind direction with a maximum tilt of 0.15° is shown as a conservative case. **(b)** height between the two PPI planes, due to positioning of the devices on transition piece, scanning with elevation angles (1.35° and 1.96°) without induced tilt.

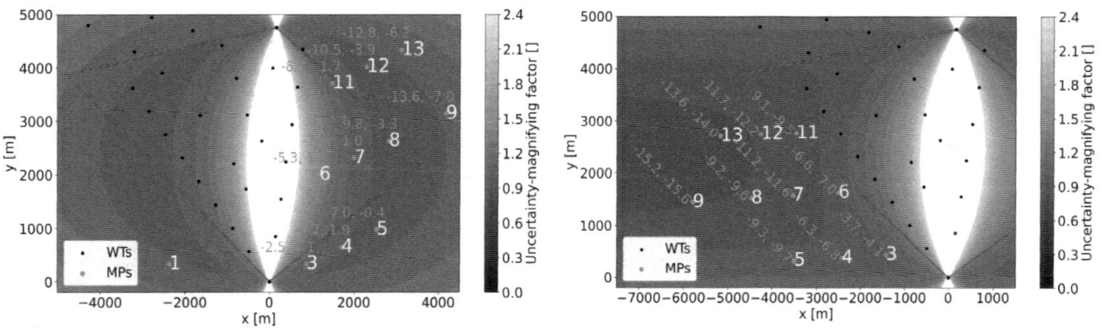

(a) Wind direction of 240° (b) Wind direction of 90°

Figure 4: Dual-Doppler scan uncertainty magnifying factor depending on measurement location, plot also displays the height deviations in meters caused by platform tilt from both scanners (scanner 1: first value; scanner 2: second value)

to derive the average or typical uncertainties, in waked flows this approach is not applicable. Here vertical profiles strongly vary with distance, position and number of the upstream turbines. We therefore suggest to define areas of varying levels of confidence for the wind measurements based on the spatial separation of the lidar beams. In figure 3b we define these areas as ≤15 m, ≤45 m and ≤75 m height difference between the two lidar beams. While this is somewhat arbitrary, it offers a tool to classify the confidence in the PPI retrievals.

Estimates of uncertainty of the reconstructed wind velocity of the dual-Doppler measurement caused by the azimuth angle difference between the lidars are shown in figure 4. The calculations are based on the error propagation of the radial velocity uncertainty for the dual-Doppler wind vector reconstruction. The calculations assume the same uncertainties for the radial velocities for both scanners and those uncertainties to be independent. In the white area, the uncertainty tends to infinity as the beams become more and more parallel. Therefore, all grid points with an intersection angle ≥ 150° are excluded in further analysis (section 3.2) due to high measurement uncertainty.

It is evident that the majority of the overlapping PPI scan area (s. figure 4a) is within areas with rather low intersection angle-induced uncertainty. As a reference, typical uncertainties of radial velocities are given as approximately 0.1 ms^{-1} (Vasiljević et al., 2020). This indicates that except from the area directly between the two devices low uncertainties can be achieved. As a consequence, the influence of the height differences between the measurement lidar beams is likely to be the more important factor, when planning a dual-Doppler PPI setup. For the step-stare measurement the effect from turbine tilt on the measurement height is visualized by orange numbers in the figure 4 and shows that deviations even for conservative assumptions of 0.15° are resulting in values ranging from -0.4 m to -16 m. The height deviations for scanner 1 (first value) and for scanner 2 (second value) are depicted for 240° in figure 4a and for 90° in 4b. The thrust induced turbine tilt mostly results in the beams measuring below the hub height, because of the measurement points being positioned downstream of the wind farm. Besides the distance from the device the largest height deviations occur, when the beam is parallel to the wind direction.

3.2 Wake effects measured by dual-Doppler scans

The first part of the measurement campaign was performed using a dual-Doppler PPI setup as outlined in section 2.2.1. Figure 5 depicts the horizontal wind speed and wind direction derived from a PPI measurement (10-minute mean). The fine scale structure of the wakes are clearly visible. The spatial evolution looks as expected and the wakes of the individual turbines are clearly identifiable. However, from the discussion in section 3.1, it is clear that the quality of the measurements strongly varies with locations within the scanned transect.

To provide an indication of how quantitatively the measurements can be used, the black lines from figure 3b are inserted into figure 5a. These lines show that high quality measurements can mainly be obtained from the scale of the individual wake up to a few turbines. Here, the dual-Doppler approach has a significantly reduced uncertainty when compared to a single scanning lidar, because much weaker assumptions regarding the homogeneity of the wind conditions have to be made. Over the scale of (a larger part of) the wind farm, the information on wake evolution becomes increasingly more qualitative. In these areas the advantages, when compared to a single-lidar PPI scan, are significantly reduced.

On the very small scale around the turbine, it appears as if a strong induction zone can be measured. However, this is likely to be an effect of the spatial averaging of the probe volume (150 m) and the spatial resolution of the PPI scan (several tens of meters, depending on the distance from the scanners). The white spots except the area between the scanners, which is also displayed in figure 4, are caused by contaminated lidar signals, due to hitting the structure of the wind turbine within the measurement volume or generally a blocked line of sight. Therefore, the data were removed during quality control.

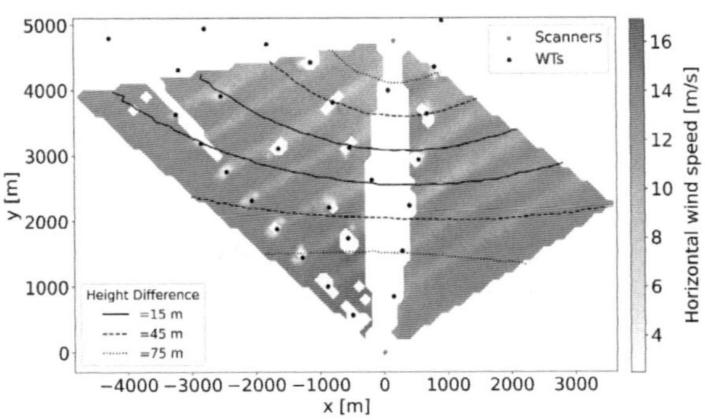

(a)

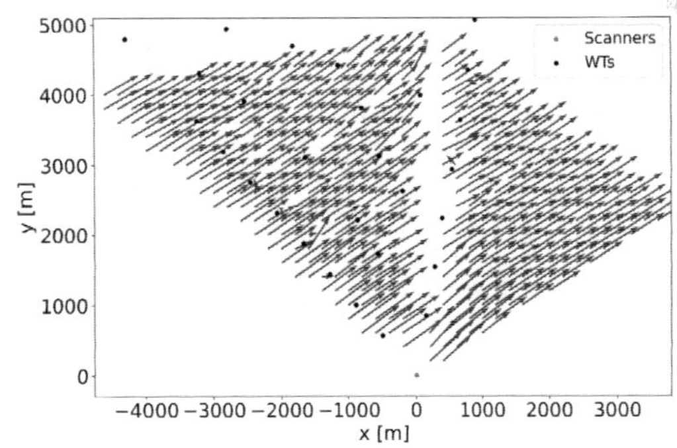

(b)

Figure 5: Wind measurements of dual-Doppler PPI scan; **(a)** horizontal wind speeds as reconstructed from the dual-Doppler PPI measurements at 25.03.2023 10:30-10:40; contour lines indicate the height between PPI planes (see also 3b for explanation); **(b)** horizontal wind directions as reconstructed from the dual-Doppler PPI measurements at 25.03.2023 10:30-10:40.

The second measurement strategy used during the experimental campaign is based on the dual-Doppler step-stare mode. During this period, the lidars were controlled using ArgusScan to automatically adapt the scan layout to measure downwind of the wind farm (see also section 2.3). This allowed optimising the time measured in the wake, while at the same time increasing the sampling rates at each individual point. Westerly wind directions were observed approximately 75 % of the time in that measurement mode, easterly wind directions occurred approximately 25 % of the time. During the step-stare campaign, time synchronisation of the lidars was better than 1 s in 85 % and better than 1.5 s in 96 % of the cases. This observation reconfirms the robustness of the scheduler-based approach for time synchronisations.

Journal of Physics: Conference Series **3025** (2025) 012005 doi:10.1088/1742-6596/3025/1/012005

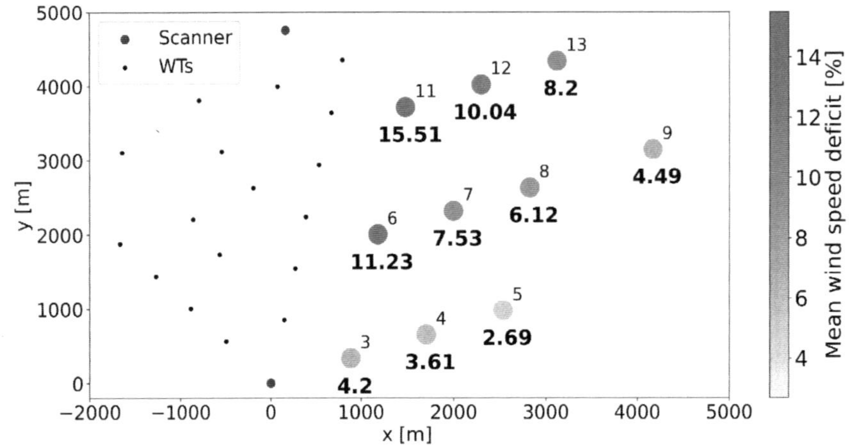

(a)

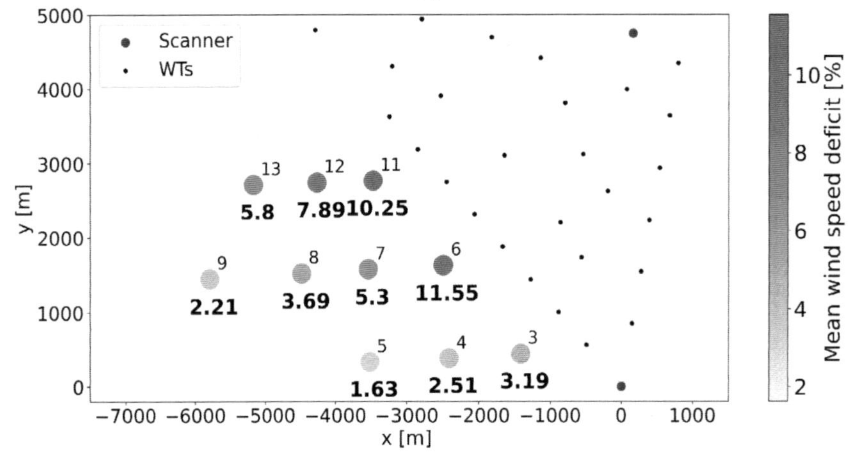

(b)

Figure 6: Wind speed deficits [%] calculated against the reference measurement point upstream of the wind farm of the dual-Doppler scanning lidar measurement campaign in step-stare mode for **(a)** westerly wind directions and **(b)** easterly wind directions.

The observed mean wake deficits during the measurement are depicted in figure 6. They range from 2.69 % to 15.51 % for westerly wind directions and are lower for easterly wind directions (1.63 % to 11.55 %). The dependency of the deficits on the number of upstream turbines can clearly be observed. A higher number of upstream turbines causes (as expected) a greater deficit behind the wind farm. Also, the typical wake recovery with increasing distance from the wind farm can be observed. In this context, it should be noted that the differences between neighbouring points are often only in the order of a few percent. A good example is the difference in the wake deficit between points 8 and 9 for westerly wind directions (figure 7a) with a difference of less than two percentage points. These relatively small differences underline the need for high accuracy measurements to be able to resolve the differences in wind speed deficits. This is especially important for all wind speeds in larger distance to the wind farm, where deficits and differences in deficits get smaller.

3.3 Wake modelling

Figure 7 compares the observed wake deficits to those measured by the step-stare approach. Trends in the results of PyWake simulations agree well with the measurements. However, the magnitude of the deficits between the dual-Doppler lidar measurements and the simulations with the TurbOPark and the

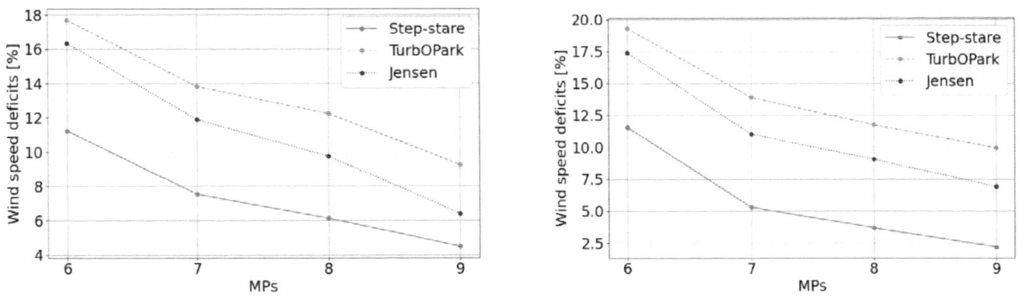

(a) West campaign (b) East campaign

Figure 7: Wind speed deficits measured by the dual-Doppler scanning lidar measurement campaign in step-stare mode for west and east campaign showing wake evolution behind the wind for measurement points on the line [6, 7, 8, 9] (s. figure 2).

Jensen model show significant differences. Both models estimate the deficits to be significantly larger. However, the used parameterisation of the models might not be optimal for this offshore wind farm setting, and also the curtailment time periods were not considered in this analysis and could be a reason that observed deficits are smaller than the modelled ones. The absolute difference is more or less constant with distance from the wind farm for TurbOPark and is around six and eight percentage points and below and around five percentage points for Jensen for westerly and easterly wind directions, respectively. Relative differences in wind speed deficits increase with distance, reaching factors of 1.6 to 2.1 (west) and 1.7 to 4.5 (east) for TurbOPark, while remaining at 1.5 (west) and increasing from 1.5 to 3.1 (east) for Jensen. While the observed differences of several percentage points are highly significant for the accurate estimation of the annual energy production, it is difficult to achieve corresponding accuracy with a single scanning lidar in wake conditions characterised by heterogeneous flows. In contrast, dual-Doppler lidar measurements significantly reduce uncertainties, allowing for a quantitative assessment of the models.

4 Conclusion

The presented results evaluate the suitability of different dual-Doppler measurement strategies to measure wake effects in wind farms at different scales. The theoretical considerations show that dual-Doppler PPI scans are an effective tool in capturing the small(er)-scale dynamics of wakes within the wind farm from the scale of single wakes to the interaction of a few turbines with high spatial coverage and resolution. The uncertainty varies strongly with location in relation to both scanning lidars. Hence, high fidelity areas should be defined. Due to the height differences inherent in the technique, measurements at larger scales become increasingly more qualitative.

If the step-stare approach is used, the beam directions are programmed in a way that the measurement point is close to the desired height at all measurement locations. Therefore, the assumptions which have to be made on the vertical wind profile are reduced. The step-stare approach can be used to capture wind speeds in the wake over a large area and in the wind farm with high accuracy. This leads to smaller uncertainties than those in the dual-PPI approach and is especially useful in the wake, since the vertical wind profile does not follow a classical shape and can vary significantly. Tracking wake evolution on the other hand is only possible to pseudo-spatially resolve the flow as the number of measurement points is limited to keep recovery rates (amount of samples in a 10-minute interval) in a reasonable range.

The use of an adaptive control software like ArgusScan can ensure a highly efficient and tailored measurement design and significantly increase the measurement time in the area of interest. This is especially interesting for the step-stare approach, where the measurement points are limited. While only demonstrated here for a simple setup with two scan patterns, more complex strategies with a higher number of scan patterns can be used to fully take advantage of the gained flexibility.

The comparison of the dual-Doppler step-stare measurement with widely used engineering wake models (TurbOPark and Jensen) shows good agreement regarding the trends. However, significant deviations could be observed regarding the magnitude of the deficits. The observed differences between the models and the measurements are in the order of a few percentage points. This underlines the need for high fidelity wind speed measurements to be able to support and validate the current intensive efforts to move towards high fidelity wake modelling. Dual-Doppler measurements are a powerful tool in this context.

Software availability
Information on the software can be found on https://www.iee.fraunhofer.de/argusscan. Licenses for academic purposes can be made available free of charge upon request.

Author contributions
TH and LP developed the concept for the overall structure and content of the study. TH, LP and FJ wrote the manuscript. TH, NG, FJ and AK performed the measurement data analysis. TH and AK performed and analysed the PyWake simulations. TH, LP and FJ developed the for the uncertainty estimation. LM and LP developed the lidar control Software ArgusScan. TH, FJ, DC, JR and LP designed the measurement campaign. TH, FJ, JR and LM were responsible for performing the measurement campaign. LP had a supervisory role for all parts of the analysis. All authors read, reviewed and approved the final manuscript and provided valuable editorial and critical input and feedback for the final document.

Competing interests
The authors declare that they have no conflict of interest.

Acknowledgements
The work for this paper was co-founded by the European Union (via the WindTwin project, Grant Agreement number: 101147377) and the German Federal Ministry for Economic Affairs and Climate Action (via the Window project, funding code: 0324159).

References
Aitken, M. L., Banta, R. M., Pichugina, Y. L., and Lundquist, J. K.: Quantifying Wind Turbine Wake Characteristics from Scanning Remote Sensor Data, Journal of Atmospheric and Oceanic Technology, 31, 765 – 787, https://doi.org/10.1175/JTECH-D-13-00104.1, 2014.

Angelou, N., Mann, J., and Dubreuil-Boisclair, C.: Revealing inflow and wake conditions of a 6 MW floating turbine, Wind Energy Science, 8, 1511–1531, https://doi.org/10.5194/wes-8-1511-2023, 2023.

Banta, R. M., Pichugina, Y. L., Brewer, W. A., Lundquist, J. K., Kelley, N. D., Sandberg, S. P., II, R. J. A., Hardesty, R. M., and Weickmann, A. M.: 3D Volumetric Analysis of Wind Turbine Wake Properties in the Atmosphere Using High-Resolution Doppler Lidar, Journal of Atmospheric and Oceanic Technology, 32, 904 – 914, https://doi.org/10.1175/JTECH-D-14-00078.1, 2015.

Barthelmie, R. J., Hansen, K., Frandsen, S. T., Rathmann, O., Schepers, J. G., Schlez, W., Phillips, J., Rados, K., Zervos, A., Politis, E. S., and Chaviaropoulos, P. K.: Modelling and measuring flow and wind turbine wakes in large wind farms offshore, Wind Energy, 12, 431–444, https://doi.org/10.1002/we.348, 2009.

Cañadillas, B., Beckenbauer, M., Trujillo, J. J., Dörenkämper, M., Foreman, R., Neumann, T., and Lampert, A.: Offshore wind farm cluster wakes as observed by long-range-scanning wind lidar measurements and mesoscale modeling, Wind Energy Science, 7, 1241–1262, https://doi.org/10.5194/wes-7-1241-2022, 2022.

Coimbra, I. L., Mann, J., Palma, J. M. L. M., and Batista, V. T. P.: Exploring dual-lidar mean and turbulence measurements over Perdigão's complex terrain, Atmospheric Measurement Techniques, 18, 287–303, https://doi.org/10.5194/amt-18-287-2025, 2025.

Goit, J. P., Yamaguchi, A., and Ishihara, T.: Measurement and Prediction of Wind Fields at an Offshore Site by Scanning Doppler LiDAR and WRF, Atmosphere, 11, 442, https://doi.org/10.3390/atmos11050442, 2020.

Gramitzky, K., Jäger, F., Hildebrand, T., Gloria, N., Riechert, J., Steger, M., and Pauscher, L.: Alignment calibration and correction for offshore wind measurements using scanning lidars, Journal of Physics: Conference Series, 2767, 042 014, https://doi.org/10.1088/1742-6596/2767/4/042014, 2024.

Hildebrand, T., Thorsen, G. R., Simon, E., Courtney, M., Jäger, F., Callies, D., Riechert, J., Schmitt, C., and Pauscher, L.: The effects of turbine tilt on offshore wind speed measurements with scanning lidar, in preparation.

Hill, M., Calhoun, R., Fernando, H. J. S., Wieser, A., Dörnbrack, A., Weissmann, M., Mayr, G., and Newsom, R.: Coplanar Doppler Lidar Retrieval of Rotors from T-REX, Journal of the Atmospheric Sciences, 67, 713–729, https://doi.org/10.1175/2009JAS3016.1, 2010.

Hung, L.-Y., Santos, P., and Gottschall, J.: Comparing scanning lidar configurations for wake measurements based on the reduction of associated measurement uncertainties, Journal of Physics: Conference Series, 2505, 012 031, https://doi.org/10.1088/1742-6596/2505/1/012031, 2023.

JCGM: Evaluation of Measurement Data: Guide to the Expression of Uncertainty in Measurement; Technical Report; Joint Committee for Guides in Meteorology; Bureau International des Poids et Mesures (BIPM): Sèvres, France, 2008.

Jensen, N. O.: A note on wind generator interaction, URL https://backend.orbit.dtu.dk/ws/portalfiles/portal/55857682/ris_m_2411.pdf, 1983.

Katic, I., Højstrup, J., and Jensen, N.: A Simple Model for Cluster Efficiency, in: EWEC'86. Proceedings. Vol. 1, edited by Palz, W. and Sesto, E., pp. 407–410, A. Raguzzi, european Wind Energy Association Conference and Exhibition, EWEC '86 ; Conference date: 06-10-1986 Through 08-10-1986, 1987.

Liang, Z., Cote, R., and Cariou, J.-P.: Field Study of Turbulence Intensity measurement by Nacelle Mounted Lidar (NML), Journal of Physics: Conference Series, 2265, 022 104, https://doi.org/10.1088/1742-6596/2265/2/022104, 2022.

Mann, J., Angelou, N., Arnqvist, J., Callies, D., Cantero, E., Arroyo, R. C., Courtney, M., Cuxart, J., Dellwik, E., Gottschall, J., Ivanell, S., Kühn, P., Lea, G., Matos, J. C., Palma, J. M. L. M., Pauscher, L., Peña, A., Rodrigo, J. S., Söderberg, S., Vasiljevic, N., and Rodrigues, C. V.: Complex terrain experiments in the New European Wind Atlas, Philosophical Transactions of the Royal Society A: Mathematical, Physical and Engineering Sciences, 375, 20160 101, https://doi.org/10.1098/rsta.2016.0101, 2017.

Nygaard, N. G., Steen, S. T., Poulsen, L., and Pedersen, J. G.: Modelling cluster wakes and wind farm blockage, Journal of Physics: Conference Series, 1618, 062 072, https://doi.org/10.1088/1742-6596/1618/6/062072, 2020.

Nygaard, N. G., Poulsen, L., Svensson, E., and Pedersen, J. G.: Large-scale benchmarking of wake models for offshore wind farms, Journal of Physics: Conference Series, 2265, 022 008, https://doi.org/10.1088/1742-6596/2265/2/022008, 2022.

Pauscher, L., Vasiljevic, N., Callies, D., Lea, G., Mann, J., Klaas, T., Hieronimus, J., Gottschall, J., Schwesig, A., Kühn, M., et al.: An inter-comparison study of multi-and DBS lidar measurements in complex terrain, Remote Sensing, 8, 782, https://doi.org/10.3390/rs8090782, 2016.

Pedersen, J. G., Svensson, E., Poulsen, L., and Nygaard, N. G.: Turbulence Optimized Park model with Gaussian wake profile, Journal of Physics: Conference Series, 2265, 022 063, https://doi.org/10.1088/1742-6596/2265/2/022063, 2022.

Pedersen, M. M., Meyer Forsting, A., van der Laan, P., Riva, R., Alcayaga Romàn, L. A., Criado Risco, J., Friis-Møller, M., Quick, J., Schøler Christiansen, J. P., Valotta Rodrigues, R., Olsen, B. J., and Réthoré, P.-E.: PyWake 2.6.3: An open-source wind farm simulation tool, URL https://gitlab.windenergy.dtu.dk/TOPFARM/PyWake, 2023.

Porté-Agel, F., Bastankhah, M., and Shamsoddin, S.: Wind-Turbine and Wind-Farm Flows: A Review, Boundary-layer meteorology, 174, 1–59, https://doi.org/10.1007/s10546-019-00473-0, 2020.

Rott, A., Schneemann, J., Theuer, F., Trujillo Quintero, J. J., and Kühn, M.: Alignment of scanning lidars in offshore wind farms, Wind Energy Science, 7, 283–297, https://doi.org/10.5194/wes-7-283-2022, 2022.

Schneemann, J., Rott, A., Dörenkämper, M., Steinfeld, G., and Kühn, M.: Cluster wakes impact on a far-distant offshore wind farm's power, Wind Energy Science, 5, 29–49, https://doi.org/10.5194/wes-5-29-2020, 2020.

Shimada, S., Kogaki, T., Konagaya, M., Mito, T., Araki, R., Ueda, Y., and Ohsawa, T.: Validation of near-shore wind measurements using a dual scanning light detection and ranging system, Wind Energy, 25, 1555–1572, https://doi.org/10.1002/we.2757, 2022.

Smalikho, I. N., Banakh, V. A., Pichugina, Y. L., Brewer, W. A., Banta, R. M., Lundquist, J. K., and Kelley, N. D.: Lidar Investigation of Atmosphere Effect on a Wind Turbine Wake, Journal of Atmospheric and Oceanic Technology, 30, 2554 – 2570, https://doi.org/10.1175/JTECH-D-12-00108.1, 2013.

Stawiarski, C., Träumner, K., Knigge, C., and Calhoun, R.: Scopes and Challenges of Dual-Doppler Lidar Wind Measurements—An Error Analysis, Journal of Atmospheric and Oceanic Technology, 30, 2044–2062, https://doi.org/10.1175/JTECH-D-12-00244.1, 2013.

Van Dooren, M. F., Trabucchi, D., and Kühn, M.: A Methodology for the Reconstruction of 2D Horizontal Wind Fields of Wind Turbine Wakes Based on Dual-Doppler Lidar Measurements, Remote Sensing, 8, https://doi.org/10.3390/rs8100809, 2016.

Vasiljević, N., Courtney, M., and Pedersen, A. T.: Uncertainty model for dual-Doppler retrievals of wind speed and wind direction, Atmospheric Measurement Techniques Discussions, 2020, 1–25, https://doi.org/10.5194/amt-2020-321, 2020.

Vasiljević, N., Lea, G., Courtney, M., Cariou, J.-P., Mann, J., and Mikkelsen, T.: Long-Range Wind-Scanner System, Remote Sensing, 8, https://doi.org/10.3390/rs8110896, 2016.

Wildmann, N., Vasiljevic, N., and Gerz, T.: Wind turbine wake measurements with automatically adjusting scanning trajectories in a multi-Doppler lidar setup, Atmospheric Measurement Techniques, 11, 3801–3814, https://doi.org/10.5194/amt-11-3801-2018, 2018.

Wu, S., Liu, B., Liu, J., Zhai, X., Feng, C., Wang, G., Zhang, H., Yin, J., Wang, X., Li, R., and Gallacher, D.: Wind turbine wake visualization and characteristics analysis by Doppler lidar, Opt. Express, 24, A762–A780, https://doi.org/10.1364/OE.24.00A762, 2016.

Journal of Physics: Conference Series **3025** (2025) 012006 doi:10.1088/1742-6596/3025/1/012006

Numerical evaluation of the smearing effect in a spinner-mounted lidar for wind field reconstruction

Esperanza Soto Sagredo, Jennifer M. Rinker, Ásta Hannesdóttir, Michael Courtney

Department of Wind and Energy Systems, Technical University of Denmark, Denmark

E-mail: espa@dtu.dk

Abstract. This study assesses the impact of the smearing effect on full-rotor longitudinal wind speed reconstruction for a spinner-mounted lidar, applying two reconstruction methods: Inverse Distance Weighting (IDW) interpolation and a Least-Squares fit with Proper Orthogonal Decomposition (POD-LSQ). The smearing effect arises as the spinner-mounted lidar beam, constrained to the hub, rotates with the rotor, causing the beam to sample across an arc during the sampling period. This results in a spatial and temporal integration of wind speeds, referred to in this study as the smearing effect. We evaluate a numerical single-beam pulsed lidar mounted on the hub of the DTU 10MW reference wind turbine using the HAWC2 v13.1 lidar sensor to generate synthetic lidar measurements. The lidar data is sampled from a turbulence field within an atmospheric boundary layer simulated using Large Eddy Simulation (LES), featuring a mean wind speed of 16.17 m/s and a turbulence intensity of 8%, over a 10-minute period. Reconstructed fields are analyzed with and without smearing effects across different sampling times. Contrary to our initial hypothesis that smearing would degrade reconstruction accuracy, results show that it improves accuracy by mitigating errors introduced by cross-contamination effects. Smearing reduces the Mean Absolute Error (MAE) in reconstructed wind fields, particularly for longer sampling times, with reductions of up to 2.9% for IDW and 8.8% for POD-LSQ for a sampling time of 0.5 s. Additionally, smearing improves Rotor Average Wind Speed (RAWS) estimation, reducing RAWS MAE by 25% for POD-LSQ—from 0.049 to 0.037 m/s. These findings suggest that, rather than degrading lidar-based wind field reconstruction, smearing serves as a beneficial averaging mechanism that enhances accuracy of the evaluated reconstruction techniques by mitigating errors associated with the assumption that lateral and vertical wind components are negligible in line-of-sight velocity when estimating the longitudinal wind speed for the reconstruction.

1 Introduction

Nacelle-mounted lidars are widely used for wind measurements in wind turbine applications. However, their positioning behind the rotor introduces challenges, as the rotating blades obstruct the lidar's line of sight, resulting in periodic data loss and measurement gaps. This blade blockage effect reduces data availability and increases uncertainty in wind field reconstruction, affecting the accuracy of inflow characterization [1, 2].

A novel alternative, the spinner-mounted multi-beam pulsed lidar (hereafter, hub-lidar) [3, 4], enables spatial measurements across the rotor plane while avoiding blade blockage. Since the lidar beams are fixed to the hub, sampling occurs as the rotor rotates, covering different regions of the wind field. However, this configuration introduces a new challenge: the smearing effect. As the beams sweep along an arc

Content from this work may be used under the terms of the Creative Commons Attribution 4.0 licence. Any further distribution of this work must maintain attribution to the author(s) and the title of the work, journal citation and DOI.
Published under licence by IOP Publishing Ltd

during the sampling time, the measured wind speeds represent an integration over this arc rather than discrete points, combining both the inherent weighting effect of lidar technology and an additional spatial averaging. Consequently, longer sampling times amplify the smearing effect, impacting the accuracy of line-of-sight velocity estimations.

This study evaluates the impact of smearing on wind field reconstruction accuracy by comparing two methods: Inverse Distance Weighting (IDW) interpolation [3] and least-squares fit with Proper Orthogonal Decomposition (POD-LSQ) [4]. Numerical lidar data, both with and without smearing effects, are analyzed under different sampling times to assess their influence on reconstruction accuracy.

These methods rely on precise estimation of the longitudinal wind speed across the rotor plane. Understanding the effects of smearing is therefore essential to determine their limitations and define lidar specifications necessary to achieve the accuracy required for applications such as Lidar-Assisted Control (LAC) and load validation.

We analyze reconstructed fields using a single-beam configuration with sampling times ranging from 0.1-0.5 s. All cases are compared to the reference wind field used in the aeroelastic simulations performed in HAWC2, from which the numerical lidar data were extracted.

This paper is structured as follows: Sect. 2 describes the methodology, including lidar data generation and post-processing, and the two reconstruction methods evaluated on this study. Sect. 3 presents the results for the smearing effect evaluation as a function of sampling time and its impact on wind field reconstruction estimation. A discussion on the study's assumptions and result is provided in Sect. 4, and Sect. 5 summarizes the key findings.

2 Methodology

Numerical hub-lidar data are generated using an aeroelastic simulation in HAWC2 v13.1 [5] for the DTU 10 MW reference wind turbine [6]. A rigid tower and no tilt are assumed to eliminate their influence on lidar measurements, as the sensor inherently accounts for tower motion and tilt in line-of-sight wind speed calculations. The simulation runs for 900 s, discarding the first 250 s to remove transient effects and the last 50 s to ensure the lidar probe volume remains within the turbulence box. Outputs are registered every 0.01 s on HAWC2.

The reference wind field is sampled from an atmospheric boundary layer simulated with Large Eddy Simulation (LES). The selected subsection has a hub-height average wind speed of $\bar{U} = 16.17$ m/s and a turbulence intensity of 8%. A turbulence box is extracted from this field for use in HAWC2.

The turbulence box has dimensions of 1800 × 61 × 138 (t × y × z), with grid resolutions of 5.009 m laterally and 2.190 m vertically, resulting in a total domain size of 300.523 m × 300.021 m (y × z). These dimensions provide a sufficiently large turbulence field for a 900 s simulation of a large rotor while maintaining the spatial and temporal resolution needed for the numerical lidar sensor to capture probe volume effects.

To incorporate the 4D LES field into HAWC2, the longitudinal and time dimensions are combined using Taylor's frozen turbulence hypothesis [7], assuming the inflow advects at a constant speed of $\bar{U} = 16.17$ m/s. Consequently, the spatial resolution along the x-axis is determined as: $dx = (\bar{U}T_{sim})/Nx$, where T_{sim} is the simulation time (900 s), with a time step of $dt = 0.5$ s and $Nx = 1800$.

This wind speed is chosen to evaluate the smearing effect at the rated rotational speed of $\omega_{rated} =$ 9.6 rpm for the DTU 10 MW wind turbine, corresponding to wind speeds above $\bar{U} = 11.4$ m/s.

For the lidar configuration, a single-beam lidar with a half-cone angle of 20° is used. Since the lidar is hub-mounted, it rotates with the hub, and the azimuthal angle is defined by the position of blade 1 in the wind turbine model, as illustrated in Fig. 1. At a rotor speed of $\omega_{rated} = 9.6$ rpm, the single-beam lidar completes a full scan in approximately 6.25 s. The total number of measurements per scan depends on the beam's sampling frequency.

The lidar operates using pulsed technology, enabling simultaneous measurement of multiple ranges. Ranges from 75 m to 500 m are selected at 10 m intervals, resulting in 43 simultaneous range gates.

2.1 Numerical lidar measurements

As the wind turbine rotates, the position of the lidar beam continuously shifts, allowing for spatial sampling across the rotor plane through a combination of rotational motion and multiple range settings. The hub-lidar measures line-of-sight velocities (V_{LOS}) at each scanning location by projecting the longitudinal (u), lateral (v), and vertical (w) velocity components of the turbulent wind field onto the line-of-sight (LOS) coordinate system.

The hub-lidar sensor provides two types of line-of-sight velocity measurements: (i) volume-averaged ($V_{LOS,wgh}$), incorporating a weighting function to account for the probe volume effect, and (ii) nominal

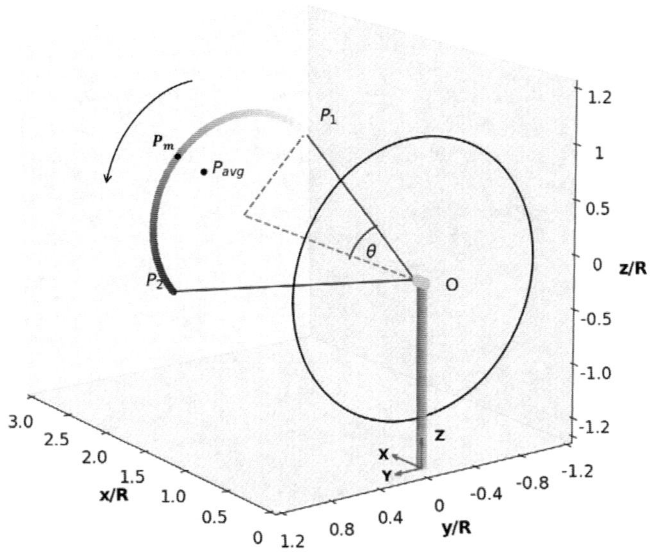

Figure 1: Schematic of a single-beam hub-lidar with a half-cone opening angle θ. The trajectory from P_1 to P_2 represents a 3 s sampling period, with outputs recorded every $dt = 0.01$ s at a rotational speed of $\omega = 9.6$ rpm. Two post-processing cases are considered: (a) **Smeared case**: All available measurements and locations (x, y, z) within the beam's sampling time are averaged, yielding point P_{avg}, representing the smeared measurement over the arc from P_1 to P_2. (b) **No smeared case**: The measurement at the midpoint of the arc, P_m, is selected as the representative value.

($V_{LOS,nom}$), which excludes volume-averaging effects. For details on the hub-lidar sensor implementation and validation, refer to [8].

To obtain V_{LOS} measurements at the desired sampling frequency without switching delays, data are filtered to match the required frequency and smearing conditions. The post-processing follows two approaches:

- **Smeared case:** All available measurements within the sampling period are averaged to determine the wind speed, $V_{LOS,smeared}$, and location (x, y, z). This process, illustrated in Fig. 1, is represented by point P_{avg}. For example, to estimate $V_{LOS,smeared}$ at $t = 250$ s with a sampling frequency of 10 Hz, ten measurements from 250.00 to 250.09 s are averaged (HAWC2 outputs data every 0.01 s), including both V_{LOS} and location (x, y, z). This results in a new measurement point incorporating the smearing effect. Consequently, the wind speed values are averaged, and the measurement point is spatially shifted compared to the no-smearing case.
- **No smeared case:** The line-of-sight wind speed, $V_{LOS,non-smeared}$, is taken directly from the lidar sensor without averaging over the sampling period. Instead, the measurement located at the midpoint of the trajectory is selected. For example, for a 10 Hz sampling frequency, considering measurements from 250.00 to 250.09 s, the value at 250.05 s is chosen, while the remaining measurements are discarded.

Once $V_{LOS,smeared}$ and $V_{LOS,non-smeared}$ are computed for different sampling frequencies, the longitudinal wind speed is estimated by scaling the line-of-sight velocity using:

$$u_{lidar} = \frac{V_{LOS}}{\cos(\theta)}, \tag{1}$$

where $\theta = 20°$ is the lidar beam's half-cone opening angle. This procedure is applied to both cases, yielding $u_{lidar,smeared}$ and $u_{lidar,non-smeared}$ for the smeared and non-smeared case respectively. This approach neglects the contributions of the v- and w-components to the LOS velocity, an effect hereafter referred to as cross-contamination.

Figure 2 provides a schematic overview of the hub-lidar data processing methodology. Subplot (a) illustrates the output measurements from HAWC2 at a 0.01 s resolution. For wind field reconstruction, hub-lidar measurements ($u_{lidar,smeared}$, $u_{lidar,non-smeared}$) are selected within a specified longitudinal range, corresponding to the time for the beam to complete a full rotor scan ($t_{fullscan}$ = 6.25 s). This introduces a spatial filtering effect, defined as $d_{span} = \bar{U} \times t_{fullscan}$, as shown in Fig. 2b, where x_{target} represents the plane where the reconstruction is performed, located at the center of d_{span}. The selected measurements are then mapped onto a fixed YZ grid, serving as input for the reconstruction methods (Fig. 2c). This process is repeated at each time step to generate reconstructed wind fields for the longitudinal component using two methods: IDW interpolation and POD-LSQ. Brief descriptions of these methods are provided in Sects. 2.2 and 2.3.

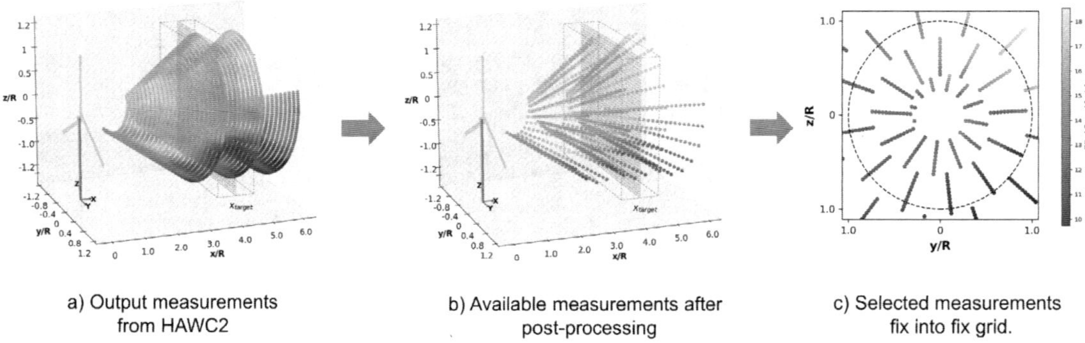

Figure 2: Schematic of 15 s hub-lidar data from the aeroelastic simulation in HAWC2: (a) Shows all available data recorded every 0.01 s over 15 s, (b) presents the data after post-processing for a sampling time of t_s = 0.5 s. The reconstruction at a target distance $x_{target} = t_{desire} \times \bar{U}$ is performed using the available measurements inside the cuboid, which represents the spatial filtering applied due to the elapsed time for a full scan (6.25 s). Finally, (c) shows the processed data mapped onto a fixed YZ grid for reconstruction.

2.2 Inverse Distance Weighting interpolation

Inverse Distance Weighting (IDW) interpolation estimates values at unmeasured locations using known values from nearby points, assigning greater influence to closer points while diminishing the impact of more distant ones. The longitudinal wind speed at each grid point in the YZ plane is computed as:

$$u(\mathbf{x}) = \frac{\sum_{i=1}^{n} w_i(\mathbf{x}) \cdot u_i}{\sum_{i=1}^{n} w_i(\mathbf{x})}, \tag{2}$$

where $w_i(\mathbf{x})$ is the weight function, defined as:

$$w_i(\mathbf{x}) = \frac{1}{d(x, x_i)^p}. \tag{3}$$

Here, $d(x, x_i)$ represents the distance between the target location $\mathbf{x}$, where the wind speed is to be estimated, and the location of the measurement point $\mathbf{x_i}$. The parameter p is a positive exponent controlling the rate at which the weight decreases with distance. A higher p results in a steeper decay, and in this study, we set $p = 3$ as it provided the best performance. If the target location coincides with a measurement point ($\mathbf{x} = \mathbf{x_i}$), then $u(\mathbf{x}) = u_i$.

To reconstruct the wind field using IDW from the available measurements (Sect. 2.1), we first remove the known vertical wind speed profile from the LES inflow, u_{shear}, before applying the interpolation. The IDW method is then used to estimate the fluctuations across the YZ plane. Once the interpolation is completed, the known vertical wind speed profile, u_{shear}, is reintroduced, yielding the full reconstructed wind field across the rotor plane. This process is repeated at each time step of the turbulence box, corresponding to $dt = 0.5$ s.

Journal of Physics: Conference Series **3025** (2025) 012006 doi:10.1088/1742-6596/3025/1/012006

2.3 Least-square fit with Proper Orthogonal Decomposition

Proper Orthogonal Decomposition (POD) is a method that breaks down turbulent flow into spatially varying modes, where the orthogonal components efficiently represent a large fraction of the flow's fluctuating kinetic energy [9, 10].

POD can be applied locally, generating a basis from a specific flow case within the parameter space, or globally, integrating multiple datasets to construct a basis that represents flows across a broader range of conditions [11, 12].

For a turbulence box of dimensions (Nx, Ny, Nz), the matrix $\mathbf{u}$ has dimensions $m \times n$, where $m = Nx$ and $n = Ny \times Nz$. Each flow field $\mathbf{u_c}$ (with c representing the number of LES inflow conditions used to estimate the global basis, ranging from 1 to 4, corresponding to the four different wind speeds, each with a duration of four hours) is first normalized by their reference wind speed, $\bar{U}_c$, and decomposed into its temporal mean and fluctuations: $\mathbf{u_c} = \hat{\mathbf{u}} + \mathbf{u'_c}$.

The fluctuations $\mathbf{u'_c}$ are arranged as row vectors in matrix $\mathbf{M}(m \times n)$, to which POD is applied. The auto-covariance matrix is computed as:

$$\mathbf{R} = \frac{1}{(n-1)}\mathbf{M}^T\mathbf{M}. \tag{4}$$

Solving the eigenvalue problem $\mathbf{RG} = \mathbf{G\Lambda}$ yields the diagonal matrix $\mathbf{\Lambda}$ containing real, positive eigenvalues and the matrix $\mathbf{G} = [g_1, ..., g_{n-1}]$ containing orthonormal eigenvectors, which represent the global spatial POD modes [11]. The modes are arranged in descending order based on variance, reflecting each mode's contribution to turbulent kinetic energy.

A reduced-order representation of the flow field is constructed as a sum of K POD modes and their corresponding temporal series $\phi_i(t)$,

$$\mathbf{u(t)} = \hat{\mathbf{u}} + \sum_{\mathbf{i=1}}^{\mathbf{K}} \mathbf{g_i}\phi_i(t) \tag{5}$$

where, in standard POD, the temporal series $\phi_i(t)$ for a given flow case are derived by projecting the fluctuating flow, $\mathbf{u}'(t)$, onto the global POD modes, $\mathbf{g_i}$, using a standard inner product:

$$\phi_i(t) =< \mathbf{u}'(t), \mathbf{g_i} > \tag{6}$$

To reconstruct the wind field, we use the available scaled lidar measurements, u_{lidar}, after removing the known LES vertical wind speed profile, u_{shear}. This step eliminates large-scale trend effects, ensuring a more accurate fit of the POD modes.

The fluctuations u'_{lidar} are stored in the matrix $M_{\text{lidar}} \in \mathbb{R}^{N_y \times N_z}$, which maps the data onto a fixed YZ grid (Fig. 2c). Grid points without available data are assigned `NaN` values in `Python` [13].

We define the index set $\mathcal{I} = \{i_1, i_2, \ldots, i_{n_{\text{lidar}}}\}$, which identifies the spatial grid positions where valid lidar measurements exist (i.e., where entries in M_{lidar} are not `NaN`). Here, n_{lidar} denotes the total number of valid measurements.

The fluctuation vector used for reconstruction is then defined as $\mathbf{D}_{\text{lidar}} = \text{vec}(M_{\text{lidar}})_{\mathcal{I}} \in \mathbb{R}^{n_{\text{lidar}}}$, where $\text{vec}(M_{\text{lidar}})$ denotes the column-wise vectorization of the matrix, and the subscript $\mathcal{I}$ selects the non-`NaN` entries. As a result, $\mathbf{D}_{\text{lidar}}$ contains only the available data u'_{lidar} in the filtered space (Sect. 2.1).

To match the locations of the available measurements, we sub-sample the global POD modes at the corresponding positions in the YZ plane. We select the first $K = 50$ modes for reconstruction, forming the matrix $\mathbf{A} = \mathbf{G}_{\mathcal{I},1:K} \in \mathbb{R}^{n_{\text{lidar}} \times K}$, where $\mathbf{G}$ contains the full set of global POD spatial modes defined over the grid.

The modal amplitudes $\mathbf{C}$ are estimated using the Moore–Penrose pseudo-inverse [14, 15] by solving a least-squares problem. This is done by projecting $\mathbf{D}_{\text{lidar}}$ onto the column space of $\mathbf{A}$ using its pseudo-inverse $\mathbf{A}^+$, as follows:

$$\mathbf{C} = (\mathbf{A}^\top\mathbf{A})^{-1}\mathbf{A}^\top\mathbf{D}_{\text{lidar}} \tag{7}$$

The reconstructed field is then computed by combining the first $K = 50$ global POD modes with the estimated modal amplitudes:

$$\mathbf{u(t)} = \bar{\mathbf{u}}_{\mathbf{shear}} + \sum_{\mathbf{i=1}}^{\mathbf{K}} \mathbf{g_i C(t)} \tag{8}$$

where $\bar{u}_{shear}$ is the known shear from the LES dataset. This method provides a low-rank approximation of the wind field based on the selected modes and estimated modal amplitudes. The reconstruction is applied at each time step ($dt = 0.5$ s) to generate the longitudinal turbulence field.

3 Results

To evaluate the impact of the smearing effect on reconstructed wind fields, we computed the Mean Absolute Error (MAE) by comparing the reconstructed field to the reference field at each time step, as:

$$MAE = \frac{1}{N_t}\sum_{i=1}^{N_t}\left[\frac{1}{N}\sum_{(y,z)\in A}\left|u_{reconstructed}(t,y,z) - u_{reference}(t,y,z)\right|\right], \quad (9)$$

where $N_t = 1200$ denotes the number of time steps over a 10-minute period, and N represents the total number of grid points on the YZ plane within the rotor area.

To perform the evaluation, we define three representations of the longitudinal wind speed: (1) the true longitudinal wind speed, $u_{\text{lidar, fw}}$, which corresponds to the actual u-component of the wind on the YZ plane at the lidar measurement locations; (2) the weighted longitudinal line-of-sight wind speed, $u_{\text{lidar, wgh}}$, obtained by scaling the volume-averaged LOS velocity, $V_{\text{LOS, wgh}}$, and (3) the nominal line-of-sight wind speed, $u_{\text{lidar, nom}}$, computed by scaling $V_{\text{LOS, nom}}$ without accounting for volume-averaging effects (Sect. 2.1).

Figure 3 compares the average MAE of wind field reconstructions using the IDW and POD-LSQ methods for the three wind speed representations, both with (solid lines) and without (dashed lines) smearing effects, across sampling times ranging from 0.1 s to 0.5 s. Subplot (a) shows the average number of available lidar measurements per time step as a function of sampling time. As the sampling time increases, the number of available measurements decreases significantly due to less frequent data acquisition, reducing the number of usable data points within each time window. This trend is further illustrated in Fig. 4, which displays hub-lidar measurement snapshots for sampling times of $t_s = 0.1$ s (a) and $t_s = 0.5$ s (b), both at time step 450 s. Subplots (b) and (c) of Fig. 3 show the average MAE over the rotor plane for IDW and POD-LSQ reconstructions, respectively, over the full 10-minute period and using the different wind speed representations.

When comparing the MAE using $u_{\text{lidar, wgh}}$ and $u_{\text{lidar, nom}}$ for reconstruction, we find that—contrary to our initial hypothesis—the smearing case (solid line) consistently yields lower errors across all sampling times compared to the non-smearing case (dashed line). In contrast, when using the true longitudinal wind speed ($u_{\text{lidar, fw}}$), the MAE curves for smearing and non-smearing cases overlap, indicating that smearing has almost no effect on the estimation with the true u-component. This observation suggests that smearing partially averages out the contributions from the lateral (v) and vertical (w) wind components, thereby mitigating "cross-contamination" effects that arise when estimating the longitudinal wind speed ($u_{\text{lidar, wgh}}$ and $u_{\text{lidar, nom}}$) under the assumption that v and w are negligible. As expected, the MAE increases with sampling time due to the reduced number of available measurements for reconstruction. Notably, the beneficial effect of smearing becomes more pronounced at longer sampling times. A summary of the average MAE values for both reconstruction method, and sampling time is presented in Table 1.

Comparing the smearing effect across both reconstruction methods, we find that POD-LSQ benefits more from error reduction than IDW, as it is more sensitive to cross-contamination effects and to the number of available measurements for accurate reconstruction [4]. Without smearing, reconstruction errors in POD-LSQ increase more rapidly than in IDW as the sampling time increases (Table 1). However, when smearing is applied, the growth in error is significantly mitigated—particularly for POD-LSQ—highlighting its greater susceptibility to both measurement scarcity and cross-contamination. The results also indicate that longer beam sampling times negatively affect reconstruction accuracy due to the reduced number of available lidar measurements. As shown in Fig. 3a, the number of available measurements is substantially lower at $t_s = 0.5$ s compared to $t_s = 0.1$ s, which limits the amount of input data for reconstruction and results in increased error.

The impact of sampling time and smearing on both measurement availability and reconstruction accuracy is further illustrated in Fig. 4 for the POD-LSQ method at time step $t = 434$ s. Subplot (a) shows the reconstruction with smearing at a sampling time of $t_s = 0.1$ s, while subplots (b) and (c) correspond to the cases with and without smearing, respectively, at $t_s = 0.5$ s. Each column, from left to right, presents: the available hub-lidar measurements used for reconstruction, the reference and reconstructed wind fields, and the difference between them across the YZ plane, including the MAE for the given time step. As shown in subplots (a) and (b), increasing the sampling time from 0.1 s to 0.5 s results in a 34% increase in MAE, primarily due to the reduced number of available measurements used to estimate the modal amplitudes. A comparison between subplots (b) and (c) demonstrates that smearing reduces reconstruction error, particularly along the rotor plane edges, where the discrepancy between the reference and reconstructed fields is visibly lower. This improvement is attributed to the temporal averaging effect introduced by smearing, which helps mitigate cross-contamination errors. It is

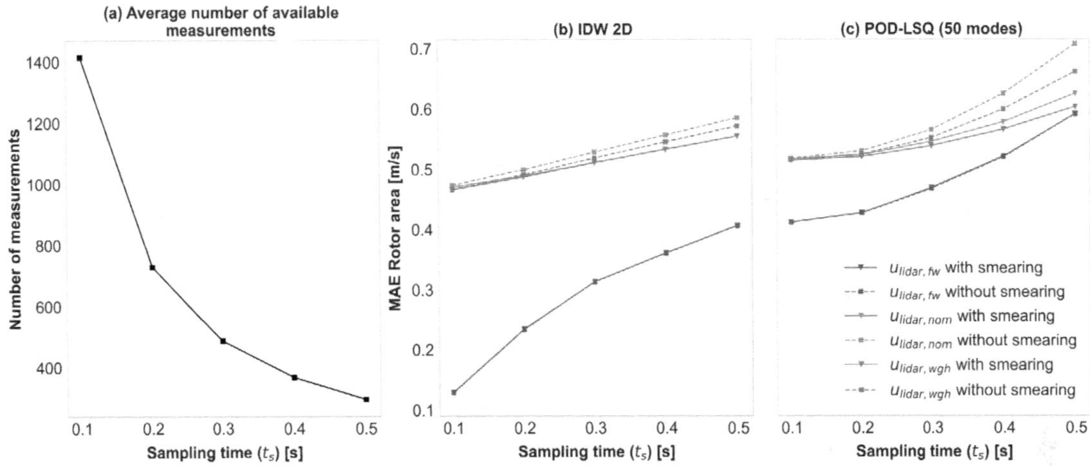

Figure 3: Evaluation of the smearing effect on wind field reconstruction using IDW and POD-LSQ methods for different wind speed representations: true longitudinal wind speed ($u_{\text{lidar, fw}}$), nominal ($u_{\text{lidar, nom}}$), and weighted ($u_{\text{lidar, wgh}}$) scaled line-of-sight lidar measurements. From left to right: (a) Average number of available measurements per time step as a function of sampling time, (b) and (c) average Mean Absolute Error (MAE) across the rotor plane for IDW and POD-LSQ reconstructions, respectively, compared to the reference LES inflow. Solid and dashed lines indicate results with and without smearing, respectively. Smearing is shown to reduce reconstruction errors, particularly for the POD-LSQ method.

also important to note that the location of available measurements may vary slightly due to the post-processing steps described in Sect. 2.1, which may influence reconstruction accuracy, albeit to a lesser extent.

To illustrate the impact of smearing on Rotor Average Wind Speed (RAWS) estimation from POD-LSQ-reconstructed wind fields, Fig. 5a presents the RAWS time series over a 10-minute period. The LES reference inflow is shown as a solid black line, while the smeared and non-smeared cases—both reconstructed using a sampling time of $t_s = 0.5$ s—are represented by dashed blue and dashed orange lines, respectively. The RAWS is computed at each time step by averaging the wind speed values across all grid points within the rotor plane. Figure 5b shows the time series of RAWS differences between the reconstructed cases and the LES reference, with seven representative time steps selected for detailed analysis in rows (c) to (f). Subplots (c) and (d) display the hub-lidar measurements used for reconstruction in the smeared and non-smeared cases, respectively. Although the number of available measurements is the same in both cases, the measured wind speeds, $u_{\text{lidar, smeared}}$ and $u_{\text{lidar, non-smeared}}$, differ due to the post-processing procedure described in Sect. 2.1. Notably, both representations are based on the weighted

Table 1: Average Mean Absolute Error (MAE) across the rotor area over a 10-minute period, showing the effect of sampling time and smearing on reconstruction accuracy. Results are presented for both IDW and POD-LSQ methods using $u_{\text{lidar, wgh}}$ as lidar measurement.

t_s [s]	MAE Smearing [m/s]		MAE No Smearing [m/s]		Error increase due to smearing		Error increase due to t_s (Smearing case)		Error increase due to t_s (No smearing case)	
	IDW	POD-LSQ	IDW	POD-LSQ	IDW	POD-LSQ	IDW	POD-LSQ	IDW	POD-LSQ
0.1	0.466	0.515	0.466	0.515	-0.05%	-0.07%	-	-	-	-
0.2	0.487	0.521	0.491	0.525	-0.8%	-0.8%	4.5%	1.2%	5.3%	1.9%
0.3	0.511	0.539	0.519	0.552	-1.5%	-2.4%	9.7%	4.6%	11.3%	7.2%
0.4	0.533	0.566	0.545	0.599	-2.2%	-5.6%	14.5%	9.9%	17.0%	16.3%
0.5	0.555	0.603	0.572	0.661	-2.9%	-8.8%	19.1%	17.2%	22.6%	28.4%

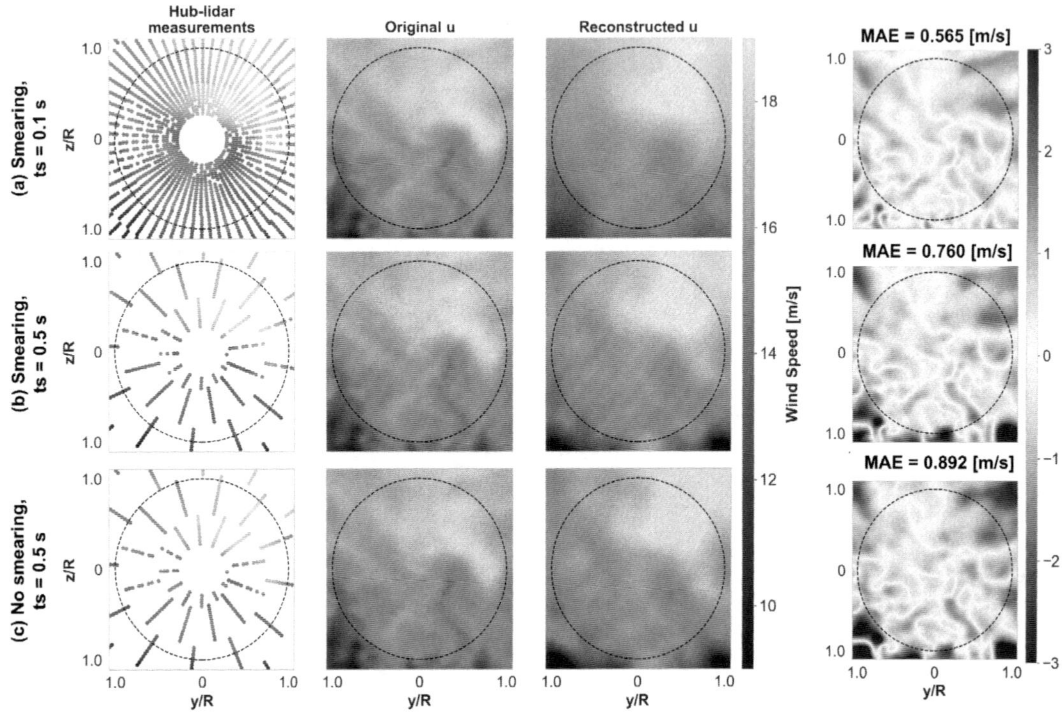

Figure 4: Wind field reconstruction using POD-LSQ at time step 434 s. Each column, from left to right, shows the available hub-lidar measurements used for reconstruction, the reference and reconstructed wind fields, and their differences across the YZ plane. Subplots illustrate the effect of sampling time and smearing: (a) $t_s = 0.1$ s with smearing, (b) $t_s = 0.5$ s with smearing, and (c) $t_s = 0.5$ s without smearing.

longitudinal wind speed, $u_{\text{lidar, wgh}}$, which accounts for the probe volume effect during reconstruction. Subplots (e) and (f) illustrate the spatial differences between the POD-LSQ-reconstructed wind fields and the LES reference for the selected time steps, with and without smearing, respectively. The MAE over the rotor plane is also indicated for each case.

The results indicate that smearing, by averaging multiple measurements over the sampling period, reduces discrepancies in the reconstructed wind fields—particularly along the rotor plane edges, where differences become less pronounced. For instance, at $t_6 = 452.5$ s, the smeared case shows smaller discrepancies on the left side of the plane (Fig. 5e) compared to the non-smeared case (Fig. 5f), reducing the MAE across the rotor plane from 0.540 to 0.425 m/s. This improvement is primarily attributed to the averaging process, which lowers the wind speed values of $u_{\text{lidar, wgh}}$ in that region (Fig. 5c). By integrating more data across the sampling interval, smearing provides a more representative estimate of u_{lidar} for reconstruction, effectively mitigating cross-contamination effects. Although the differences between $u_{\text{lidar, smeared}}$ (Fig. 5c) and $u_{\text{lidar, non-smeared}}$ (Fig. 5d) appear small, they consistently contribute to lower MAE at each time step, as shown in Fig. 5e–f. Overall, the RAWS estimation error over the 10-minute period is reduced by 25% when smearing is applied. These findings suggest that averaging over the sampling period helps reduce errors introduced by assumptions such as neglecting the v and w components during line-of-sight scaling. Furthermore, integrating multiple measurements allows each data point to reflect more representative flow characteristics of a given region, rather than relying on a single instantaneous observation.

Journal of Physics: Conference Series **3025** (2025) 012006 doi:10.1088/1742-6596/3025/1/012006

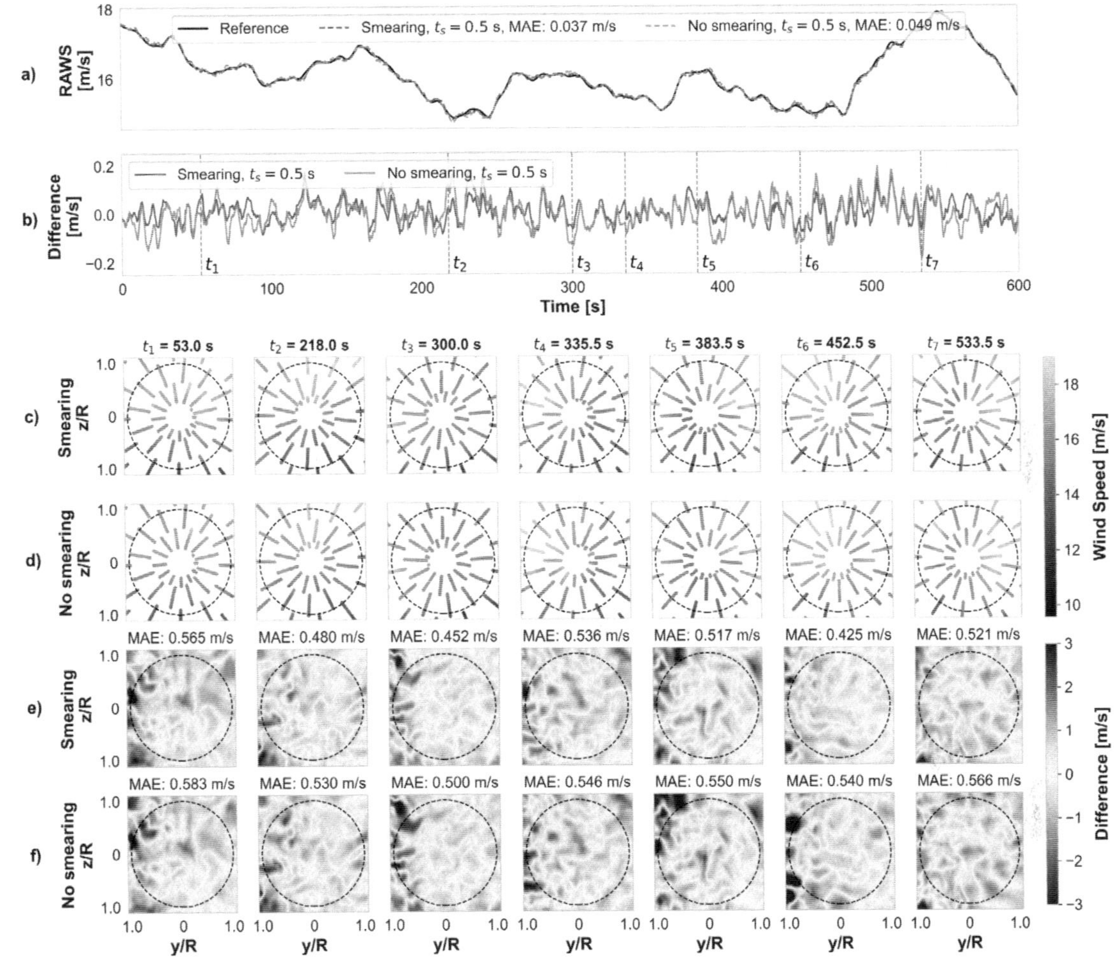

Figure 5: RAWS evaluation over a 10-minute period for POD-LSQ reconstructions with and without smearing, using a sampling time of 0.5 s. (a) RAWS time series for the reference (black), smeared (blue), and non-smeared (orange) cases. (b) RAWS differences relative to the reference; dashed gray lines indicate selected time steps shown in (c)–(f). (c) and (d) Available hub-lidar measurements in the YZ plane for smeared and non-smeared cases at selected time steps. Both have the same number of points, but values and locations differ due to post-processing (Sect. 2). (e) and (f) Wind speed differences between reconstructed and reference fields across the YZ plane, including MAE for each case.

4 Discussion

This study demonstrates that smearing significantly influences wind field reconstruction using hub-mounted lidar data for both POD-LSQ and IDW methods, primarily by mitigating cross-contamination errors in the estimation of longitudinal wind speed. However, several limitations should be considered:

- **Smearing and sampling time dependencies:** Smearing and sampling time are critical factors influencing the feasibility of lidar-based reconstruction methods. In practice, smearing arises from a combination of volume averaging along the line-of-sight and the beam's motion over an arc during the sampling period, resulting in a three-dimensional (3D) averaging process. This study simplified the smearing effect by averaging all available measurements within each sampling period, while retaining volume averaging at the individual measurement level. Real lidar systems, however, operate under more complex conditions where these averaging and weighting processes interact dynamically

- **Lidar configuration effects:** A single-beam lidar configuration was used in this study to isolate the impact of smearing. In contrast, multi-beam configurations introduce additional complexities such as switching delays between beams, which reduce the number of measurements per time step. These factors may affect the observed benefits of smearing. Future work should assess the influence of smearing in multi-beam lidar setups, where beam sequencing and data availability vary dynamically.
- **Impact of half-cone angle:** Since smearing reduces reconstruction errors by mitigating cross-contamination—introduced by the assumption that $v = w = 0$ in line-of-sight scaling—it is important to investigate its effect at higher half-cone angles (θ). Increasing θ amplifies cross-contamination effects while decreasing the number of usable measurements, as more points fall outside the rotor area. Understanding how smearing interacts with these factors is crucial to assessing its benefits.
- **Influence of tower motion:** The present analysis is based on a wind turbine model with rigid tower, thereby neglecting errors introduced by tower motion into the mounted-lidar measurements. The influence of smearing under realistic conditions, including relative tower velocity, remains an open question. Future studies should address how smearing interacts with structural motion and its effect on lidar measurements.
- **Turbulence intensity:** This study considered a single inflow case with a turbulence intensity (TI) of 8%. The accuracy of reconstruction methods is expected to vary with TI, as higher TI levels introduce greater variability in wind conditions and may exacerbate discrepancies in reconstruction. The influence of smearing under different turbulence intensities remains unexplored and warrants further investigation to assess its general applicability in varying atmospheric conditions.

Future research should focus on extending the evaluation to multi-beam lidar systems with realistic operational constraints, exploring a wider range of half-cone angles, and quantifying the effectiveness of smearing across different turbulence intensities and turbine operating conditions.

5 Conclusions

This study evaluated the impact of the smearing effect and sampling time on wind field reconstruction using a hub-mounted single-beam pulsed lidar. Two reconstruction techniques were considered: Inverse Distance Weighting (IDW) interpolation and a Least-Squares fit with Proper Orthogonal Decomposition (POD-LSQ). By comparing reconstructed wind fields with and without smearing across various sampling frequencies, we demonstrated that smearing improves reconstruction accuracy by mitigating errors associated with cross-contamination and line-of-sight scaling.

Averaging multiple measurements along the lidar beam's trajectory reduced the Mean Absolute Error (MAE), particularly at longer sampling times ($t_s = 0.5$ s), with error reductions of up to 2.9% for IDW and 8.8% for POD-LSQ (Table 1). The effectiveness of smearing increases with sampling time, as temporal averaging incorporates additional flow information.

The results also highlight the dominant influence of sampling frequency on reconstruction accuracy. As the sampling time increases, the number of available measurements decreases, leading to higher reconstruction errors. However, smearing mitigates this effect by smoothing discrepancies and improving wind speed estimates for both reconstruction methods (Fig. 3). These improvements are particularly evident in critical regions of the rotor area, where localized errors are significantly reduced (Fig. 4b and Fig. 4c).

Improved reconstruction accuracy also leads to better Rotor Average Wind Speed (RAWS) estimation. For POD-LSQ at $t_s = 0.5$ s, the RAWS MAE decreased by 25%, from 0.049 m/s to 0.037 m/s (Fig. 5a), further demonstrating the benefits of smearing.

Overall, the findings suggest that the smearing effect, rather than degrading lidar-based wind field reconstruction, serves as a beneficial averaging mechanism. It enhances accuracy by mitigating errors related to the estimation of longitudinal wind speed from line-of-sight measurements, particularly where the v and w components are neglected. These benefits become more pronounced as the lidar beam's sampling time increases, underscoring the importance of optimizing temporal resolution in wind field reconstruction.

Acknowledgments
This work is part of the CONTINUE project, which has received funding from the Danish Energy Technology Development and Demonstration Programme (EUDP), under grant agreement 640222-496980.

References

[1] Schlipf, D., et al. (2018). IEA Wind Task 32: Best Practices for the Certification of Lidar-Assisted Control Applications. https://doi.org/10.1088/1742-6596/1102/1/012010

[2] Angelou, N., & Sjöholm, M. (2022). Data Reliability Enhancement for Wind-Turbine-Mounted Lidars. Remote Sensing, 14(13). https://doi.org/10.3390/rs14133225

[3] E. Soto Sagredo, Á. Hannesdóttir, J.M. Rinker and M. Courtney (2024). Reconstructing turbulent wind-fields using inverse-distance-weighting interpolation and measurements from a pulsed mounted-hub lidar, Journal of Physics: Conference Series, Volume 2745, WindEurope Annual Event 2024 20/03/2024 - 22/03/2024 Bilbao, Spain. https://doi.org/10.1088/1742-6596/2745/1/012017

[4] E. Soto Sagredo, J.M. Rinker, S.J. Andersen and J.P. Forrest (2024). Wind-field characterization using synthetic lidar measurements and proper orthogonal decomposition, Journal of Physics: Conference Series, Volume 2767, Modeling and simulation technology. https://doi.org/10.1088/1742-6596/2767/5/052061

[5] DTU (2019). HAWC2 Horizontal Axis Wind turbine simulation Code 2nd generation, Official Webpage. http://www.hawc2.dk/hawc2-info/.

[6] Bak, C., Zahle, F., Bitsche, R., Kim, T., Yde, A., Henriksen, L. C., Hansen, M. H., Blasques, J. P. A. A., Gaunaa, M., & Natarajan, A. (2013). The DTU 10-MW Reference Wind Turbine. Sound/Visual production (digital)

[7] Taylor G.I. (1938). The spectrum of turbulence, Proceedings of the Royal Society of London. Series A-Mathematical and Physical Sciences 164 (919) 476-490. https://doi.org/10.1098/rspa.1938.0032

[8] Soto Sagredo, E., Rinker J.M and Lund, R.S. (2023). Verification of numerical lidars in HAWC2: Analysis of nacelle- and hub-mounted lidars, Technical report, DTU Wind Energy, Roskilde, Denmark. https://doi.org/10.11581/DTU.00000302

[9] Lumley, J. L.: The Structure of Inhomogeneous Turbulence, in: Atmospheric Turbulence and Wave Propagation, pp. 166–178, Nauka, Moscow, ISBN 9783937655239, 1967.

[10] Berkooz, G., Holmes, P., and Lumley, J. L.: The proper orthogonal decomposition in the analysis of turbulent flows, Annual Review of Fluid Mechanics, 25, 539–575, https://doi.org/10.1146/annurev.fl.25.010193.002543, 1993.

[11] Andersen, S. J., & Murcia Leon, J. P. (2022). Predictive and stochastic reduced-order modeling of wind turbine wake dynamics. Wind Energy Science, 7(5), 2117–2133. https://doi.org/10.5194/wes-7-2117-2022

[12] Céspedes Moreno, J. F., Murcia León, J. P., and Andersen, S. J.: Convergence and efficiency of global bases using proper orthogonal decomposition for capturing wind turbine wake aerodynamics, Wind Energ. Sci., 10, 597–611, https://doi.org/10.5194/wes-10-597-2025, 2025

Céspedes Moreno, J. F., Murcia León, J. P., and Andersen, S. J.: Convergence and Efficiency of Global Bases Using Proper Orthogonal Decomposition for Capturing Wind Turbine Wake Aerodynamics (Preprint), Wind Energy Science, https://doi.org/10.5194/wes-2024-81, 2024

[13] Python Software Foundation. `Python`, Version 3.10, 2023. Available at: `https://www.python.org`.

[14] Moore, H., (1920). On the reciprocal of the general algebraic matrix. Bulletin of the American Mathematical Society 26, 394–395.

[15] Penrose, R., (1955). A generalized inverse for matrices, Proceedings of the Cambridge Philosophical Society 51, 406–413.

doi:10.1088/1742-6596/3025/1/012007

EDiGaPS: A novel machine learning-based approach to model the influence of turbulence intensity on wind turbine power curves

Fernando Saraiva Filho[1*], Nathianne Andrade[1], Rodrigo Queiroz[1], Guido Santos[1], Samuel Lima[1] and Gustavo Carvalho[1]

[1] Delfos Energy, Barcelona, Spain

[*]E-mail: fernando.filho@delfos.energy

Abstract. Wind turbines are installed in various wind climates, each with unique conditions, but they are often developed and tested in locations with different wind characteristics. This creates a need for normalization across different wind conditions, including factors like air density and turbulence intensity. Turbulence intensity, in particular, impacts wind turbine power output in multiple ways. Some effects are physical, such as the aerodynamic and inertia-related forces on the turbine, but another effect is the bias induced by averaging measured power output and wind speed over 10-minute intervals. Annex M of the IEC 61400 12-1 standard offers a complex methodology that only addresses the correction for the 10-minute averaging bias, covering allegedly 50% of the total turbulence effect. This paper introduces the Equivalent Difference in Gaussian Process Surface (EDiGaPS), a novel machine-learning approach for correcting turbulence effects on wind turbine power curves using data from the SCADA system of wind turbines. The corrections provided by EDiGaPS were compared with the ones obtained from the methodology proposed by the IEC standard. On average, the new procedure corrected 60.2% of the total turbulence influence in the tests performed, which is good evidence that it can handle not only the effect of 10-minute time-averaging, but also other possible effects of turbulence. Furthermore, implementing and comparing both methods, we noticed an already mentioned effect in the literature: the normalization method tends to cause an overcompensation that leads to too high AEPs; the novel proposed approach does not present such an effect. To the best of our knowledge, our work represents an important contribution, due to the fact that, although the normalization method has existed for some years, there are very few works in the literature addressing the topic, possibly due to the limited adoption by the community.

Content from this work may be used under the terms of the Creative Commons Attribution 4.0 licence. Any further distribution of this work must maintain attribution to the author(s) and the title of the work, journal citation and DOI.
Published under licence by IOP Publishing Ltd

1. Introduction

Wind turbines are deployed in diverse wind climates, each with unique conditions. However, they are often developed and tested in locations with different or limited wind characteristics [1]. The IEC Standard 61400-12-2 provides methods for standardizing turbine performance across varying environments, including density and turbulence normalization [2]. Yet, only density normalization is widely adopted due to the complexities of implementing turbulence normalization [3]. Additionally, the proposed method for turbulence normalization, which adjusts only the effect of 10-minute power curve averaging, overlooks other factors on the power curve, such as: the aerodynamic and inertia effects of the wind turbine [4].

The literature provides examples of the application of the methodology described in Annex M of the IEC standard [2]. Bardal and Sætran [4] employed measurements from a Lidar and a wind turbine in order to investigate the impact of turbulence intensity on wind turbine power curves and to evaluate a turbulence normalization method proposed in the draft of the IEC 61400-12-1 standard. Using experimental data collected over a 10-month period at a test site on the Norwegian coast, variations in annual energy production (AEP) under different turbulence conditions were analyzed. The study found that turbulence significantly influences the power curve and that applying the normalization method reduces the estimated AEP difference by 50% compared to traditional power curves. However, the method does not directly account for the aerodynamic effects of turbulence on the turbine rotor.

Clifton and Wagner [1] compares three methods for modeling wind turbine performance while accounting for the influence of turbulence conducted experiments using simulation data: (1) the traditional binning method, (2) turbulence normalization, and (3) a machine learning model based on random forests. Using simulations of a 1.5 MW turbine, the results indicate that the binning method fails to capture the effects of turbulence, whereas turbulence normalization improves accuracy but tends to either overestimate, when adjusting from high to low turbulence, or underestimate, when adjusting from low to high turbulence, power output depending on the conditions. The machine learning model demonstrated the highest predictive capability, adapting to various atmospheric variables and reducing errors compared to the other methods. and showed that the corrections tended to overcompensate for turbulence effects.

Tissot et al. [3] proposed an implementation of turbulence normalization that is well-suited for large fleets, integrating the zero-turbulence power curve with a Gaussian wind speed distribution after air density normalization. The study examines turbulence intensity effects on wind turbine power curve assessments using SCADA-based nacelle anemometry instead of met mast or lidar data. The method reduced power standard deviation by 6.9% for a single turbine and 5% fleet-wide, decreasing data scatter and uncertainty.

Given the primary limitation identified in the existing literature— the compensation of only the 10-minute averaging bias and the application on operational data — this paper seeks to propose a novel turbulence correction method and test it on operational data. To account for a broader range of turbulence effects, the approach will employ machine learning (ML) techniques to model the turbulence-related phenomena. Furthermore, to ensure its applicability to large fleet operations, the method will exclusively utilize data from Supervisory Control and Data Acquisition (SCADA) systems. Lastly, the performance of this new method will be evaluated in comparison to the approach outlined in the IEC standard [2], using SCADA system data from a 1.8 MW wind turbine.

2. Materials and Methods

2.1 Turbulence normalization

The procedure proposed in the IEC standard addresses the effects of averaging measured power output and wind speed over 10-minute intervals. The method is based on the application of the following equation:

$$\overline{P_{I_{ref}}(v)} = \overline{P(v)} + (\overline{P_{sim,I_{ref}}(v)} - \overline{P_{sim,I}(v)}) \tag{1}$$

where the normalized power output $P_{I_{ref}}(v)$ is determined by the average measured power output $\overline{P(v)}$ and the simulated average power output for the measured turbulence $\overline{P_{sim,I}(v)}$ and for the reference turbulence $\overline{P_{sim,I_{ref}}(v)}$. The simulated values $\overline{P_{sim}(v)}$ are computed as:

$$\overline{P_{sim}(v)} = \int_{v=0}^{\infty} \overline{P_{I=0}(v)}.f(v).dv \tag{2}$$

where $\overline{P_{I=0}(v)}$ is the so-called Zero Turbulence Power Curve (ZTPC) and $f(v)$ denotes the wind speed distribution within a 10-minute period. According to [2], $f(v)$ can be approximated using a gaussian distribution, which is fully determined by the average wind speed and the standard deviation of the wind speed as recorded within the 10 min period. Further details on the procedure are available in [1], [2] and [5].

2.2 EDiGaPS

This study proposes a new method, called Equivalent Difference in Gaussian Process Surface (EDiGaPS), that has some similarities with the one proposed by IEC. The IEC standard proposes the obtention of a simulated curve for each turbulence value so that corrections in new measured power values must be based on the differences between these simulated curves, as shown in Equation (1). The essence of our algorithm is the same: applying corrections based on the differences between the curves obtained for each turbulence. The fundamental difference is the way we obtain such curves. Instead of using the procedure involving the ZTPC, we use a training dataset to model the relation between the output and the inputs using a Gaussian Process (GP), that is described below.

A GP is a regression machine learning algorithm widely used in the literature, being a bayesian, non-parametric and non-linear approach. GPs have been used in many applications related to wind energy studies, due to their construction flexibility and ease of operation [6, 7].

Mathematically, a GP model is a continuous stochastic process $\{f(x)\}_{x \in L}$, $L \subseteq \mathbb{R}^D$, such that for any finite set of N inputs $x_i \in L$, $i = 1, 2, ..., N$, the function follows a N-variate normal distribution. The first and second order statistics of a GP are the mean function $\mu: L \rightarrow \mathbb{R}$ and the covariance function $\kappa: L \times L \rightarrow \mathbb{R}$ are, respectively, and they fully describe a GP model:

$$\mu(x) = E[f(x)] \tag{3}$$

$$\kappa(x, x') = E[(f(x) - \mu(x))(f(x') - \mu(x'))^T] \tag{4}$$

$$f(x) \sim GP(\mu(x), \kappa(x, x')) \tag{5}$$

In our problem, the output that we are modelling is the power of the wind turbine and inputs are the wind speed and the turbulence measured every 10 minutes. It is important to mention that, in our work, we do not directly use the gaussian process algorithm. As explained in [6], the S-shape of the power curve can be represented considering the gaussian process mean function as a logistic function. We use the technique in our work and more details about it can be found in [6]. Furthermore, following the approach in [8], we use two chained gaussian processes to model both the mean and variance of the power output, which enables us to account for the heteroscedasticity, capturing uncertainty across different wind conditions.

In our work, we employ a GP to model the training dataset. The features used as inputs for the ML algorithm are wind speed and turbulence and the output is the power. Using a seven-month training dataset, we construct a three-dimensional reference surface (wind speed, turbulence, and power) to enable power value correction, as described below. Since the correction method, described below, relies on calculating euclidean distances, the values of turbulence intensity and wind speed were normalized, according to equations (6) and (7):

$$t_{norm} = \frac{(t-t_{min})}{(t_{max}-t_{min})} \tag{6}$$

$$v_{norm} = \frac{(v-v_{min})}{(v_{max}-v_{min})} \tag{7}$$

Given a measurement of wind speed, turbulence and power (v, I, P), we want to obtain a new value of power P' that would be associated with the same wind speed value but with a new reference turbulence value I_{ref}, so that we can obtain a new curve in which all the power values are associated with the same turbulence intensity. The difference between P' and P is due only to the difference in turbulence and its value is unknown at first. We call it ΔP.

Consider two points, A and B, on the gaussian surface. Point A, with coordinates (v_a, I_a, P_a), represents the point on the surface where the Euclidean distance from (v_a, I_a) to a to a given measurement (v, I) is minimum. Similarly, Point B, with coordinates (v_b, I_b, P_b), corresponds to the point on the surface where the Euclidean distance from (v_b, I_b) to (v, I_{ref}) is minimum, where I_{ref} denotes the reference turbulence intensity for which the power curve is to be analyzed. The proposed method consists of considering that the difference ΔP, due only to the turbulence difference effect, can be considered equivalent and approximately equal to the difference between P_a and P_b.

Furthermore, in order to obtain the closest equivalents in the surface, it was computationally unfeasible to calculate the distance of every new measurement to every point on the surface due to the huge number of points. Therefore, for this task, it was used the NearestNeighbors algorithm of the library Scikit-learn [9].

Using the obtained surface, we apply corrections to the values of power using an idea very similar to the one proposed in the IEC standard, expressed in the equation:

$$\overline{P_{I_{ref}}(v)} = \overline{P(v)} + (P_a - P_b) \tag{8}$$

where P_a and P_b are power on the surface at the point for which the coordinates of turbulence and velocity are as close as possible to (I, v) and (I_{ref}, v), respectively.

Journal of Physics: Conference Series **3025** (2025) 012007 doi:10.1088/1742-6596/3025/1/012007

2.3 Experimental data

Seven months of 10-minute average SCADA data from a subpark of seven identical 1.8MW wind turbines were used to fit the reference surface through the gaussian process training algorithm. The original unfiltered dataset has 213703 points, however, after applying the filters this amount is reduced to 135060 points. Before the training process, two additional procedures are applied that do not alter the number of data points: normalization to a reference air density and the application of a nacelle transfer function (NTF).

After the reference surface is obtained during the training process, the effectiveness of the proposed method is assessed in a test dataset consisting of fifteen months of 10-minute average SCADA data from a single wind turbine of the same subpark. The assessment procedure is described below.

2.4 Assessment method

In order to assess the effectiveness of the method, the test dataset is divided into various subsets, each one associated with a value of turbulence intensity, namely: 10%, 16% and 22%. Each one of these subsets has an associated Annual Energy Production (AEP), that is one of the main key indicators of performance, because it represents an expectancy of how much energy the WTG would produce in a year, considering that the power curve would remain constant and taking into account the local wind speed distribution.

Different subsets of points of the same wind turbine in the same period but with different values of turbulence intensities should have different associated values of AEP because the turbulence influences the performance of the turbine. However, if we apply a reference turbulence correction in the values of power, these differences between the values of AEP should disappear. Therefore, we adopt the reduction of this difference as our metric of effectiveness of the correction method, in order to compare the turbulence normalization method and our new proposed method.

3. Results

Initially, we show in Figure 1 that our dataset behaves as described in the literature, that is, measurements associated with higher turbulence values tend to generate lower powers in high winds and higher powers in low winds.

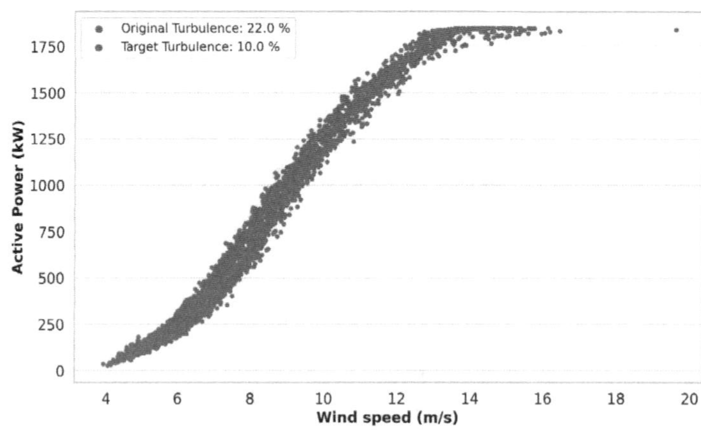

Figure 1. Power versus wind speed for specific values of turbulence.

Therefore, a straightforward manner to check whether the methods are making reasonable corrections is to test corrections from lower to higher turbulence values. In this scenario, in high winds, the power should decrease and in low winds the power should increase.

Using the normalization procedure, we obtained the ZTPC, shown in Figure 2, which is the main tool to perform the normalization corrections. Using the GP to model the training dataset, we obtained the surface shown in Figure 3, which is the main tool to perform EDiGaPS corrections.

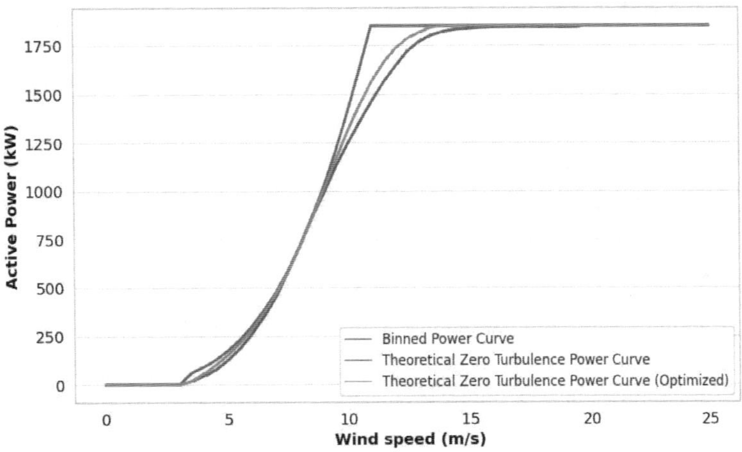

Figure 2. ZTPC from the normalization procedure.

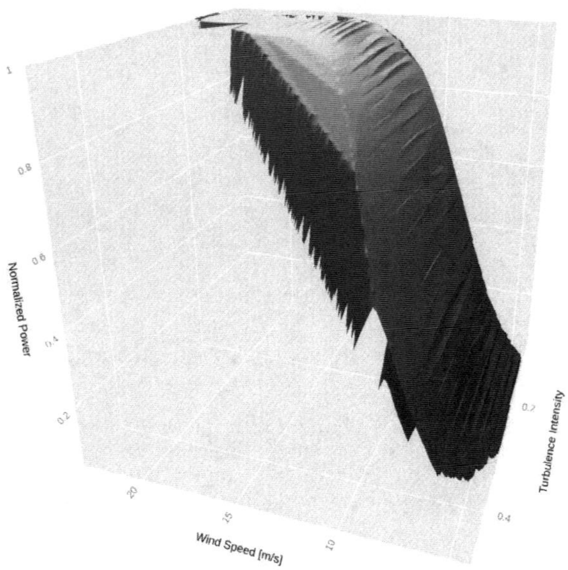

Figure 3. Surface of GP

In Figures 4a and 4b, we can see that both methods perform reasonable corrections.

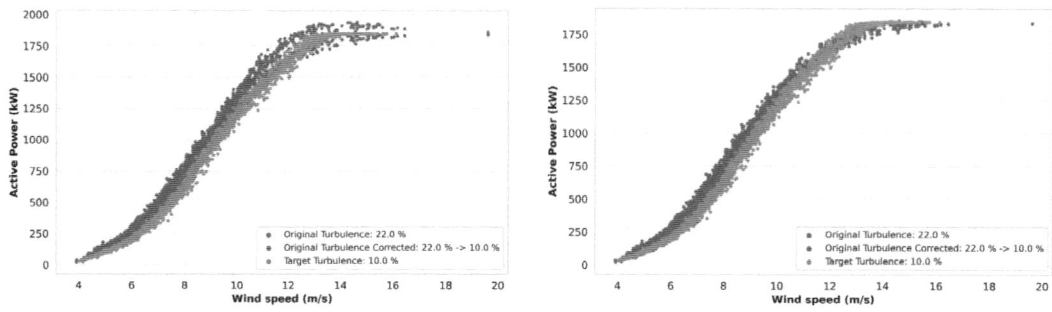

(a) Normalization (b) EDiGaPS

Figure 4. Effect of correction of turbulence from 22% to 10% on the points

However, the normalization procedure causes an overcompensation, as the corrected points in Figure 4a exceed even the nominal power. This kind of effect was already observed in other works in literature in [1] and [5]. On the other hand, the novel method EDiGaPS did not present such an effect.

For enhanced visualization, the data points can be replaced with curves derived using the binning method, as illustrated in Figures 5a and 5b. The binning method applied follows the standard outlined in IEC 61400-12-1 [2], which involves grouping wind speed measurements into discrete intervals of 0.5 m/s and calculating the average power for each bin.

(a) Normalization (b) EDiGaPS

Figure 5. Effect of correction of turbulence from 22% to 10% on the curves

The corrections in the opposite direction, that is, from 10% to 22%, may be seen in Figures 6a and 6b.

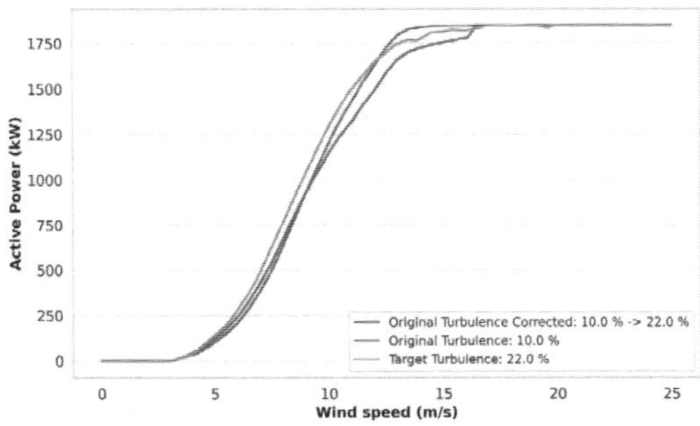

(a) Normalization

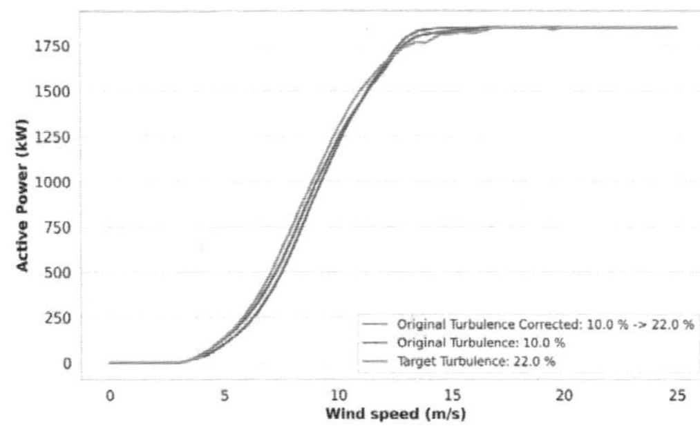

(b) EDiGaPS

Figure 6. Effect of correction of turbulence from 10% to 22% on the curves

From Figures 5 and 6, it can be seen that both methods perform corrections in the expected direction. To determine which of the two methods makes the most assertive corrections, we chose to employ a comparison similar to the one performed in [4]. The principle is to calculate the AEP for two datasets with different turbulence values and to check how much this difference is reduced after performing the correction. With a perfect correction, this difference should disappear almost completely, assuming that the difference is mainly due to the effect of turbulence. The evaluation of the AEP difference reduction was done using the proposed metric:

$$Score = 1 - \frac{|AEP_{corrected} - AEP_{target}|}{|AEP_{original} - AEP_{target}|} \quad (4)$$

The proposed metric range is (-∞, 1]. A perfectly effective correction receives a score of 1 (or 100%), while a completely ineffective correction is assigned a score of 0. Conversely, a correction that results in an increased difference in AEPs yields a negative score. We chose three values of turbulence (10%, 16%, 22%) and performed multiple experiments of corrections, as shown in Table 1.

Table 1. AEP correction with normalization and with EDiGaPS

$TI_{original}$	TI_{target}	$AEP_{original}$	AEP_{target}	$AEP_{corrected-IEC}$	$AEP_{corrected-EDiGaPS}$
10%	16%	712.3	753.7	704.8	727.1
10%	22%	712.3	764.3	696.7	737.4
16%	10%	753.7	712.3	761.3	738.7
16%	22%	753.7	764.3	744.8	763.8
22%	10%	764.3	712.3	780.1	739.1
22%	16%	764.3	753.7	772.6	753.9

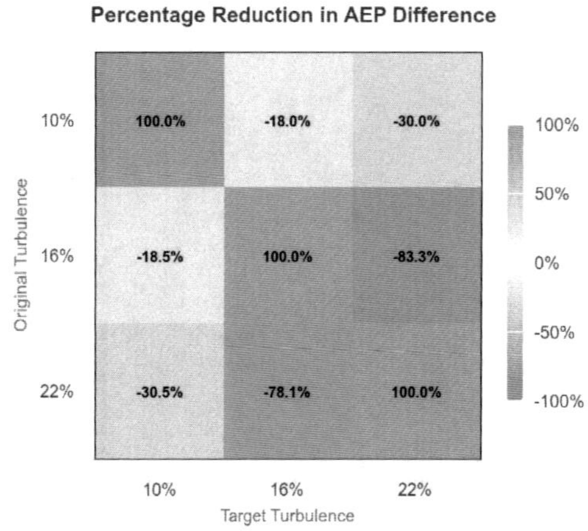

(a) Normalization

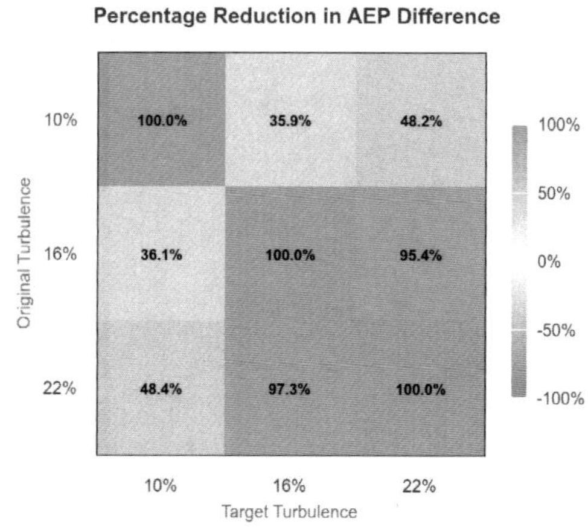

(b) EDiGaPS

Figure 7. Percentage Reduction in AEP difference after turbulence correction

From Table 1 and Figure 7, we can see that, although both methods provide corrections in the right direction, we have strong evidence that the new model is capable of generating more assertive corrections than normalization, given that, while in these examples the IEC method actually increases the difference between the AEPs, the EDiGaPS method decreases such difference very significantly, with an average reduction of 60.2% in our tests and almost 100% in some cases.

4. Conclusions and Future Work

In this work, we proposed a new approach for turbulence effect correction in WTPC. Through a real-world data case study, we demonstrated that our method yields more precise corrections compared to those proposed in the IEC standard, which provides strong evidence that this new data-driven approach is correcting not only the time averaging effect, as the normalization procedure proposes, but also other physical effects associated with turbulence.

Our findings confirm that the renormalization method overcompensates for turbulence effects, a phenomenon previously documented in the literature [1]. With our new method, on the other hand, we did not observe this effect. The proposed EDiGaPS method demonstrated a significant improvement over the normalization method, reducing AEP difference by an average of 60.2%, with some cases approaching near-total correction.

A limitation of the present study is that only one kind of wind turbine was evaluated. In order to have a more comprehensive comparison between the turbulence normalization method and our method, new studies should be conducted with different kinds of turbines of different manufacturers.

References

[1] Clifton A and Wagner R 2014 Accounting for the effect of turbulence on wind turbine power curves. *J. Phys.: Conf. Ser.* 524 012109
[2] International Electrotechnical Commission 2017 Power Performance Measurements of Electricity Producing Wind Turbines (IEC 61400-12-1) (Geneva: International Electrotechnical Commission)

[3] Tissot J, Schlacher A and Gray C 2022 Turbulence intensity normalisation of power curves and large-scale fleet validation. *Proc. Wind Europe Annu. Event* 2022
[4] Bardal L M and Sætran L R 2017 Influence of turbulence intensity on wind turbine power curves. *Energy Procedia* **137** 553–58
[5] Albers A 2010 Turbulence and shear normalisation of wind turbine power curve. *Proc. Eur. Wind Energy Conf.* Exhib. 2010 6 4116–23
[6] Virgolino G C, Mattos C L, Magalhães J A F and Barreto G A 2020 Gaussian processes with logistic mean function for modeling wind turbine power curves. *Renew. Energy* **162** 458–65
[7] Pandit R K, Infield D and Kolios A 2020 Gaussian process power curve models incorporating wind turbine operational variables. *Energy Rep.* **6** 1658–69
[8] Saul A D, Hensman J, Vehtari A and Lawrence N D 2016 Chained Gaussian processes. *Proc. Artif. Intell. Stat.* (AISTATS) (PMLR) pp 1431–40
[9] Pedregosa F et al 2011 Scikit-learn: Machine learning in Python. *J. Mach. Learn.* Res. **12** 2825–30

Journal of Physics: Conference Series **3025** (2025) 012008 doi:10.1088/1742-6596/3025/1/012008

Physics-Based, Data-Augmented Model for Wind Turbine Control Design

Jalal Heidari[1,2], Nezmin Kayedpour[1,2], Lieven Vandevelde[1,2], and Guillaume Crevecoeur[1,2]

[1]Department of Electromechanical, Systems and Metal Engineering, Ghent University, 9052 Ghent, Belgium
[2]FlandersMake@UGent - Corelab MIRO, Belgium

E-mail: jalal.heidari@ugent.be

Abstract. Accurate modeling of wind turbine dynamics is essential, as it enables engineers to design optimized control systems that enhance the performance and reliability of wind turbines. This study demonstrates how a hybrid model, integrating physics-based equations and a data-driven neural network, outperforms both low-fidelity physics models and purely data-driven approaches in estimating rotational speed with higher accuracy. By leveraging fundamental physical principles to capture key aerodynamic and mechanical behaviors and incorporating machine learning to estimate effective wind speed, the hybrid approach addresses the limitations of traditional methods. Purely physics-based models often struggle with uncertainties and variability in real-world conditions, while data-driven models lack generalizability and fail to incorporate physical constraints. In contrast, the hybrid model combines the strengths of both approaches, ensuring consistency with physical laws while adapting to operational variability. Validation using the high-fidelity OpenFAST simulation environment highlights the potential of the hybrid model not only for accurate dynamic modeling but also as a foundation for future model-based optimal control designs.

1 Introduction

The growing impact of the global greenhouse effect has led to a significant increase in global efforts to eliminate greenhouse gas emissions by 2050 [1]. Historically, fossil fuels have been the primary energy source worldwide, but their overuse has raised serious environmental concerns. In response, there has been a consistent move towards adopting clean and sustainable energy alternatives. Among these alternatives, wind energy stands out due to its vast availability, cost-effectiveness, and minimal environmental impact, playing a crucial role in the transition to a sustainable energy landscape [2]. Wind power technology has experienced rapid advancements in the 21st century, leading to considerable growth in both onshore and offshore wind capacities. According to the World Wind Energy Association's 2023 Annual Report, the total installed global wind power capacity has surpassed 1,047 GW, with an additional 116 GW added in 2023, signifying a robust growth rate of 12.5% [3]. These developments highlight the crucial role of wind energy in achieving sustainable global energy goals and highlight the pressing need for innovative strategies to improve the efficiency, reliability, and operational performance of wind energy technologies, particularly wind turbines [4].

The ability to effectively model wind turbines plays a fundamental role in ensuring their optimal performance and long-term reliability. From a control systems standpoint, reliable models enable operators to execute more effective control strategies [5]. Moreover, modeling supports the forecasting of

Content from this work may be used under the terms of the Creative Commons Attribution 4.0 licence. Any further distribution of this work must maintain attribution to the author(s) and the title of the work, journal citation and DOI.
Published under licence by IOP Publishing Ltd

system states and the continuous monitoring of operational parameters, which are vital for proactive maintenance and performance optimization [6]. Research on wind turbine modeling mostly focuses on two main approaches. The first approach uses physical equations derived from aerodynamic and mechanical theories to simulate turbine behavior. The problem with these physical models is that they can be inaccurate due to system uncertainties. The second approach relies on data-driven models that utilize machine learning algorithms to analyze operational data. Although these algorithms are effective in modeling complex patterns, they might lack the ability to generalize to new, unseen data, limiting their usefulness in diverse operational contexts [7]. To overcome these challenges, hybrid models have been introduced, combining the strengths of both physical and data-driven techniques.

In the literature, there are various ways to combine machine learning algorithms with physical equations. The most well-known approach is physics-informed neural networks, where cost functions and constraints derived from physics ensure that the neural network finds a solution consistent with physical laws [8]. For example, in [9], a physics-informed neural network was developed to integrate the physical constraints of wind turbine systems into the data-driven model. By integrating structural physical constraints into a deep residual recurrent neural network, the model reduces computational costs and improves accuracy compared to purely data-driven methods. Another approach is using a corrective term for a physical model with a data-driven model. For example, in [10], a hybrid two-stage deep learning framework is proposed to reconstruct urban wind fields. Significant accuracy improvements in the prediction of complex wind fields are achieved by integrating a computational fluid dynamics-based physical feature extraction model with sparse measurement-based error correction. Furthermore, incorporating input from physical equations into a data-driven model is another technique. For instance, in [11], physical equations were used to estimate wind turbine control actions by adding additional inputs to a data-driven model, thereby enhancing the prediction accuracy.

The main contribution of this study is the design of a novel hybrid framework—a physics-based approach that integrates a data-driven component specifically aimed at estimating rotor-effective wind speed. The data-driven component employs a simple neural network with low execution time, making it suitable for real-time applications within a control system. By combining physical equations with machine learning methods, the hybrid model compensates for uncertainties inherent in physical models and enhances its ability to generalize across various operating conditions.

The paper is organized as follows. Section 2 discuss the wind turbine under study. Section 3 outlines the methodology, encompassing the physical modeling, data-driven modeling, and the development of the hybrid physics-based, data-augmented model. Section 4 presents the results. Finally, in Section 5, the conclusion is presented along with a discussion of potential future research directions.

2 System under study

In this study, the wind turbine is modeled using OpenFAST, a software developed by the National Renewable Energy Laboratory (NREL) [12]. OpenFAST employs a modular framework, with each module utilizing high-fidelity physical equations to simulate turbine dynamics in a coordinated manner [12]. The selected wind turbine for analysis in this software is the 5 MW NREL reference wind turbine. This wind turbine serves as a benchmark for numerous research applications due to its detailed design specifications, making it ideal for validating models and developing control systems. This turbine features a three-blade, upwind rotor system with a rotor diameter of 126 meters. The hub is positioned 90 meters above the ground, supported by a hollow cylindrical tower, providing structural stability and flexibility under various loading conditions [13]. Table 1 presents the key parameters of this wind turbine.

To ensure the efficient operation of the 5 MW NREL wind turbine, it is essential to implement effective controllers that can manage the turbine's behavior across different operating regions. An effective controller for the OpenFAST simulation environment is the NREL Reference Open-Source Controller (ROSCO) [14]. ROSCO is an advanced control system developed for wind turbines, designed to simplify controller tuning while aligning with standard industry practices. It is designed to maximize power production in below-rated wind speeds and regulate rotor speed in above-rated conditions, ensuring optimal performance across different operating regions. Figure 1 shows the steady-state responses of this wind turbine in different operating regions. In Region 1, when the wind speed is below the cut-in speed of 3 m/s, the turbine remains inactive. In Region $1\frac{1}{2}$, between the cut-in and rated wind speeds, the turbine operates in variable-speed mode, increasing power output as wind speed rises from 3 m/s to 11.4 m/s. In Region 2, the turbine continues in variable-speed operation, optimizing power capture by adjusting the rotor speed to match the wind speed. As the wind nears the rated speed in Region $2\frac{1}{2}$, the system transitions to pitch control to maintain a consistent power output. Finally, in Region 3, when wind speeds exceed the rated level, the turbine operates at its rated power, with pitch control regulating aerodynamic power capture and controlling rotor speed [13, 14].

Table 1. *Parameters of 5 MW NREL Wind Turbine*

Parameter	**5 MW NREL**
Rated Power	5,000 kW
Rated Wind Speed	11.4 m/s
Cut-in Wind Speed	4 m/s
Cut-off Wind Speed	30 m/s
Rated Rotor Speed	12.1 rpm
Rotor Diameter	126 m
Hub Height	90 m
Gear box ratio	97:1
Generator Inertia	534.116 kg·m^2
Rotor Inertia	38,677,056 kg·m^2
Drivetrain Torsional-Spring Constant	867,637,000 N·m/rad
Drivetrain Torsional-Damping Constant	6,215,000 N·m/(rad/s)

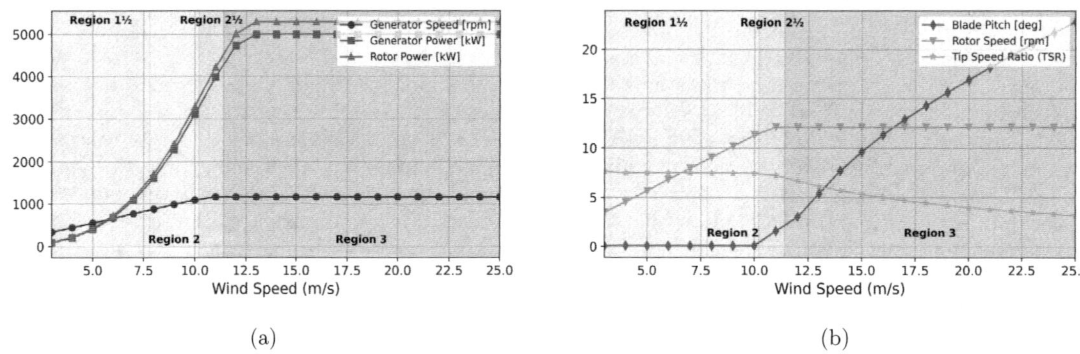

Figure 1: Steady-state response of 5 MW NREL wind turbine using ROSCO controller: (a) Generator characteristics and rotor power output, (b) Aerodynamic control response

3 Methodology

3.1 Physical Modeling

One approach to modeling a wind turbine is to rely solely on fundamental physics. Accordingly, the aerodynamic torque of a wind turbine can be determined by the following equation:

$$T_r = \frac{1}{2}\rho A v^3 C_p(\lambda, \beta)/\omega_r \tag{1}$$

$$\lambda = \frac{R \cdot \omega_r}{v} \tag{2}$$

where ρ is air density, v is wind speed, A is the swept area of the wind turbine, ω_r is rotational speed, and C_p is the power coefficient. It is clear that controlling β directly regulates T_r and can also be influenced indirectly by generator torque, yaw, etc. This torque is calculated using the AeroDyn module in OpenFAST by integrating drag and lift forces across the blade, taking into account wind speed and direction at different nodes. After that, the calculated aerodynamic torque will be an input for the ElastoDyn module, where it will be used to simulate the dynamic response of the wind turbine. Depending on the degrees of freedom considered, the rotor speed can be calculated using different physical equations. For instance, consider a simple model of the wind turbine (one inertia model), in which the rotor is treated as a rigid body:

$$J_{eq}\dot{\omega}_r = T_r - T_g \cdot N \tag{3}$$

$$J_{eq} = J_r + J_g \cdot N^2 \tag{4}$$

where J_{eq} is the equivalent moment of inertia of the rotor and generator, ω_r is the rotor speed, T_r is the aerodynamic torque, and T_g is the generator torque [15]. It should be noted that in this study, the term 'rotor' specifically refers to the 'turbine rotor'. In addition, considering drivetrain rotational-flexibility (two inertia model), the following equations can be used [16]:

$$(J_r + J_g N^2)\ddot{\theta}_r - J_g N^2 \ddot{\nu} = T_r - T_g \cdot N \quad (5)$$

$$-J_g N^2 \ddot{\theta}_r + J_g N^2 \ddot{\nu} + K\nu + B\dot{\nu} = T_g \cdot N \quad (6)$$

where J_r is the inertia of the rotor, J_g is the inertia of the generator, N is the gear ratio between the rotor and generator, θ_r is the rotor angle, θ_g is the generator angle, T_r is the torque applied to the rotor, T_g is the torque applied to the generator, K is the drivetrain torsional spring, B is the drivetrain torsional damper, and ν is the torsion angle between the two angles:

$$\nu = \frac{\theta_r - \theta_g}{N} \quad (7)$$

Differentiating ν with respect to time, we get:

$$\dot{\nu} = \dot{\theta}_r - \frac{\dot{\theta}_g}{N}; \ddot{\nu} = \ddot{\theta}_r - \frac{\ddot{\theta}_g}{N} \quad (8)$$

Substituting $\ddot{\nu}$ into the original system:

$$(J_r + J_g N^2)\ddot{\theta}_r - J_g N^2 \left(\ddot{\theta}_r - \frac{\ddot{\theta}_g}{N}\right) = T_r - T_g \cdot N \quad (9)$$

$$-J_g N^2 \ddot{\theta}_r + J_g N^2 \left(\ddot{\theta}_r - \frac{\ddot{\theta}_g}{N}\right) + K\left(\theta_r - \frac{\theta_g}{N}\right) + B\left(\dot{\theta}_r - \frac{\dot{\theta}_g}{N}\right) = T_g \cdot N \quad (10)$$

Simplifying these equations:

$$J_r \ddot{\theta}_r + J_g N \ddot{\theta}_g = T_r - T_g \cdot N \quad (11)$$

$$-J_g N \ddot{\theta}_g + K\theta_r - \frac{K\theta_g}{N} + B\dot{\theta}_r - \frac{B\dot{\theta}_g}{N} = T_g \cdot N \quad (12)$$

Solving for the angular accelerations $\ddot{\theta}_r$ and $\ddot{\theta}_g$:

$$\ddot{\theta}_g = \dot{\omega}_g = \frac{K(\theta_r - \frac{\theta_g}{N}) + B\left(\dot{\theta}_r - \frac{\dot{\theta}_g}{N}\right) - T_g \cdot N}{J_g N} \quad (13)$$

$$\ddot{\theta}_r = \dot{\omega}_r = \frac{T_r - T_g \cdot N - J_g N \ddot{\theta}_g}{J_r} \quad (14)$$

It is evident that in both the one-inertia and two-inertia models, adjusting T_g will result in changes to the rotor speed. In this study, a two-inertia model was used to achieve a more realistic representation of the drivetrain's behavior and dynamics.

3.2 Data-driven Modeling

In this section, the structure of the implemented data-driven model is presented. To generate data, OpenFAST with the ROSCO controller needs to be tested across a wide range of wind profiles. To achieve this, a code for data extraction, as shown in Code 1, is developed. Here, wind profiles are generated using TurbSim [17]. These wind profiles are then fed into OpenFAST, and the outputs of interest are recorded. In the simulations, 100 wind profiles are generated using the IEC 61400 standard with turbulence intensity category B. In this analysis, it was presumed that the wind was perfectly aligned with the turbine's hub axis, with no yaw misalignment or divergence between the turbine's orientation and the wind direction. Each simulation lasts 600 seconds with a time step of 0.01 seconds. A total of 80 operating points are used for the training dataset, while 20 are reserved for the test dataset. Figure 2 shows the collected data for the test dataset.

Journal of Physics: Conference Series **3025** (2025) 012008 doi:10.1088/1742-6596/3025/1/012008

Code 1: Data extraction for different wind profiles using OpenFAST

1. Define constants, including file paths and parameters (U_{Ref}, Rand, and the number of operating points to be generated).
2. Generate random U_{Ref} values within the specified range of 5 to 25 m/s.
3. For each U_{Ref} value, generate a random Rand value.
4. Update the TurbSim input file with the current U_{Ref} and Rand values.
5. Run TurbSim and capture the output.
6. Prepare the `InflowWind` file for each run by updating the wind file reference.
7. Update the `.fst` file to reference the new `InflowWind` file.
8. Run OpenFAST with the updated `.fst` file and capture the output.
9. Rename the OpenFAST output file for each run.
10. Clean up by deleting temporary files.

Note: U_{Ref} is the mean of wind speed, and `Rand` is the seed value used to create a unique wind profile in TurbSim. Additionally, `.fst` files serve as the primary input files in OpenFAST.

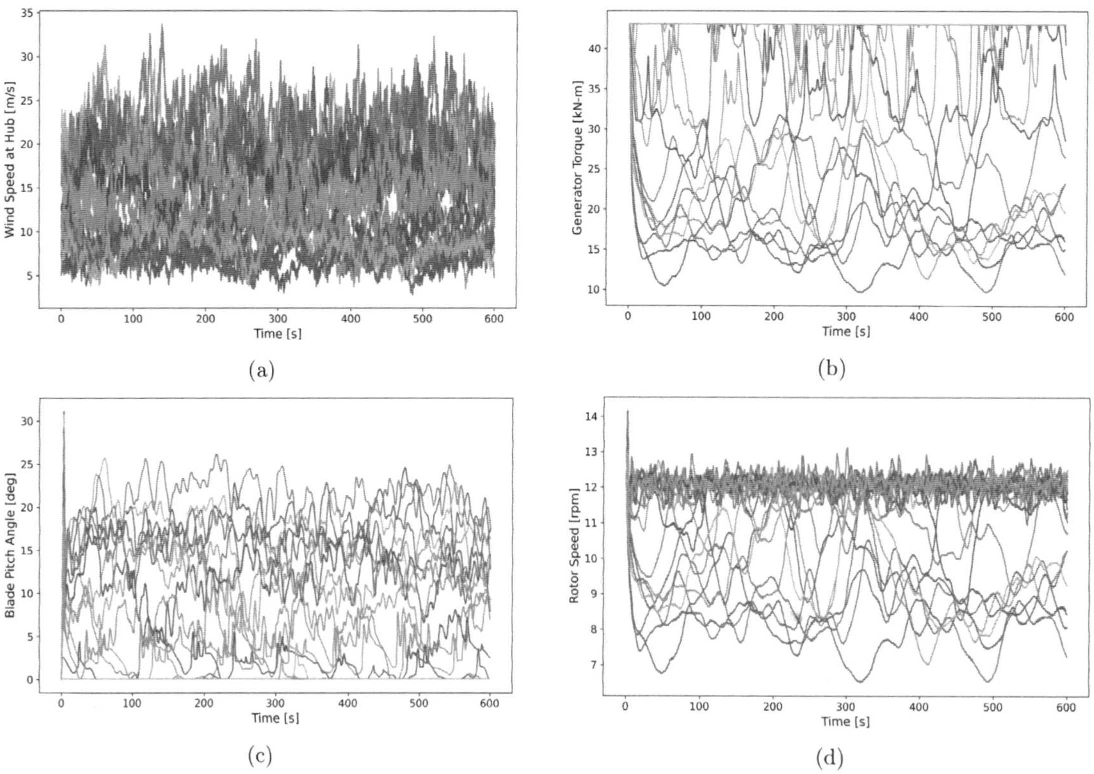

Figure 2: Generated data for test dataset: (a) Wind speed at hub, (b) Generator torque, (c) Blade pitch angle, (d) Rotor speed.

3.3 Hybrid Physics-Based, Data-Augmented Model

In this section, the structure of the proposed hybrid physics-based, data-augmented model is discussed. This model is designed to combine the strengths of both physics-based models and data-driven techniques to enhance the accuracy and robustness of wind turbine control strategies.

Here, we assume that the physical parameters of the wind turbine represent the ground truth values, as shown in Table 1. Reconsider Equation 1. The issue with the physical model is that v, the wind speed, should represent the average rotor-effective wind speed across entire swept area, rather than the hub wind speed or any other measured wind speed. However, the average rotor-effective wind speed cannot be directly measured; it is a response derived from the entire wind turbine system. Therefore, we first calculate the average effective rotor wind speed at different operating points, relying on known physical parameters. We will then apply a data-driven component to estimate the effective wind speed for unknown operating points.

The approach begins by using physical parameters such as turbine inertia, gearbox ratio, and blade radius of the NREL 5 MW reference wind turbine. With these known values and the fundamental physical equations, an inverse data extraction methodology is applied to calculate the effective wind speed. Figure 3 shows the flowchart of the process for developing a hybrid model by integrating data-driven and physics-based approaches, beginning with data extraction using OpenFAST and ending with the integration of neural network-based components with physical equations.

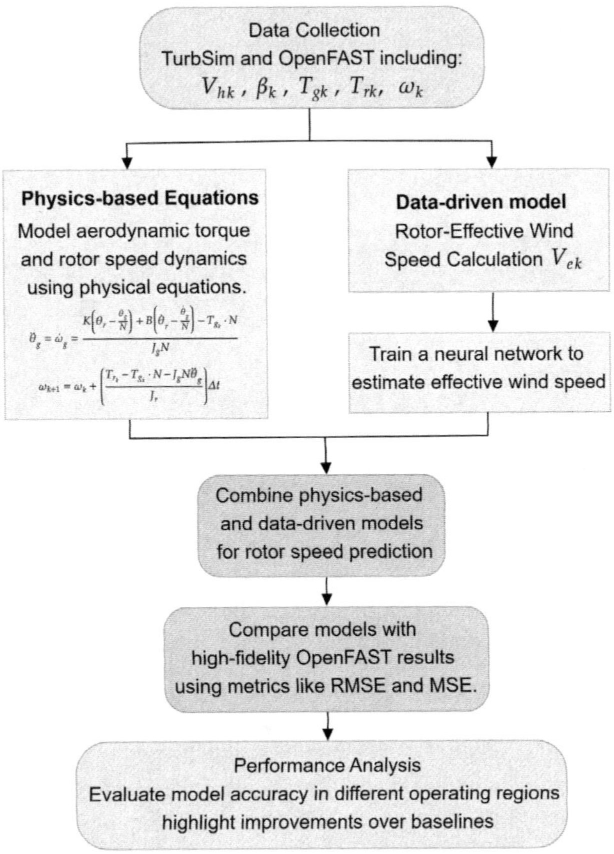

Figure 3: Flowchart of the hybrid model development process.

The following steps outline the process:

1. Aerodynamic Torque Calculation: The aerodynamic torque is calculated using a numerical solver. The aerodynamic torque can be obtained from measurements of rotor speed and generator torque. Here, it is assumed that inertia values are known parameters.

2. Effective Wind Speed Estimation: Once the aerodynamic torque is determined, the model applies the well-established relationships between aerodynamic forces, rotor speed, and wind speed to estimate effective wind speed. The power coefficient (C_p) is a key factor here, as it defines the efficiency of the wind turbine in converting wind energy into mechanical energy. These values are calculated based on predefined C_p lookup tables of 5MW NREL wind turbine. The effective wind speed is calculated for each operating point using Equation 1, with the goal of determining the rotor-effective wind speed, V_{e_k}.

3. Neural Network Integration: Instead of directly predicting the rotor speed as in traditional data-driven models, the hybrid model takes a more nuanced approach by configuring the neural network to predict the effective wind speed. This step allows the model to incorporate physical constraints while using data-driven methods to enhance its prediction capabilities. The neural network, trained on operating data from multiple scenarios, uses the predicted effective wind speed to estimate the rotor speed indirectly through the physical model. The rationale behind this approach is that effective wind speed can capture dynamic changes more accurately than direct rotor speed predictions.

Figure 4 shows the structure of the final three models.

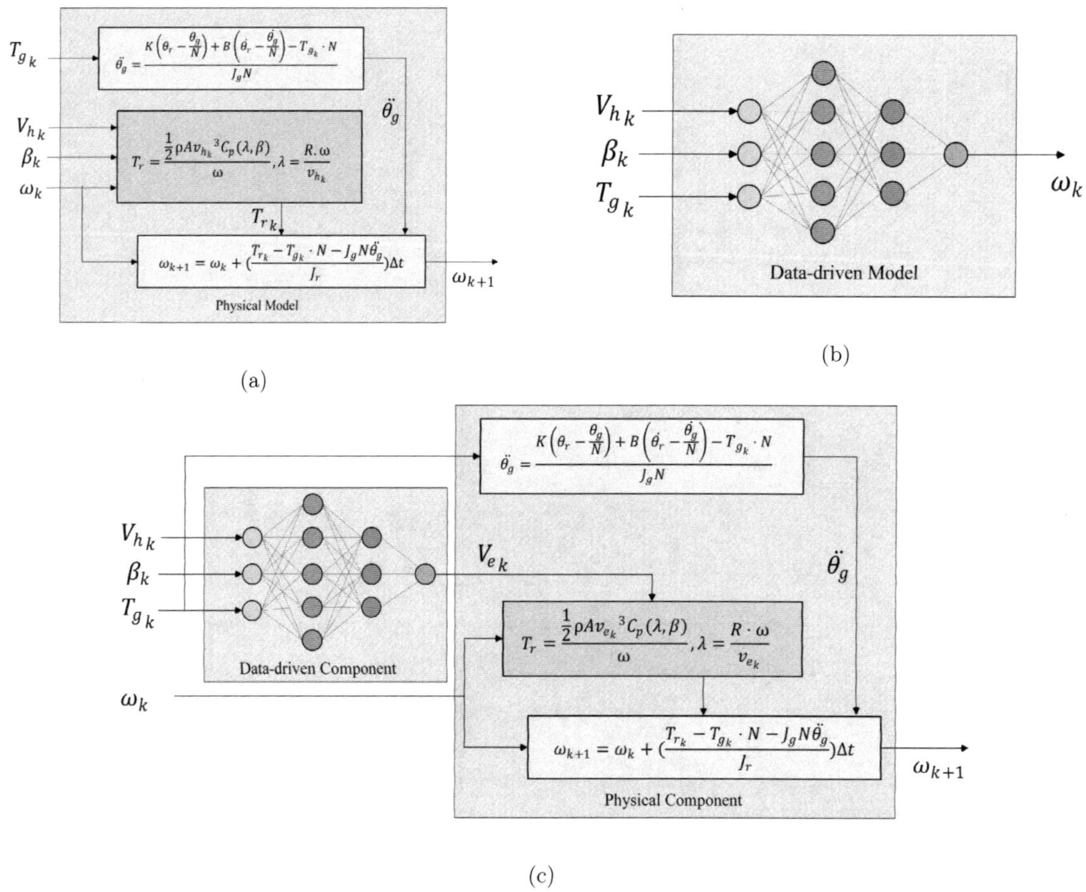

Figure 4: Wind turbine Models: (a) Physical Model, (b) Data-driven Model, (c) Hybrid Physics-based, Data-augmented Model.

4 Results

In this section, the performance of the developed hybrid physics-based data-augmented model is compared with a low-fidelity physics-based model, a purely data-driven model, and the high-fidelity OpenFAST model. Figure 5 shows the performance of the data-driven model and the data-driven component of the hybrid model. Since we employed a simple neural network with two hidden layers, the training time was only 170 seconds on a PC with a 13th Gen Intel® Core™ i7-1365U 1.80 GHz processor. Furthermore, processing the entire test dataset took just 0.15 seconds, demonstrating that the model can be integrated into a control framework for real-time operation. In the data-driven model, the rotor speed is directly predicted, while in the hybrid model, the effective average wind speed is predicted, which is then used to estimate rotor speed through the physical approach. The figures indicate that the neural network can more effectively follow the trends in the effective wind speed, compared to predicting the rotor speed directly.

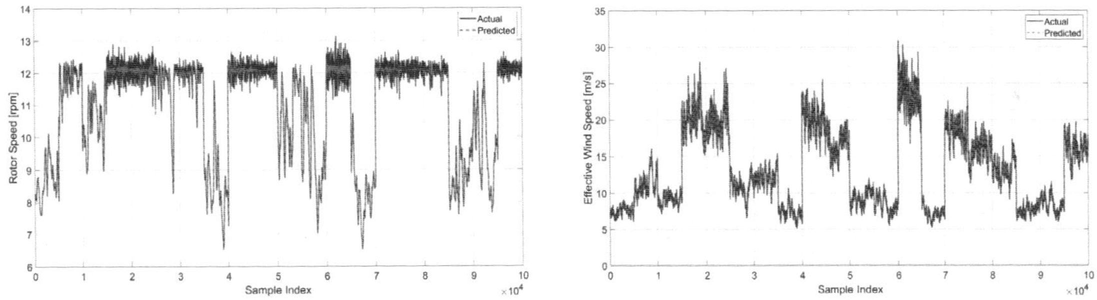

(a) (b)

Figure 5: Comparison of neural network prediction with actual values for test data: (a) Rotor speed prediction or data-driven model (b) Wind speed prediction or data-driven component of hybrid model.

To enable a more accurate comparison between different models, two different operating points were selected among all operating points in test data set: one in Region 2, where the mean wind speed at the hub is 10 m/s, and one in Region 3, where the mean wind speed at the hub is 15 m/s. Figures 6 and 7 show the effective wind speed, predicted effective wind speed, and measured wind speed at the hub, as well as a comparison of different models for Regions 2 and 3, respectively. Moreover, Figure 8 shows the probability density function for both regions. Finally, Table 2 shows the MSE and RMSE values for rotor speed prediction in both regions using the following equations:

$$\mathrm{MSE} = \frac{1}{n}\sum_{i=1}^{n}(\omega_i - \hat{\omega}_i)^2 \tag{15}$$

$$\mathrm{RMSE} = \sqrt{\frac{1}{n}\sum_{i=1}^{n}(\omega_i - \hat{\omega}_i)^2} \tag{16}$$

where ω_i is the actual rotor speed from the high-fidelity OpenFAST model, and $\hat{\omega}_i$ is the predicted rotor speed using different models. It can be seen that the proposed hybrid model consistently outperformed the low-fidelity model and data-driven model in both regions. In addition to the MSE and RMSE comparisons, the probability density function provides further insight into model performance. The results indicate that the hybrid model yields a narrower error distribution, suggesting that it captures the rotor speed trends more effectively.

Journal of Physics: Conference Series **3025** (2025) 012008 doi:10.1088/1742-6596/3025/1/012008

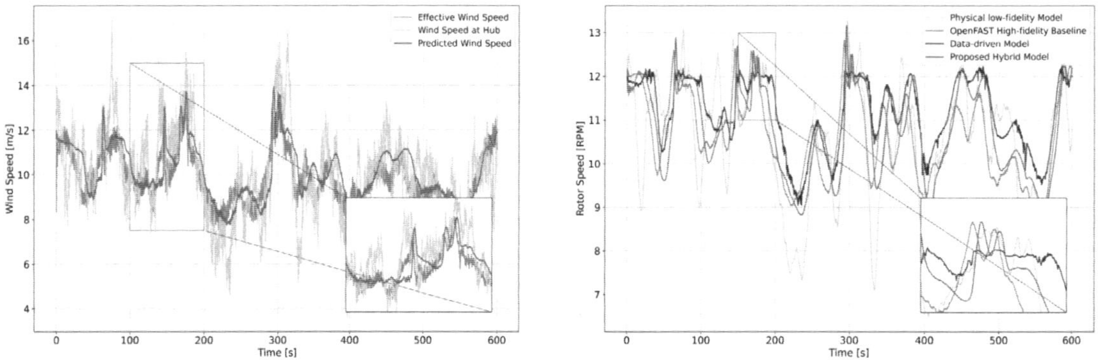

(a) (b)

Figure 6: Region 2: (a) Wind speed prediction, (b) Comparison of rotor speed predictions across models

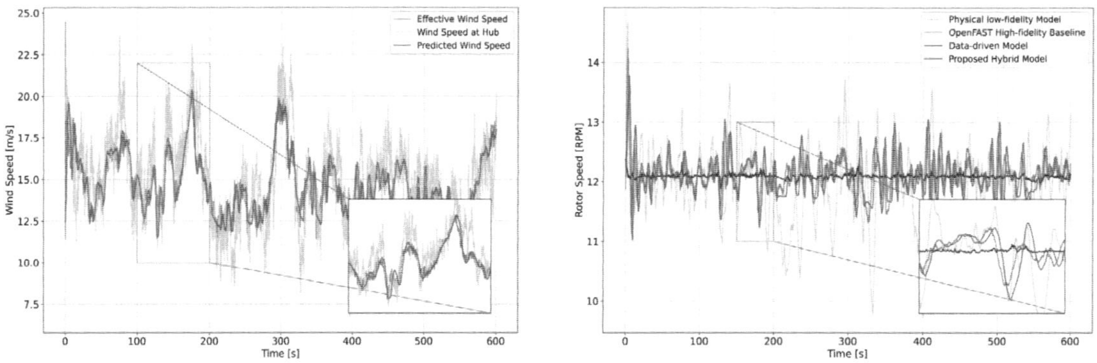

(a) (b)

Figure 7: Region 3: (a) Wind speed prediction, (b) Comparison of rotor speed predictions across models

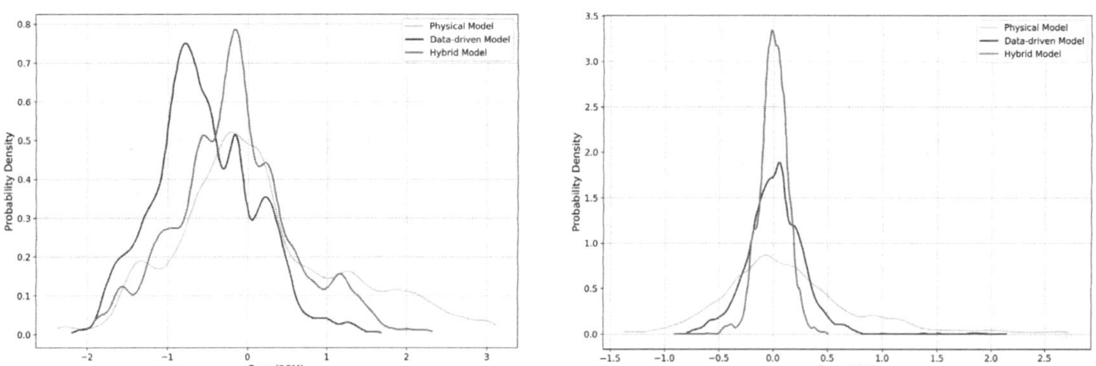

(a) (b)

Figure 8: Comparison of probability density function: (a) Region 2, (b) Region 3

Table 2. *Comparison of MSE and RMSE for two chosen operating points across different models compared to High-fidelity OpenFAST*

Model	Region 2		Region 3	
	MSE	RMSE	MSE	RMSE
Low-fidelity Model	0.0654	0.2558	0.0509	0.2255
Data-driven Model	0.0398	0.1995	0.0088	0.0938
Hybrid Model	0.0326	0.1807	0.0021	0.0456

5 Conclusion

In conclusion, this study has developed a hybrid model for predicting rotational speed that can be incorporated into predictive control approaches by integrating physics-based and data-driven methodologies. The hybrid model leverages the strengths of both approaches, addressing the limitations of purely physics-based models in capturing real-world complexities and uncertainties, while avoiding the generalization issues of purely data-driven models. Using this combined approach, the model is better equipped to estimate the rotor speed and handle unmodeled dynamics, particularly in highly variable wind conditions. Although the current implementation relies on a multilayer perceptron neural network configuration, future work could explore more advanced data-driven techniques to estimate the unknown uncertainties within the physical equations. These techniques have the potential to further enhance model predictions by complementing and building upon the current approach. More accurate models can lead to optimized control actions, which are crucial for efficient and reliable wind turbine operations, particularly in model predictive control frameworks. In addition, incorporating a physical parameter estimation framework into the model can enhance its robustness against uncertainties in the physical parameters.

Acknowledgment

This work was conducted as part of the project "Smart, Aware, Integrated Wind Farm Control Interacting with Digital Twins (ICONIC)" and has received funding from the European Union's Horizon Europe Research and Innovation Actions under the grant agreement No 101122329 and the UKRI Horizon Europe Guarantee scheme under the grant No 10095874.

References

[1] Stéphanie Bouckaert, Araceli Fernandez Pales, Christophe McGlade, Uwe Remme, Brent Wanner, Laszlo Varro, Davide D'Ambrosio, and Thomas Spencer. Net zero by 2050: A roadmap for the global energy sector. 2021.

[2] Han Shao, Rui Henriques, Hugo Morais, and Elisabetta Tedeschi. Power quality monitoring in electric grid integrating offshore wind energy: A review. *Renewable and Sustainable Energy Reviews*, 191:114094, 2024.

[3] World Wind Energy Association. WWEA Annual Report 2023: Record Year for Wind Power, 2023. [Accessed 20 November 2024].

[4] Pankaj Ravindra Gode and Arild Aspelund. Addressing heterogeneity in the development of circular economy strategies in the offshore wind industry: A review. *Heliyon*, 2024.

[5] Jingjie Xie, Hongyang Dong, and Xiaowei Zhao. Data-driven torque and pitch control of wind turbines via reinforcement learning. *Renewable Energy*, 215:118893, 2023.

[6] Jiayang Liu, Xiaosun Wang, Shijing Wu, Liang Wan, and Fuqi Xie. Wind turbine fault detection based on deep residual networks. *Expert Systems with Applications*, 213:119102, 2023.

[7] Muhammad Fahim, Vishal Sharma, Tuan-Vu Cao, Berk Canberk, and Trung Q Duong. Machine learning-based digital twin for predictive modeling in wind turbines. *IEEE Access*, 10:14184–14194, 2022.

[8] Maziar Raissi, Paris Perdikaris, and George E Karniadakis. Physics-informed neural networks: A deep learning framework for solving forward and inverse problems involving nonlinear partial differential equations. *Journal of Computational physics*, 378:686–707, 2019.

[9] Xuan Li and Wei Zhang. Physics-informed deep learning model in wind turbine response prediction. *Renewable Energy*, 185:932–944, 2022.

[10] Yi Liu, Ranpeng Wang, Yin Gu, Congjian Li, and Gangqiao Wang. Physics-inspired and data-driven two-stage deep learning approach for wind field reconstruction with experimental validation. *Energy*, 298:131230, 2024.

[11] Nezmin Kayedpour, Jixiang Qing, Jolan Wauters, Jeroen DM De Kooning, Ivo Couckuyt, and Guillaume Crevecoeur. Wind turbine hybrid physics-based deep learning model for a health monitoring approach considering provision of ancillary services. *IEEE Transactions on Instrumentation and Measurement*, 2024.

[12] National Renewable Energy Laboratory. OpenFAST: Open-Source Wind Turbine Simulation Software, 2024. Accessed: 10 December 2024.

[13] J Jonkman. Definition of a 5-mw reference wind turbine for offshore system development. Technical report, National Renewable Energy Laboratory, 2009.

[14] Nikhar J Abbas, Daniel S Zalkind, Lucy Pao, and Alan Wright. A reference open-source controller for fixed and floating offshore wind turbines. *Wind Energy Science*, 7(1):53–73, 2022.

[15] Babak Mehdizadeh Gavgani, Arash Farnam, Jeroen DM De Kooning, and Guillaume Crevecoeur. Efficiency enhancements of wind energy conversion systems using soft switching multiple model predictive control. *IEEE Transactions on Energy Conversion*, 37(2):1187–1199, 2021.

[16] Irving P Girsang, Jaspreet S Dhupia, Eduard Muljadi, Mohit Singh, and Lucy Y Pao. Gearbox and drivetrain models to study dynamic effects of modern wind turbines. *IEEE Transactions on Industry Applications*, 50(6):3777–3786, 2014.

[17] National Renewable Energy Laboratory. TurbSim: Wind Turbine Inflow Turbulence Simulator, 2024. Accessed: 10 December 2024.

Journal of Physics: Conference Series **3025** (2025) 012009 doi:10.1088/1742-6596/3025/1/012009

A parameter study of leading edge protection in a wind tunnel

Christian Bak[1], Anders S Olsen[1], Jimmie Beckerlee[1], Sigurd Ildvedsen[1], Andreas Fischer[1], Oliver Lylloff[1], Michael Drachmann Haag[2], Jonas R. Langebæk[2], Stefan Weber[2]

[1] DTU Wind and Energy Systems, Frederiksborgvej 399, 4000 Roskilde, Denmark
[2] Polytech A/S, Industrivej 75, 6740 Bramming, Denmark

E-mail: chba@dtu.dk

Abstract. This paper describes the influence of leading edge protection (LEP) on the aerodynamic performance of an airfoil based on wind tunnel tests of a NACA 63_3-418 airfoil. Measurements were carried out at a Reynolds number of $Re = 5$ million representing modern wind turbines. Different configurations of LEP were tested to carry out a parameter study. The parameters were the width and the height of the LEP but also the chamfer at the trailing edge of the LEP. It was seen how the aerodynamic performance was decreasing as the position of the trailing edge of the LEP was moving towards the leading edge of the airfoil. Also, if the LEP had a constant thickness it was seen that the aerodynamic losses would be smaller, the thinner the LEP was. Further to this, tests with in-build chamfer in the LEP indicated that it was the thickness of the trailing edge of the LEP together with the width of the LEP that determined the loss in aerodynamic performance.

1 Introduction

Leading Edge Erosion (LEE) and contaminated leading edges of blades cause a significant uncertainty for the annual energy production (AEP), see e.g. [1]. The power and the energy output from full scale wind turbines are measured, but it is a big challenge to make any final conclusions on the energy losses. This is especially due to the stochastic nature of the wind, which results in uncertainties in the relation between wind speed and power, see e.g. [2]. This challenge is discussed intensively in these years and much effort is put into providing methods to determine the power losses in a precise and reliable manner. Due to the impact on the performance from LEE, methods to reduce LEE have been developed. One method is to equip the rotor blades with leading edge protection (LEP). Since LEP is used on many both new and existing wind turbine blades, it is important to provide knowledge about how it affects the aerodynamic performance. This is because LEE cause a drop in aerodynamic performance, but if LEP is also causing a drop in aerodynamic performance, it is important to find the right balance between LEE and the use of LEP. Investigations of how Leading Edge Roughness (LER) and LEE influence the aerodynamic performance of wind turbine rotors and thereby the power output have been carried out ([3], [4], [5], [6], [7]), but also analysis of losses as measured in wind tunnels have been made ([8], [9], [10], [11], [12]).

Different overviews of the problem are given and solutions are proposed ([13], [14], [15], [16]). Some investigations of the aerodynamic impact of the LEP solutions have been carried out, where investigations using either Computational Fluid Dynamics (CFD) and/or wind tunnel tests have been made. Using both CFD and wind tunnel tests, [17], it was investigated how protection tapes affected the aerodynamic

Content from this work may be used under the terms of the Creative Commons Attribution 4.0 licence. Any further distribution of this work must maintain attribution to the author(s) and the title of the work, journal citation and DOI.
Published under licence by IOP Publishing Ltd

performance. The protection tapes had maximum thicknesses between 350μm and 500μm, tapered to 75μm at the ends and was 152.4mm wide. The investigations showed that the tape increased the drag significantly but also that tapering the tape was efficient. Tape with tapering showed an increase in drag of around 30% at 0 degrees angle of attack, whereas tape without tapering showed a drag increase of around 80% when tested at a Reynolds number of 3 million. Another study, [18], where LEP was tested in a wind tunnel showed the same trends where LEP with no tapering/chamfer showed a significant increase in drag at 0 degrees angle of attack around 80%, but where careful chamfering of the LEP reduced the drag increase to around 30%. Finally, in wind tunnel tests of different roughness/erosion types, [19], also a glossy paper with a thickness of 0.8mm was tested corresponding to LEP. Here it was again found that the drag increased with around 80% but also that if the LEP was not correctly mounted the drag could increase even further.

In this work the wind turbine sub-supplier Polytech together with DTU have carried out a systematic investigation of how LEP influences the aerodynamic performance using wind tunnel tests. The investigation was carried out in the *Poul la Cour Tunnel* at Reynolds numbers between 3 and 7 million corresponding to wind turbines of between in the order of 1 and 7 MW, respectively. In the tests the aerodynamic performance was measured where different widths and thicknesses of LEP were mounted.

2 Method

Wind tunnel tests were carried out in the *Poul la Cour Tunnel (PLCT)* on a NACA $63_3 - 418$ airfoil with a chord length of 1.0m and with different types of LEP.

2.1 The wind tunnel

A sketch of *PLCT* is seen in Figure 1. With reference to the figure, (1) shows the test section where the model is mounted, (2, 3, 6 and 7) shows the corners, where guide vanes with noise absorption are mounted, (4) is the cooling surface that ensures a constant temperature of the flow, (5) is the fan with a diameter of 4.7m ensuring a flow speed up to 105m/s in the test section, (8) includes one surface of honeycomb to straighten the flow and three screens to reduce the vortices and (9) is the contraction that accelerates the flow into the test section.

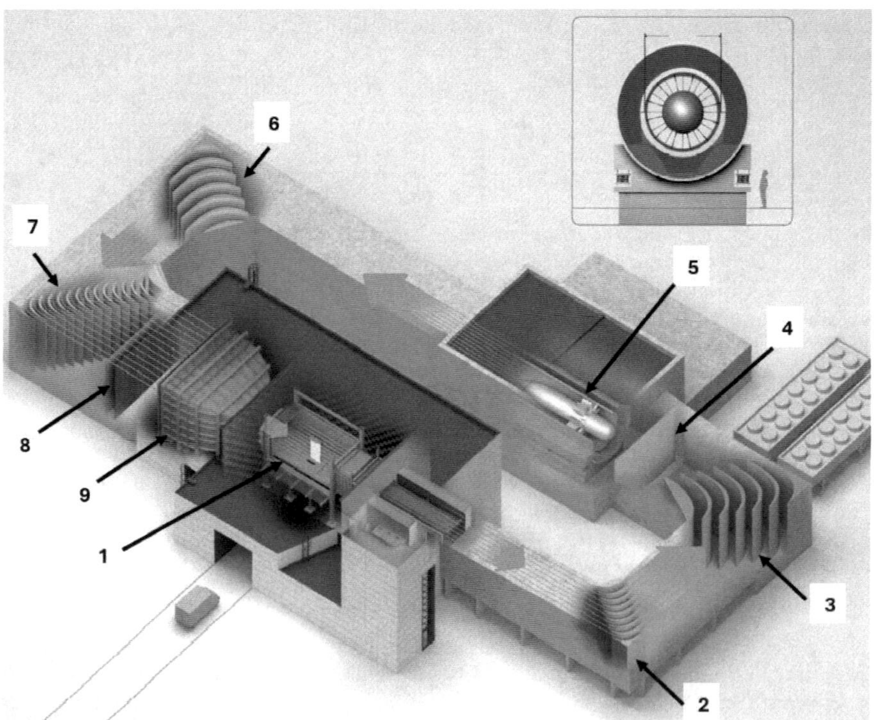

Figure 1: Sketch of the Poul la Cour Tunnel. The air is flowing anti-clockwise in the closed return wind tunnel with the closed test section

PLCT has a test section with the size width x height x length = 3.0m x 2.0m x 9.0m and the turbulence level is $TI < 0.1\%$. With velocities up to 105m/s, Reynolds numbers (Re) up to 7 million can be obtained with a chord length of 1.0m. Two different configurations of the wind tunnel exist: The aerodynamic and the aeroacoustic. The walls of the test section are hard in the aerodynamic configuration and they are made of Kevlar in the aeroacoustic configuration. In this study the aerodynamic configuration was used.

The airfoil was mounted in a turntable. Lift was measured at the airfoil surface with 100 pressure taps and at the test section walls with 96 pressure taps and the drag was measured with a wake rake with 96 pressure tubes. A photo of an airfoil in the aerodynamic configuration of the test section is seen in Figure 2. The configurations were measured at Re = 3, 5 and 7 million, however only results for Re=5 million are shown in this paper.

Figure 2: A photo of the airfoil with 1.0m chord in the test section. The airfoil is mounted in turntables in the floor and in the ceiling

2.2 The airfoil configurations

Three different types of configurations were tested: 1) A clean airfoil, 2) an airfoil with zigzag tapes ensuring tripping from laminar to turbulent flow in the boundary layer and 3) the leading edge mounted with different LEP dimensions.

A common way of simulating LER in wind tunnel tests is to mount zigzag tape close to the leading edge on both the suction side and the pressure side to ensure a tripping of the boundary layer from laminar to turbulent flow. In this study zigzag tape with a height of 0.4mm was mounted at x/c=0.05 at the suction side and x/c=0.10 at the pressure side.

LEP was tested as a 500mm wide patch that was mounted around the leading edge. This 500mm wide patch was during the tests narrowed down with 100mm between each test to reduce the width until it had a width of 100mm. Thus, five different widths of the LEP patch were tested, see Table 1. Furthermore, two thicknesses of the LEP was tested: 0.55mm and 0.85mm. At a thickness of 0.55mm the largest width was 400mm. Photos of the configuration are seen in Figure 3.

Table 1: Dimensions and position of the tested LEP

LEP width [mm]	LEP-TE: $x/c_{Pressureside}$ [%]	LEP-TE: $x/c_{Suctionside}$ [%]
100mm	3.0	3.4
200mm	8.0	8.0
300mm	12.6	12.6
400mm	17.5	17.5
500mm	22.5	22.5

Figure 3: Photo of LEP mounted at the leading edge. Three different widths are seen: 500mm, 400mm and 200mm

Furthermore, two different LEP's with build-in chamfer were tested. They were thicker at the mid part of the LEP (0.85mm) corresponding to the very leading edge of the airfoil and the thickness was reduced towards the trailing edge of the LEP. They had a width of 300mm and 270mm with thicknesses at the trailing edge of the LEP of 0.40mm and 0.45mm, respectively.

3 Results

In this section results from the wind tunnel tests are shown. The tests with LEP are compared to an airfoil with clean surface and with zigzag tape.

In Figure 4 is shown how LEP at the leading edge of the airfoil influence the aerodynamic characteristics. The LEP had a thickness of 0.85mm, no chamfer at the trailing edge part of the LEP and had different widths.

Journal of Physics: Conference Series **3025** (2025) 012009 doi:10.1088/1742-6596/3025/1/012009

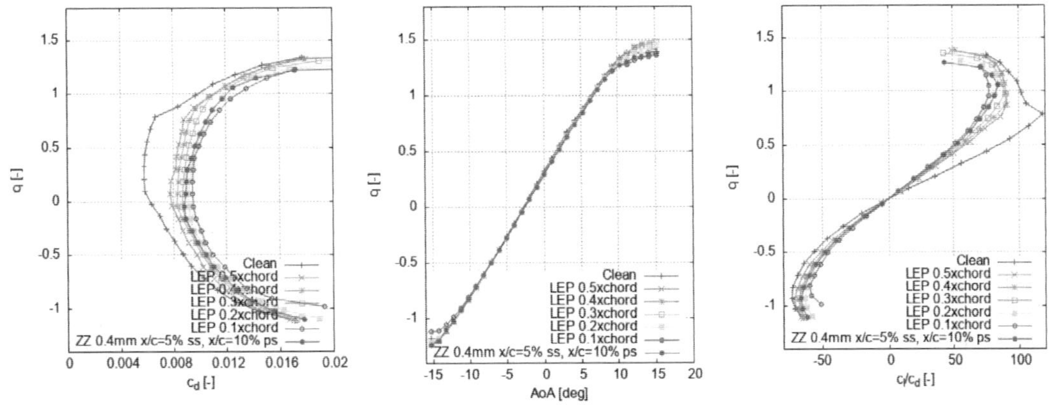

Figure 4: Aerodynamic performance of the NACA $63_3 - 418$ airfoil with 0.85mm thick LEP mounted at the leading edge. Left: c_l as a function of c_d, Mid: c_l as a function of angle of attack, AoA, Right: c_l as a function of $c_l - c_d$ ratio with several widths of the LEP. Apart from the performance with LEP, the performance of the clean airfoil is seen

It is seen how maximum lift coeffcient, $c_{l,max}$ was influenced slightly by the position of the trailing edge of the LEP (x_{TE-LEP}) with only a small decrease as the x_{TE-LEP} moved towards the leading edge of the airfoil. However, the drag coefficient, c_d, was clearly influenced. At an angle of attack of e.g. $AoA = 5.2^o$ representing the design/operating lift of the airfoil, c_d increased from about 0.0096 to 0.0115 if the TE-LEP was moved from 22.5% at the suction side to 3.4%. The corresponding c_l/c_d showed a drop from 91.1 to 73.5. It was also seen that maximum c_l/c_d changed only slightly for LEP widths of 500mm, 400mm and 300mm, but it started to decrease at 200mm and decreased further at 100mm. Comparing c_d for the 100mm LEP to c_d with zigzag tape showed that c_d with the LEP was higher. Since the zigzag tape was mounted at $x/c = 5\%$ and the trailing edge of the LEP was positioned at x/c=3.4% this indicated that the increase in c_d was controlled by the transition from laminar to turbulent flow. The closer to the leading edge that the transition would appear the higher c_d would be.

In Figure 5, LEP at the leading edge of the airfoil and with a thickness of 0.55mm are seen. The LEP had different widths and no chamfer at the trailing edge part of the LEP.

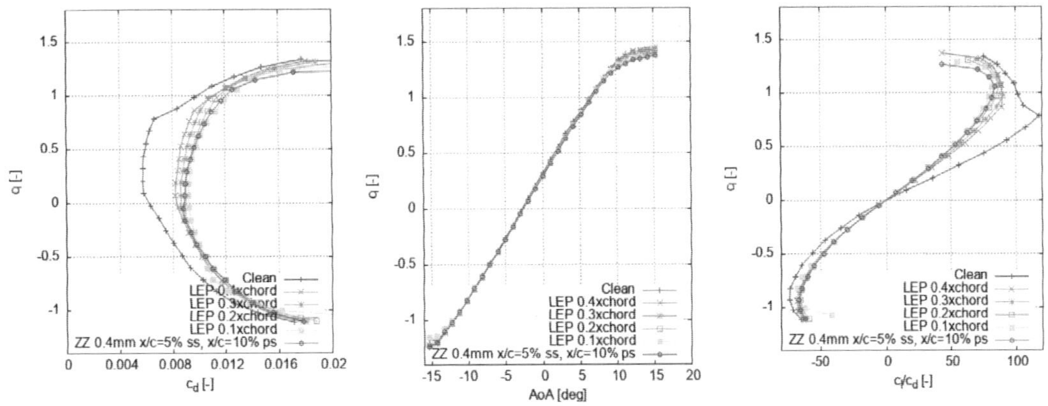

Figure 5: Aerodynamic performance of the NACA $63_3 - 418$ airfoil with 0.55mm thick LEP mounted at the leading edge. Left: c_l as a function of c_d, Mid: c_l as a function of angle of attack, AoA, Right: c_l as a function of $c_l - c_d$ ratio with several widths of the LEP. Apart from the performance with LEP, the performance of the clean airfoil is seen

The results for this thickness of LEP (0.55mm) were very similar to the results of the thicker LEP (0.85mm), see Figure 4. With the thinner LEP thickness it was also seen how $c_{l,max}$ was influenced

Journal of Physics: Conference Series **3025** (2025) 012009 doi:10.1088/1742-6596/3025/1/012009

slightly by the x_{TE-LEP} with only a small decrease as the x_{TE-LEP} moved towards the leading edge of the airfoil. However, c_d was also in this case clearly influenced by the x_{TE-LEP}. At an angle of attack of $AoA = 5.2^o$, c_d increased from about 0.0097 to 0.0113 if the TE-LEP was moved from 17.5% at the suction side to 3.4%. The corresponding c_l/c_d showed a drop from 89.3 to 75.7. It was also seen that maximum c_l/c_d changed only slightly for LEP widths of 400mm and 300mm, but it started to decrease at 200mm and decrease further at 100mm. As was the case for the 0.85mm thick LEP the 100mm long LEP with 0.55mm thickness showed higher c_d than the airfoil with zigzag tape.

In Figure 6 LEP with build-in chamfer is shown.

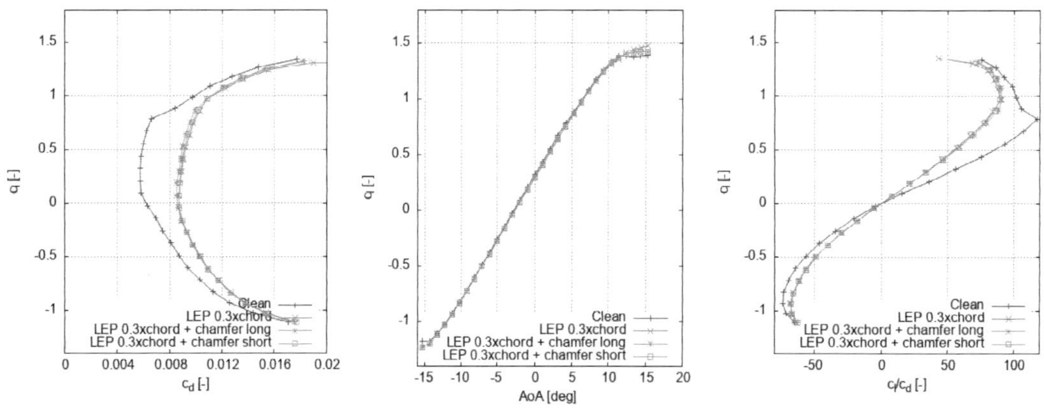

Figure 6: Aerodynamic performance of the NACA $63_3 - 418$ airfoil with 0.85mm thick LEP with two different chamfers mounted at the leading edge. Left: c_l as a function of c_d, Mid: c_l as a function of angle of attack, AoA, Right: c_l as a function of $c_l - c_d$ ratio. Apart from the performance with LEP, the performance of the clean airfoil is seen

The results for the thick LEP with a thickness of 0.85mm with chamfer showed small changes in the aerodynamic performance. It was seen how c_l and c_d were influenced slightly by the chamfer. Thus, with a 300mm long LEP the $c_l - c_d$ ratio increased from 83.7 (0.85mm thick), to 85.7 (0.55mm thick), and further to 87.9 (0.85mm thick with chamfer). If the LEP with chamfer was made slightly shorter (270mm) then $c_l - c_d$ ratio was the same as the thin LEP (0.55mm) that is 300mm long. The tests with chamfer indicated that the width and the trailing edge thickness of the LEP were the main parameters that determined the aerodynamic losses.

An overview of the loss in aerodynamic performance at $AoA = 5.2^o$, e.g. lift, $c_{l@AoA5.2}$, drag, $c_{d@AoA5.2}$ and lift-drag ratio, $(c_l/c_d)_{@AoA5.2}$, is shown in Table 2.

Table 2: Lift, $c_l@AoA5.2^o$, drag, $c_d@AoA5.2^o$ and lift-drag ratio, $(c_l/c_d)@AoA5.2^o$, for the different configurations

Configuration	LEP-TE [%]	$c_l@AoA5.2$	$c_l@AoA5.2/c_{l,clean}$	$c_d@AoA5.2$	$c_d@AoA5.2/c_{d,clean}$	$(c_l/c_d)@AoA5.2$	$(c_l/c_d)@AoA5.2/(c_l/c_d)_{clean}$
Clean	-	0.886	1	0.0084	1	105.6	1
ZZ	5	0.849	0.958	0.0110	1.310	77.2	0.730
h=0.85mm							
100mm	3.4	0.848	0.957	0.0115	1.376	73.5	0.695
200mm	8	0.855	0.965	0.0110	1.317	77.4	0.733
300mm	12.6	0.862	0.972	0.0103	1.227	83.7	0.792
400mm	17.5	0.869	0.981	0.0097	1.157	89.6	0.848
500mm	22.5	0.871	0.983	0.0096	1.141	91.1	0.862
h=0.55mm							
100mm	3.4	0.853	0.963	0.0113	1.344	75.7	0.717
200mm	8	0.861	0.972	0.0107	1.282	80.1	0.758
300mm	12.6	0.869	0.980	0.0101	1.209	85.7	0.811
400mm	17.5	0.870	0.982	0.0097	1.162	89.3	0.846
h=0.85mm chamfer							
300mm (chamfer)	12.6	0.868	0.979	0.0099	1.177	87.9	0.831
270mm (chamfer)	11.1	0.866	0.978	0.0101	1.210	85.4	0.808

Also, in Figures 7 and 8 the key parameters in the table are plotted.

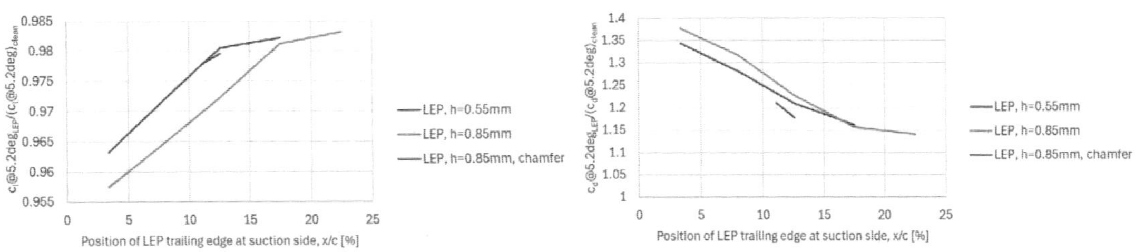

Figure 7: Lift, c_l, and drag, c_d as a function of x_{TE-LEP} at the suction side of the airfoil. Three different configurations of LEP are shown

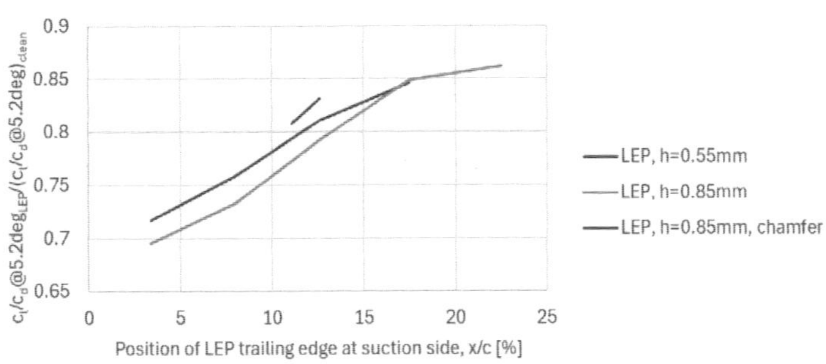

Figure 8: Lift-drag ratio, c_l/c_d, as a function of x_{TE-LEP}. Three different configurations of LEP are shown.

It was clear that the decrease in c_l at $AoA = 5.2^o$ compared to a perfectly clean airfoil was smallest for the thinnest LEP, where the slope of the the loss in c_l was steepest for x_{TE-LEP} for $x/c = 12.6\%$ or less, resulting in a reduction of 3.7%, when x_{TE-LEP} was at $x/c = 3.4\%$ against a reduction of 2.0% when x_{TE-LEP} was at $x/c = 12.6\%$. For the thickest LEP, the decrease was bigger and the slope of the the loss in c_l was steepest for x_{TE-LEP} for $x/c = 17.5\%$ or less, resulting in a reduction of 4.3%, when the x_{TE-LEP} was at $x/c = 3.4\%$ against a reduction of 1.8% when the x_{TE-LEP} was at $x/c = 17.5\%$. Comparing the thin and thick LEP it was clear that the thinner the LEP was the more narrow the LEP can be for the same reduction in c_l.

Analyzing c_d the same conclusion was reached. c_d was highest for the most narrow LEP but the thin LEP results in a smaller increase in c_d. Here, the slope in the increase in c_d was steepest for x_{TE-LEP} for $x/c = 12.6\%$ or less for the thinnest LEP, resulting in an increase of 34.4%, when x_{TE-LEP} was at $x/c = 3.4\%$ against an increase of 20.9% when x_{TE-LEP} was at $x/c = 12.6\%$. For the thickest LEP, the increase was bigger and the slope was steepest for x_{TE-LEP} for $x/c = 17.5\%$ or less, resulting in an increase of 37.6%, when the x_{TE-LEP} was at $x/c = 3.4\%$ against a reduction of 15.7% when x_{TE-LEP} was at $x/c = 17.5\%$. In the studies referred to in the introduction, [17] and [18], the drag increased up to 80% which is somewhat more than observed in this study. However, the drag reported from the other studies were observed at 0° angle of attack and not at the design/operating point around 5° to 8° angle of attack. In this study we observed a maximum increase in drag at 0° angle of attack of around 70% which is close to the reported 80% increase found in the other investigations. It is therefore important at which angle of attack the drag increase is evaluated.

However, with in-build chamfer in the LEP the plots reflect that this LEP with a maximum thickness of 0.85mm had the same or even less impact on the aerodynamic performance as the thinnest LEP with a thickness of 0.55mm. Thus, the loss in c_l was the same as the thinnest LEP but the loss in c_d was smaller.

Studying the lift-drag ratio, c_l/c_d, in Figure 8, showed that the losses were smallest for the LEP with chamfer. Thus, the loss in c_l/c_d with a 300mm long LEP with chamfer showed a loss of 16.9% compared to a perfectly clean airfoil. This was almost the same as a thick LEP that was 400mm long where a loss of 15.2% was observed.

4 Discussion

In the above comparison the reference was a perfectly clean airfoil with $Ra = 0.8\mu m$. Such a surface is likely not to be obtained for real wind turbine blades. E.g. some blade manufacturers uses a roller to apply the paint, which results in a surface roughness somewhat greater than $Ra = 0.8\mu m$. Therefore, the data shown in this paper as "Clean" is expected to be too optimistic. Thus, airfoil characteristics for a new blade is likely to be found between "Clean" data and "ZZ" data with a weighting towards the "Clean" data. Therefore, the losses described in this paper, e.g. 16.9% loss in c_l/c_d for an LEP with chamfer that is 300mm long, is probably somewhat too high compared to the losses that will be seen for real blades.

This work was based on one airfoil with a very smooth surface, where LEP with two different thicknesses and LEP with chamfer were mounted at the leading edge to be tested. In future work, more airfoils with varying surface roughness and further variation of the LEP thickness could be tested. This would create further insight into how the aerodynamic airfoil performance depends on more parameters.

5 Conclusions

A parameter study of the aerodynamic performance as a function of the widths of leading edge protection was shown. The presented measurements were carried out at Reynolds numbers of 5 million representing modern wind turbines with a rated power of about 3MW. It was seen how the aerodynamic performance was decreasing as the position of the trailing edge of the LEP was moved towards the leading edge of the airfoil. If the LEP had a constant thickness it was seen that the aerodynamic losses would be smaller the thinner the LEP was. Further to this, LEP that was 300mm long with in-build chamfer showed less aerodynamic loss than LEP with constant thickness. This indicated that it was the thickness of the trailing edge of the LEP together with the width of the LEP that determined the loss in aerodynamic performance.

References

[1] Bak C, Forsting AM, Sørensen NN. The influence of leading edge roughness, rotor control and wind climate on the loss in energy production. Journal of Physics: Conference Series. 2020;1618(5):052050 doi: 10.1088/1742-6596/1618/5/052050

[2] Malik T, Bak C, 2025 Challenges in detecting wind turbine power loss: the effects of blade erosion, turbulence, and time averaging, Wind Energy Science, Volume 10, issue 1, pp. 227–243, 2025

[3] Sareen A, Sapre C A and Selig M S 2014 Effects of leading edge erosion on wind turbine blade performance Wind Energy, 17, Issue 10, pp. 1531-1542.

[4] Standish K, Rimmington P, Laursen J and Paulsen H N 2010 Computational Prediction of Airfoil Roughness Sensitivity. In: 48th AIAA Aerospace Sciences Meeting and Exhibit, AIAA, paper no. 2010-0460

[5] Kruse E K, Sørensen N N and Bak C 2018 Predicting the Influence of Surface Protuberance on the Aerodynamic Characteristics of a NACA 633-418 J Physics: Conference Series, 1037, Issue 2, pp. 022008

[6] Bak C, Gaunaa M, Olsen A S and Kruse E K 2016 What is the critical height of leading edge roughness for aerodynamics? J Physics: Conference Series, 753, Issue 2, pp. 022023

[7] Kruse E K 2019 A Method for Quantifying Wind Turbine Leading Edge Roughness and its Influence on Energy Production. LER2AEP PhD Thesis, DTU Wind Energy

[8] Ehrmann R S, Wilcox B and White E B 2017 Effect of Surface Roughness on Wind Turbine Performance In: Sandia National Laboratories, SAND2017-10669

[9] Maniaci D C, White E B, Wilcox B, Langel C M, van Dam C P and Paquette J A 2016 Experimental Measurement and CFD Model Development of Thick Wind Turbine Airfoils with Leading Edge Erosion J Physics: Conference Series, 753, Issue 2, pp.022013 The Science of Making Torque from Wind (TORQUE 2020) Journal of Physics: Conference Series 1618 (2020) 052050 IOP Publishing doi:10.1088/1742-6596/1618/5/052050 10

[10] Gaudern N 2014 A practical study of the aerodynamic impact of wind turbine blade leading edge erosion J Physics: Conference Series, 524, Issue 1, pp. 012031

[11] Veraart M L W 2017 Deterioration in aerodynamic performance due to leading edge rain erosion Master's thesis. DTU Wind Energy and Wind Energy Research Group Faculty of Aerospace Engineering Delft University of Technology

[12] Kruse EK, Bak C, Olsen AS. Wind tunnel experiments on a NACA 633-418 airfoil with different types of leading edge roughness. Wind Energy. 2021;24(11):1263-1274. https://doi.org/10.1002/we.2630

[13] Sagol E, Reggio M and Ilinca A 2013 Issues concerning roughness on wind turbine blades Renewable and Sustainable Energy Reviews, 23, pp. 514–525

[14] Zidane I F, Saqr K M, Swadener G, Ma X and Shehadeh M F 2016 On the role of surface roughness in the aerodynamic performance and energy conversion of horizontal wind turbine blades: a review Int J Energy Research, 40, Issue 15, pp 2054-2077

[15] Bech J I, Hasager C B and Bak C 2018 Extending the life of wind turbine blade leading edges by reducing the tip speed during extreme precipitation events Wind Energy Science, 3, Issue 2, pp. 729-748

[16] Herring R, Dyer K, Martin F, Ward C, The increasing importance of leading edge erosion and a review of existing protection solutions, Renewable and Sustainable Energy Reviews journal, Volume 115, November 2019, 109382, https://doi.org/10.1016/j.rser.2019.109382

[17] Major D, Palacios J, Maughmer M, Schmitz S., Aerodynamics of leading-edge protection tapes for wind turbine blades. Wind Engineering. 2020;45(5):1296-1316. doi:10.1177/0309524X20975446

[18] Grasso F, Bak C, Smærup A, Ravnkilde L, Johansen RK, Leading Edge Protection Impact on Airfoil Performance: An Experimental Investigation, AIAA 2022-0277, AIAA SCITECH 2022 Forum, January 3-7, 2022, San Diego, CA & Virtual, https://doi.org/10.2514/6.2022-0277

[19] Bak, C and Olsen, A S and Forsting, A M and Bjerge, M H and Handberg, M E and Shkalov, H, Wind tunnel test of airfoil with erosion and leading edge protection, Journal of Physics: Conference Series, IOP Publishing, vol. 2507, 2023, 10.1088/1742-6596/2507/1/012022

Journal of Physics: Conference Series **3025** (2025) 012010 doi:10.1088/1742-6596/3025/1/012010

Reduction of crosstalk effects in multicomponent force and moment transducers for wind turbine test benches via geometrical design optimisation

Lukas Schmitz[1*], Georg Jacobs[1], Christian Hollas[1], Maximilian Zweiffel[1], Julian Röder[1], Fabian Renger[2], Ernst Manner[3]

[1] Chair for Wind Power Drives (CWD), RWTH Aachen University, Aachen, Germany
[2] RENK Test System GmbH, Augsburg, Germany
[3] MANNER Sensortelemetrie GmbH, Spaichingen, Germany

[*]E-mail: lukas.schmitz@cwd.rwth-aachen.de

Abstract. Wind turbine system test benches are essential for optimising the availability and reliability of wind turbines, thereby reducing development costs. The precise measurement of loads applied at the rotor hub is crucial; however, measuring all load components at a single location can lead to crosstalk effects that need to be minimized. Therefore, a highly precise multicomponent force and moment transducer (MCFMT) is required. This paper presents a method to systematically optimise the geometry of the MCFMT with the objective of reducing crosstalk effects and enhancing measurement accuracy. The optimisation process is based on finite element simulations and the creation of surrogate models that accurately map the relationships between geometrical design parameters, measurement signals and material strength utilisation. The optimised MCFMT design enables precise measurement of all six load components with minimal crosstalk effects. The crosstalk in the measurement of bending moments and torque was reduced to less than 0.09 %. However, adapting the geometry to measure thrust and shear forces proved to be a greater challenge. Thus, there is a need of post measurement mathematical compensation of crosstalk effects.

1. Introduction

Wind turbines (WT) are one of the important pillars of a climate-neutral energy supply and a key element in achieving the European Union's climate neutrality targets for 2050 [1]. To promote the expansion of wind energy, it is essential to improve the competitiveness compared to other energy sources, in particular by further reducing the levelized cost of electricity (LCOE). To decrease the LCOE, it is necessary to optimise the availability and reliability of WTs [2]. WT system test benches enable the investigation of the behaviour of the entire drivetrain, its components and control system under controlled and reproducible operating conditions. The data obtained from these tests is fed back into the design process to optimise the drivetrain (e.g. regarding robustness). This ultimately results in improved availability and reliability.

Content from this work may be used under the terms of the Creative Commons Attribution 4.0 licence. Any further distribution of this work must maintain attribution to the author(s) and the title of the work, journal citation and DOI.
Published under licence by IOP Publishing Ltd

WT system test benches in general consist of a prime mover to generate drive torque on the main shaft and a hydraulic load application system (LAS) to apply forces and bending moments (non-torque loads, NTL). Dynamic loads are applied simultaneously in six degrees of freedom to emulate a realistic load situation. Torque and NTL are applied at different positions of the test bench (prime mover and LAS). The applied loads differ from actual hub loads on the flange of the device under test (DUT) due to spatial offsets. For various applications, such as determining the efficiency of the WT [3], validating load calculations or characterising the drivetrain behaviour in critical operating conditions [4], precise application and therefore accurate measurements of the applied hub loads are required. Figure 1 shows the common measurement locations for measuring torque and NTL on the 4 MW WT system test bench of CWD [5].

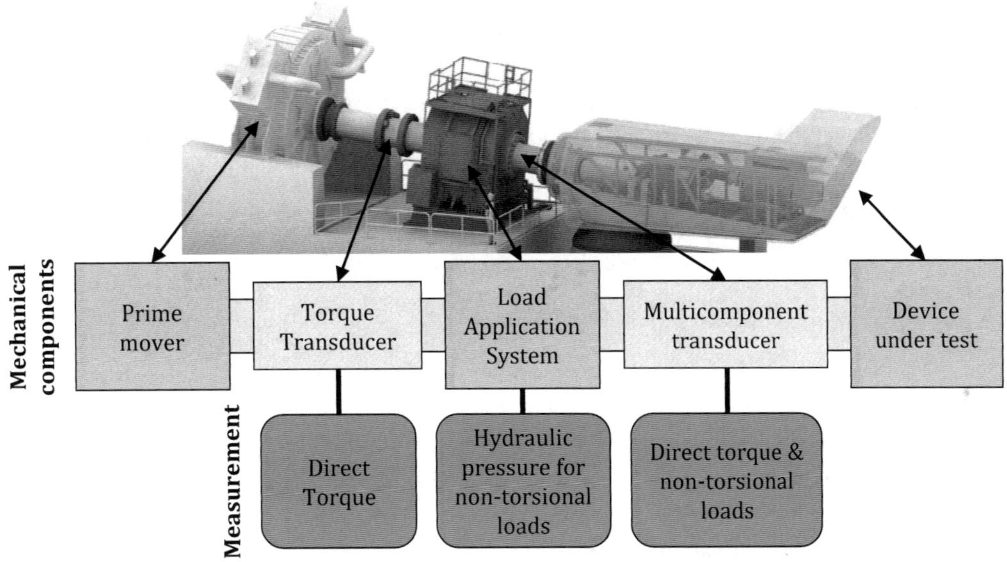

Figure 1. Structure and measurement locations of a 4 MW WT system test bench.

On this test bench, torque is measured using a torque transducer near the prime mover to prevent interference from other loads. The torque is transferred via the curved-tooth coupling and the plain bearings of the LAS to the DUT. The transfer path introduces additional torque losses due to gear and bearing friction torque [6], which are significantly influenced by the applied NTL and temperature. These torque losses cause the measured torque to deviate from the torque at the flange of the DUT. The LAS comprises four axial actuator sets and two radial actuator sets that, in combination, enable the application of NTL in five degrees of freedom. The applied NTL are usually measured indirectly using the forces acting on the hydraulic actuators of the LAS, which in turn are obtained by the hydraulic pressure in each actuator. However, this indirect measurement results in inaccuracies because influencing factors (e. g. speed-dependent friction within the cylinders) are ignored.

Thus, various direct measurement methods were investigated. Some of the measurement methods are based on the force-lever principle [7, 8], while others rely on fibre optic strain sensors [9] or strain gauges [10–12]. All of these methods are capable of measuring torque but have relatively low sensitivity for measuring the other load components and are therefore not suitable for measuring NTL.

In order to improve the measurement accuracy of both torque and NTL applied to the DUT, it is necessary to directly measure all load components between LAS and DUT flange using a multicomponent force and moment transducer (MCFMT). However, when measuring a specific load component, the measurements of the other load components can interfere, leading to inaccuracies known as crosstalk effects. Crosstalk effects are especially relevant when measuring multiple load components at the same position, where load components cannot be separated through the measurement geometry. There are various designs of MCFMT based on cylindrical shapes, tubular shapes and a combination of bending beams or double bending beams [13–15]. Although these designs can provide high accuracy and good crosstalk characteristics, the disadvantage is that the measurement capacity cannot be customised individually for each load component. An approach using a basic hollow shaft instrumented with eight strain gauges positioned around the circumference has shown that torque can be measured with high accuracy [16]. Additionally, NTL, except thrust, can be measured independently, although the exact uncertainties need to be further investigated. To ensure accurate measurement of the applied hub loads and to minimise crosstalk effects, previous approaches have relied solely on mathematical compensations. In contrast, a previous paper proposed a concept demonstrating the possibility of adapting the geometry to reduce crosstalk effects [17]. This concept is used as a baseline for the investigation presented in this paper.

2. Geometrical design optimisation

2.1 Development of the design concept

The concept development of the MCFMT is carried out according to the design methodology of VDI 2221 [18]. First, the boundary conditions and requirements for the MCFMT are defined, such as fixed screw flange connections to ensure compatibility with the system test bench. Next, a functional structure of the MCFMT is created based on the requirements. The main function of the MCFMT, "measurement of all six hub load components", can be subdivided into sub-functions for each load component. Each sub-function is realised by a principle solution, combining a physical effect with an effective structure (e.g. geometry, material). Through spatially separating the solutions, each geometry can be adjusted to achieve precise and highly sensitive measurements of the corresponding load component while minimizing crosstalk. Therefore, the geometry of the MCFMT is divided into segments, each individually designed and optimised to account for the specific characteristics of a particular load component. The overall design of the MCFMT is then assembled by combining these segments. The loads are measured via strain gauges (physical effect). The final design concept is shown in Figure 2.

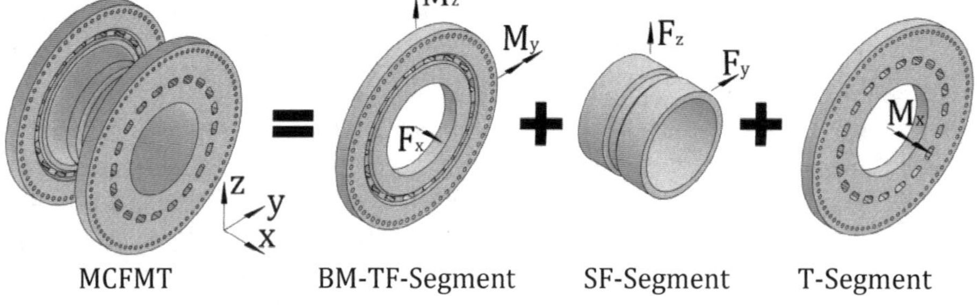

Figure 2. Segmented design concept for the MCFMT.

As illustrated from left to right, the bending-moment-thrust-force-segment (BM-TF-segment) is utilised for the detection of M_y, M_z and F_x, the shear-force-segment (SF-segment) is employed for the detection of F_y and F_z, and the torsion-segment (T-segment) is used for the detection of M_x. The measurement of bending moment and thrust force are combined into one flange, as both solutions have similar effective structures. The concept of an internal shear beam, as outlined in the preliminary work, is not pursued further in the present study. The design is considered excessively complex and highly susceptible to uncertainties regarding the attachment of the shear beam in the hollow shaft.

2.2 Bending moment and thrust force measurement

The BM-TF-segment is composed of a spoke wheel-like structure that contains multiple bending beams (struts). Strain gauges are fixed to these struts with the objective of measuring the resulting strains. The struts are designed to respond sensitively to bending moment and thrust forces, while remaining unresponsive to shear forces and torque. Therefore, the struts have a low beam section modulus in the axial direction and a high beam section modulus in the tangential direction (i.e. beams being axially thin and tangential thick). The subsequent separation of the bending moment and thrust force is achieved by selective interconnection of the strain gauges. This process involves the application of four strain gauges to two 180 degrees offset struts, to form a Wheatstone bridge. The measured values for the thrust force are averaged across all struts, while the measured values for the bending moments follow a sine distribution over all struts.

The BM-TF-segment is defined by six design parameters describing the struts: The outer diameter of the strut (D_{BM}), the inner diameter of the strut (d_{BM}), the number of struts (n_{BM}), the width of the struts (w_{BM}), and the thickness of the struts (t_{BM}) and the rounding radius to the strut (r_{BM}, constant). The hollow shaft is defined by four design parameters: The outer diameter of the shaft (D_{Shaft}), the wall thickness of the shaft (t_{Shaft}), the rounding radius to the shaft (R_{Shaft}, constant) and inner diameter of the shaft (d_{Shaft}, constant). Figure 3a) shows the geometric parameterisation of the segment for the measurement of bending moments and thrust force. Constants, which are calculated from other design parameters are marked in green, while variable design parameters for the optimization are marked in blue.

2.3 Shear force measurement

The SF-segment consists of a hollow shaft with an integrated measuring zone. Strain gauges are attached to the measuring zone in order to measure the resulting strains of the shear forces. By selectively connecting the strain gauges to Wheatstone bridges, the measurement of the shear force is separated from the other load components. In general, the shear forces in a typical load spectrum of a WT are benign compared to the dominant torsion and bending moments. The measurement of shear forces cannot be integrated into either MCFMT flange, as the flanges exhibit significant rigidity and resistance to shear forces due to their spoked wheel-like structural design. As adding internal shear beams are excluded due to mounting/manufacturing issues, the hollow shaft approach, incorporating an integrated measuring zone, is the only viable method for the measurement of shear forces.

The shear force segment is defined by a total of six design parameters: The outer diameter of the shaft (D_{Shaft}), the wall thickness of the shaft (t_{Shaft}), the inner diameter of the hollow shaft (d_{Shaft}, constant), the width of the measuring zone (w_{SF}), the wall thickness of the measuring zone (t_{SF}) and the rounding radius to the measuring zone (r_{SF}, constant). The geometric parameterisation of the segment for measuring shear forces is illustrated in Figure 3c).

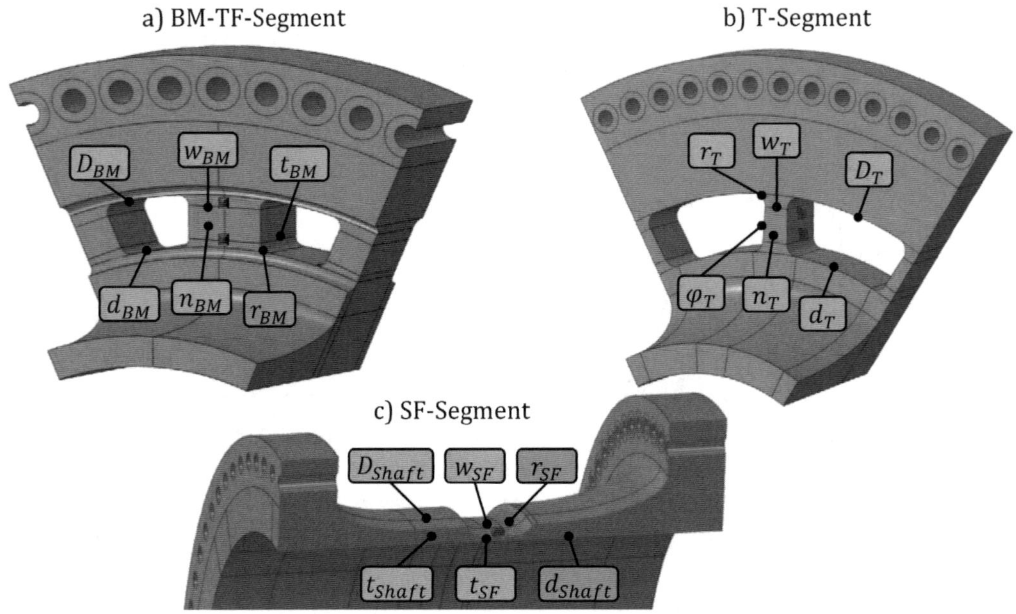

Figure 3. Geometrical design parameters of each segment. a) BM-TF-segment for measuring bending moments and thrust forces. b) T-segment for measuring torque. c) SF-segment for measuring shear forces. Possible strain gauge positions shown in red.

2.4 Torque measurement

To measure torque, a spoked wheel composed of multiple struts is selected for the T-segment. Strain gauges are attached to the struts to measure the resulting strain. The struts are designed to be sensitive to torque and insensitive to shear/thrust forces or bending moments by minimizing the tangential beam section modulus and maximizing the axial beam section modulus (i. e. beams being tangentially thin and axially thick = reversed compared to BM-TF-structs). A total of four strain gauges are applied to each strut and interconnected via a Wheatstone bridge. The measured values are averaged across all struts to compensate crosstalk effects.

The T-segment is defined by six struts design parameters: The outer diameter of the strut (D_T), the inner diameter of the strut (d_T), the number of struts (n_T), and the width of the strut (w_T), the rounding radius of the struts (r_T) and the strut angle (φ_T). The hollow shaft is described by four design parameters: The outer diameter of the shaft (D_{Shaft}), the wall thickness of the shaft (t_{Shaft}), the rounding radius to the shaft (R_{Shaft}, constant) and the inner diameter of the shaft (d_{Shaft}, constant).The parameters r_T and φ_T are exclusive to the T-segment, while the other four can be found similarly at the BM-TF-segment. The strut angle changes the thickness of the strut over the diameter of the strut, making the strut thinner towards the outside and thus more sensitive to the torque strain. Figure 3b) shows the geometric parameterisation of the segment for measuring the torques.

2.5 Systematic optimisation process

The geometrical design of the three segments and in particular the struts has a major influence on the measuring accuracy and structural integrity of the MCFMT. Manually optimising all design parameters is impractical. Therefore, a systematic optimisation method is used to identify the optimal geometrical design parameters of the MCFMT. The defined design concept and defined parameters provide the basis for the following systematic optimisation process based on [19]. Adapted to the MCFMT, the method leads to the approach shown in Figure 4, consisting of seven steps:

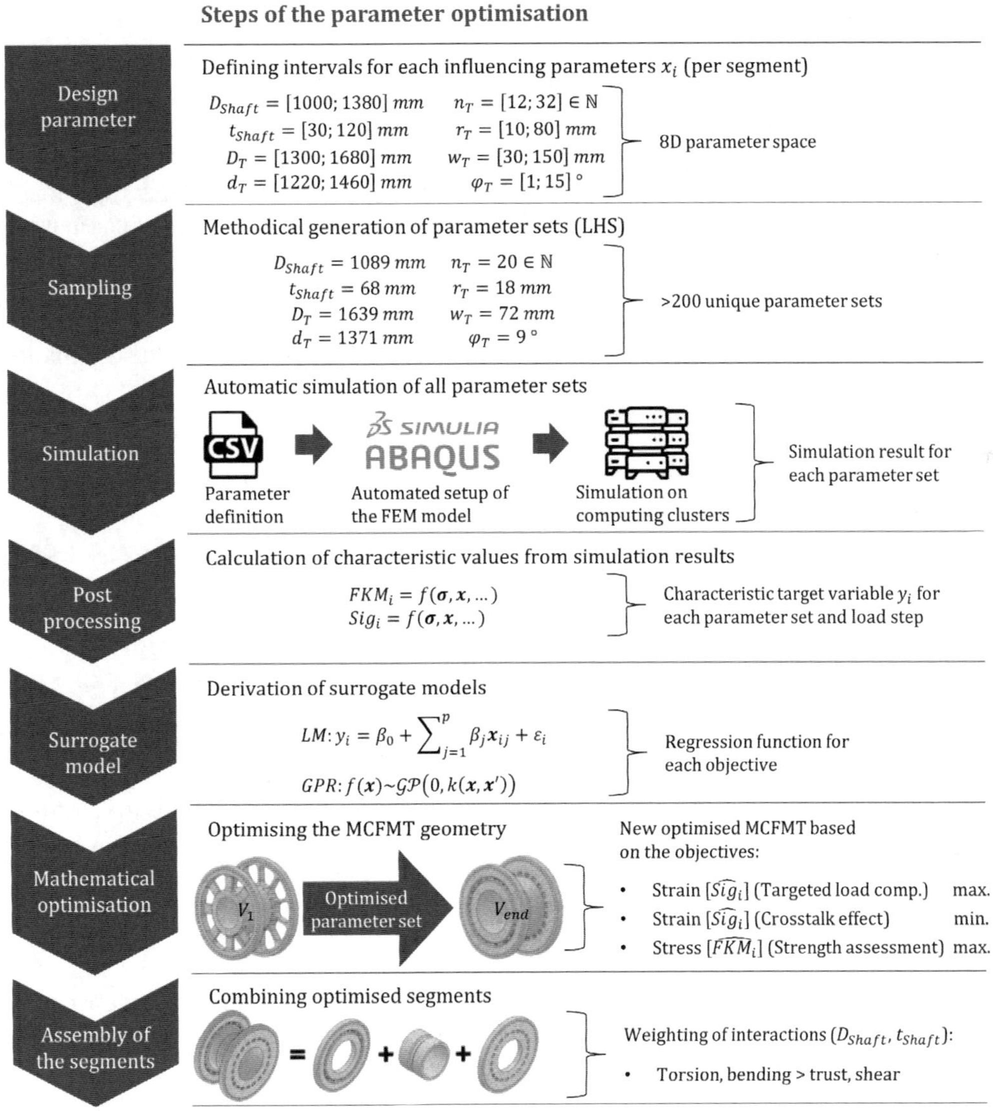

Figure 4. Steps of the geometrical design optimisation of the MCFMT.

1. <u>Design parameter</u>
   Each parameter is allowed to vary within a specified range. These ranges must be defined in advance by appropriate limits based on preliminary studies, experience etc.

2. Sampling
   Discrete and unique parameter sets are defined within the defined parameter space. To ensure efficient scanning of the parameter space, the parameter sets can be methodically generated using a sampling algorithm, such as Latin Hypercube Sampling (LHS). Furthermore, by applying multiple constraints during parameter sampling, geometrically invalid parameter sets are eliminated. (e.g. $D_T > d_T$). By defining parameters recursively (e.g. $r_T = 0 \dots 1 \cdot (D_T - d_T)/4$), the elimination rate can be greatly reduced compared to explicit definition.
3. Simulation
   The feasible sampled parameter sets are used to automatically generate several design variations of the MCFMT in the form of finite element (FE) simulation models, using an inhouse toolchain developed for Abaqus. FE simulations enable a detailed analysis of the MCFMT including stress and strain distributions. The FE simulation models are investigated under multi-dimensional load cases in order to analyse the crosstalk behaviour of the design variants. The investigation considers the elastic environment of the adapters and the nacelle, as well as the influence of gravity and centrifugal loads. To ensure a realistic behaviour, the FE simulation model includes the bolted connections of the adaption. The strain gauges are represented by substitute models, which facilitate the recording and analysis of surface strains with minimal effort and consequently accelerate post-processing. The reduction of the entire test bench model to the described simulation model enables the calculation of multiple simulation results in a time-efficient manner.
4. Post processing
   Relevant characteristic values are derived from the simulation results. The outputs of the strain gauges are combined to calculate the signals of the Wheatstone bridges (Sig_i). As the bolt connections induce loads on the flange, the Wheatstone bridge signals are offset by bolt load measurement. Furthermore, an automated strength assessment of all surface nodes of the FE simulation model is performed based on the FKM guideline [20], deriving a material utilisation rate (FKM_i) from the stresses. To ensure no plastic deformation at the struts, the strength assessment does not consider plastic load carrying capacities, resulting in higher utilisation rate (must be ≤ 1). The processed data provides the basis for evaluating measurement accuracy, sensitivity to crosstalk effects and strength assessment.
5. Surrogate models
   The relevant characteristic values (Sig_i, FKM_i) are used to derive surrogate models for the derived characteristic values, mapping them to the geometrical design parameters. Both linear regression functions (LM) with first and second order terms and Gaussian process regression (GPR) functions are utilised to derive surrogate models. In the following, the surrogate models replace the computationally intensive FE simulation model.
6. Optimisation
   The surrogate models are utilised to efficiently determine the optimal geometrical design parameters of the MCFMT segments that most effectively fulfil the design objectives. The design objectives specified in table 1 are optimised for the MCFMT segments.
7. Assembly of the segments
   In the final step of the process, the overall design of the MCFMT is composed from the segments optimised for each load component. A compromise must be made between the overlapping geometrical design parameters, depending on the relevance of the load components to be measured. The actual measuring points of each segment are only marginally affected by this compromise, as it mainly concerns the diameter of the hollow shaft that connects the segments to each other. As a result, a geometrically optimised

Journal of Physics: Conference Series **3025** (2025) 012010 doi:10.1088/1742-6596/3025/1/012010

MCFMT is generated. The assembled MCFMT is capable of measuring loads from six degrees of freedom with minimal crosstalk effects, while fulfilling the strength assessment.

Table 1. Design objectives of the optimisation process.

Objective	Target	Explanation
Strain (Targeted load comp.)	Maximize	The strain sensitivity of the measuring point of the segment needs to be maximised in order to enhance the effective detection of the desired load component of the MCFMT.
Strain (Crosstalk effect)	Minimize	The strain sensitivity of the measuring point of the segment for parasitic load components must be minimised in order to reduce crosstalk effects between the load components of the MCFMT.
Stress (Strength assessment)	Maximize to limit	The stress in the structure needs to be maximised to a limit in order to ensure the structural integrity of the MCFMT, while the MCFMT itself is required to remain sensitive enough to still detect strains.

3. Results

The systematic optimisation process of the geometrical design parameters depends primarily on the performance of the surrogate models. To ensure the reliability and validity of the optimisation process, it is crucial that the surrogate models represent the behaviour of the MCFMT of the FE simulation.

3.1 Validation of the surrogate models

To ensure a robust assessment of model quality, a K-fold cross-validation was performed. The models were evaluated using various performance metrics, including the Root Mean Square Error ($RMSE$), the coefficient of determination r^2 and the adjusted coefficient of determination r^2_{adj}. The definitions of these metrics are as follows:

$$RMSE = \sqrt{\frac{1}{n}\sum_{i=1}^{n}(y_i - \hat{y}_i)^2};\ r^2 = 1 - \frac{\sum_{i=1}^{n}(y_i - \hat{y}_i)^2}{\sum_{i=1}^{n}(y_i - \bar{y})^2};\ r^2_{adj} = 1 - (1 - r^2) \cdot \frac{n-1}{n-p-1}$$

Where y_i are the actual values, $\hat{y}_i$are the predicted values, $\bar{y}$ is the mean of the actual values, n is the number of observations, and p is the number of predictors in the model. By comparing these metrics, the surrogate models with the best predictive accuracy were identified and selected for predicting the target variables. The surrogate models were validated by comparing their predictions with the simulation results. Figure 5 illustrates the validation of the surrogate models for the prediction of the Wheatstone bridge signal of the bending moment (Sig_{BM}) and the static (FKM_{ST}) and dynamic (FKM_{DY}) material utilisation.

The low $RMSE$ values, coupled with the high r^2 and r^2_{adj} values, demonstrate that the developed surrogate models are highly precise and effectively capture the relationships in the data. The small discrepancies between the surrogate model predictions and the simulation results ($< 8\ \%$) confirm the validity of the surrogate models. The surrogate models of the other load components deliver comparable results. Therefore, it can be concluded that the surrogate models are reliable and can be successfully utilised in place of computationally intensive FE-models to

Journal of Physics: Conference Series **3025** (2025) 012010 doi:10.1088/1742-6596/3025/1/012010

predict the signals of the load components and material utilisation. In consequence, the computing effort can be reduced significantly.

	Sig_{BM}			FKM_{ST}			FKM_{DY}		
	Performance metrics								
	$RMSE$	r^2	r^2_{adj}	$RMSE$	r^2	r^2_{adj}	$RMSE$	r^2	r^2_{adj}
Surrogate Models (GPR)	0.0137 1.2 %	0.993	0.993	0.0383 1.5 %	0.978	0.977	0.0419 1.1 %	0.987	0.987
	Characteristic values								
Prediction	0.4346 mV/V			0.690			1.006		
Simulation	0.4254 mV/V			0.656			0.931		
Delta	0.0092			0.034			0.075		
Delta [%]	2.12 %			4.93 %			7.46 %		

Figure 5. Validation of the surrogate models for the bending moment and the static and dynamic material utilisation.

3.2 Optimised design of the MCFMT

The presented method was used to systematically optimise the geometrical design parameters of the segments of the MCFMT. A controlled, elitist genetic algorithm (a variant of NSGA-II [21]) was utilised. The optimised design was then analysed using a detailed FE simulation. The composed design and the resulting signals of the load components are shown in Figure 6.

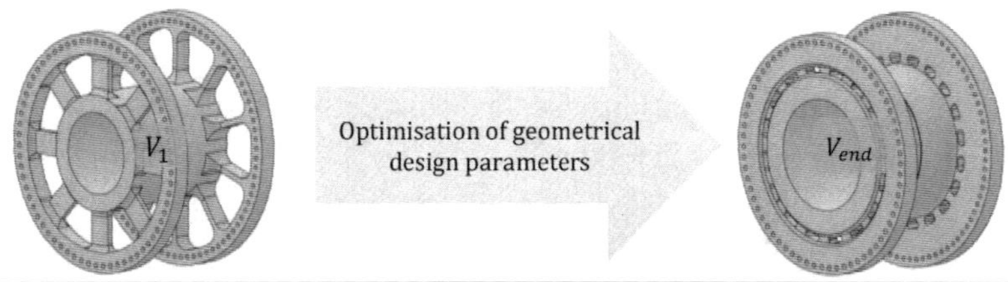

	Sig_{TF}			
Loadcase	F_x	F_z	M_x	M_y
Simulation	0.077 mV/V	0.000 mV/V	0.000 mV/V	0.001 mV/V
Crosstalk	–	0.010 %	0.134 %	1.529 %

	Sig_{SF}			
Loadcase	F_x	F_z	M_x	M_y
Simulation	0.000 mV/V	0.099 mV/V	0.000 mV/V	0.011 mV/V
Crosstalk	0.025 %	–	0.389 %	11.105 %

	Sig_{T}			
Loadcase	F_x	F_z	M_x	M_y
Simulation	0.000 mV/V	0.000 mV/V	0.890 mV/V	0.000 mV/V
Crosstalk	0.000 %	0.001 %	–	0.001 %

	Sig_{BM}			
Loadcase	F_x	F_z	M_x	M_y
Simulation	0.000 mV/V	0.001 mV/V	0.000 mV/V	0.702 mV/V
Crosstalk	0.001 %	0.085 %	0.003 %	–

Figure 6. Measurement signals and crosstalk effects of the FE simulation of the optimised geometrical design of the MCFMT.

The optimised design includes two spoked wheels and a measuring zone situated at the centre of the MCFMT. The struts of the spoked wheels and the measuring zone are equipped with strain gauges at various positions to measure strains. The geometry of the struts of the two spoked wheels has been specifically optimised for the measurement of thrust force, bending moment and torque. Shear forces are measured via the measuring zone situated at the centre of the MCFMT. Crosstalk effects in the measurement of bending moment and torque can be reduced to an absolute minimum (crosstalk under 0.09 %). However, optimising the measurement of thrust and shear forces proved to be a more complex process due to the smaller load ranges ($Sig_{TF} = 0.077\frac{mV}{V}$ and $Sig_{SF} = 0.099\frac{mV}{V}$). Within the specified load ratio, the bending moment exhibits a significantly higher magnitude compared to the thrust and shear force. The detailed analysis of the measurement signals reveals that the bending moment still influences the measurement of the thrust forces (crosstalk of 1.529 %) and shear forces (crosstalk of 11.105 %). This indicates a potential limit for the reduction of crosstalk effects via the concept of geometrical optimisation of the MCFMT, unless more complex geometries are taken into consideration.

4. Conclusion

The demand for WT system test benches is primarily driven by the necessity to optimise the availability and reliability of WTs, thereby reducing WT developing costs. A fundamental element in this process is the precise measurement of loads and moments applied at the rotor hub. Thus, a highly precise MCFMT is required. This study presented a method to systematically optimise the geometry of the MCFMT.

In a first step, a suitable design concept was parameterised. The geometry of the MCFMT was divided into specific segments, each individually designed for measuring certain load components and geometrical design parameters were defined. Subsequently, an optimisation algorithm was presented, which generates surrogate models based on simulation results. The use of surrogate models based on FE simulations led to a significant acceleration and increased efficiency of the optimisation process, with basic computational hardware reducing the computational time from hours to a few seconds.

The results of the optimisation process led to an optimised design of the MCFMT, comprising two spoked wheels and a measuring zone. The specific design of the struts and the measuring zone enables accurate measurement of all six load components with minimal crosstalk effects. While crosstalk in the measurement of bending moments and torque could be reduced to a minimum of 0.09 %, the analysis indicated that there is still room for improvement regarding the reduction of crosstalk effects when measuring thrust and shear forces. Consequently, it is recommended to consider a combination of the systematic optimisation of the geometry of the MCFMT illustrated herein, and a post mathematical compensation of the residual crosstalk effects, with an aim to differentiate the measurement of the load components as far as possible.

Acknowledgements

This work was funded by the German Federal Ministry for Economic Affairs and Climate Action (grant number: 03EE3073A).

References

[1] The European Parliament and the council of the European Union. *Regulation (EU) 2021/1119 of the European Parliament and of the Council of 30 June 2021 establishing the framework for achieving climate neutrality and amending Regulations (EC) No 401/2009 and (EU) 2018/1999 ('European Climate Law')* http://data.europa.eu/eli/reg/2021/1119/oj (accessed 23 Feb 2024)

[2] BMWK - Bundesministerium für Wirtschaft und Klimaschutz *8. Energieforschungsprogramm zur angewandten Energieforschung - Forschungsmissionen für die Energiewende* https://www.bmwk.de/Redaktion/DE/Publikationen/Energie/8-energieforschungsprogramm-zur-angewandten-energieforschung.html (accessed 14 Jan 2025)

[3] Song Z, Weidinger P, Zweiffel M, Dubowik A, Mester C, Yogal N and Oliveira R S 2023 Traceable efficiency determination of a 2.75 MW nacelle on a test bench *Forsch Ingenieurwes* **87** 259–73, DOI: 10.1007/s10010-023-00650-1

[4] Foyer G, Kock S and Weidinger P 2019 Influences on torque measurement in nacelle test benches and their effect on the measurement uncertainty and consequences of a torque calibration *ACTA IMEKO* **8** 59, DOI: 10.21014/acta_imeko.v8i3.658

[5] Averous N R, Stieneker M, Kock S, Andrei C, Helmedag A, Doncker R W de, Hameyer K, Jacobs G and Monti A 2017 Development of a 4 MW Full-Size Wind-Turbine Test Bench *IEEE J. Emerg. Sel. Topics Power Electron.* **5** 600–9, DOI: 10.1109/JESTPE.2017.2667399

[6] Kock S, Jacobs G, Bosse D and Sharma A 2018 Friction as a major uncertainty factor on torque measurement in wind turbine test benches *J. Phys.: Conf. Ser.* **1037** 62001, DOI: 10.1088/1742-6596/1037/6/062001

[7] Gnauert J, Jacobs G, Kock S and Bosse D 2018 Measurement uncertainty estimation of a novel torque transducer for wind turbine test benches *J. Phys.: Conf. Ser.* **1065** 42050, DOI: 10.1088/1742-6596/1065/4/042050

[8] Lorente-Pedreille R M, Sebastian M A, Saenz-Nuno M A and Medina-Martin M N 2019 A Metrological Characterization Approximation for the New Torque Measurement System in Wind Turbines Test Benches *IEEE Access* **7** 73469–79, DOI: 10.1109/ACCESS.2019.2920261

[9] Gutierrez Santiago U, Fernández Sisón A, Polinder H and van Wingerden J-W 2022 Input torque measurements for wind turbine gearboxes using fiber-optic strain sensors *Wind Energ. Sci.* **7** 505–21, DOI: 10.5194/wes-7-505-2022

[10] 19ENG08 WindEFCY 2025 *Traceable mechanical and electrical power measurement for efficiency determination of wind turbines* https://www.ptb.de/empir2020/windefcy/home

[11] Weidinger P, Foyer G, Kock S, Gnauert J and Kumme R 2018 Development of a torque calibration procedure under rotation for nacelle test benches *J. Phys.: Conf. Ser.* **1037** 52030, DOI: 10.1088/1742-6596/1037/5/052030

[12] Weidinger P, Foyer G, Kock S, Gnauert J and Kumme R 2019 Calibration of torque measurement under constant rotation in a wind turbine test bench *J. Sens. Sens. Syst.* **8** 149–59, DOI: 10.5194/jsss-8-149-2019

[13] Lietz S, Tegtmeier F, Kumme R, Röske D, Kolwinski U and Zöller K 2007 A new six-component force vector sensor – first investigations *IMEKO TC3*

[14] Park Y-K, Kumme R and Kang D-I 2002 Dynamic investigation of a binocular six-component force-moment sensor *Meas. Sci. Technol.* **13** 1311–8, DOI: 10.1088/0957-0233/13/8/320

[15] Park Y-K, Kumme R, Roeske D and Kang D-I 2008 Column-type multi-component force transducers and their evaluation for dynamic measurement *Meas. Sci. Technol.* **19** 115205, DOI: 10.1088/0957-0233/19/11/115205

[16] Zhang H, Pieper S and Heller M 2023 Direct measurement of input loads for the wind turbine drivetrain under test on a nacelle test bench *Forsch Ingenieurwes* **87** 93–105, DOI: 10.1007/s10010-023-00628-z

[17] Gnauert J, Jacobs G, Kock S, Bosse D and Janik B 2020 Design study for a multicomponent transducer for wind turbine test benches *J. Sens. Sens. Syst.* **9** 239–49, DOI: 10.5194/jsss-9-239-2020

[18] VDI – Verein Deutscher Ingenieure 1993 Systematic approach to the development and design of technical systems and products *(2221)* (Berlin: Beuth)

[19] Jakobs T, Jacobs G, Rolink A, Euler J, Hollas C and Röder J 2024 Designing Large Segmented Flexible Conical Plain Bearings as Wind Turbine Main Bearings *J. Phys.: Conf. Ser.* **2745** 12022, DOI: 10.1088/1742-6596/2745/1/012022

[20] FKM Forschungskuratorium Maschinenbau e.V. 2020 Rechnerischer Festigkeitsnachweis für Maschinenbauteile aus Stahl, Eisenguss- und Aluminiumwerkstoffen: FKM-Richtlinie 7th edn (Frankfurt am Main: VDMA)

[21] Deb K, Pratap A, Agarwal S and Meyarivan T 2002 A fast and elitist multiobjective genetic algorithm: NSGA-II *IEEE Trans. Evol. Computat.* **6** 182–97, DOI: 10.1109/4235.996017

Journal of Physics: Conference Series **3025** (2025) 012011 doi:10.1088/1742-6596/3025/1/012011

Virtual load sensors based on calibrated wind turbine strain sensors for damage accumulation estimation: a gap-filling technique

Bruno Rodrigues Faria[1], **Nikolay Dimitrov**[1], **Victor Perez**[2], **Athanasios Kolios**[1], **Asger Bech Abrahamsen**[1]

[1]DTU Wind And Energy Systems, Technical University of Denmark, Roskilde, Denmark

[2]GE Vernova Offshore Wind, Barcelona, Spain

E-mail: `brofa@dtu.dk`

Abstract. This study addresses the challenge of reliable lifetime assessments of the structural components of wind turbines through virtual load sensors trained and deployed on different combinations of measurement and SCADA data from a research turbine. Strain gauge measurements are introduced to improve the accuracy of the virtual sensor models while considering the impact of time-dependent drift of the strain sensor calibrations. A virtual duplicate of the turbine tower bending moment is made from the blade root strain gauge and similarly then a virtual blade root bending moment is provided from the tower bottom strain gauge combined with the SCADA data of the turbine. The virtual sensors were trained on both normal production and start-up/shutdown events resulting in a virtual tower fatigue damage equivalent load estimation with errors within 3.4 % compared to measurements. The two duplicate virtual sensors can be used to fill the gap of missing load sensor signal in periods, where the load sensors fail and it is demonstrated that the error of the continuous tower accumulated damage is in the order of 5 % for 6 months of missing sensor signal.

1. Introduction

Lifetime extension strategies are crucial for enhancing the asset value of wind turbines. In this context, the DNV lifetime extension of wind turbine standard [1] recommends the usage of load measurements and counted fatigue damage accumulation for updated reliability assessment of operating turbines. But once the cost of installing and maintaining load sensors becomes a barrier, virtual load sensors based on machine learning models are a feasible solution, as these can identify complex patterns with limited information [2]. A significant amount of research has been conducted in this area, as detailed in [3]. In particular, the usage of virtual sensors for high-frequency time-series load estimation has shown promising results from aeroelastic simulations [4, 5, 6, 7]. Literature on virtual sensors validated on high-frequency measurements from real-world operation scenarios remains limited and is the topic of this paper.

Content from this work may be used under the terms of the Creative Commons Attribution 4.0 licence. Any further distribution of this work must maintain attribution to the author(s) and the title of the work, journal citation and DOI.
Published under licence by IOP Publishing Ltd

2. Approach

The main objective of this work is to propose a health condition monitoring framework for reliable lifetime fatigue counting, where high-frequency virtual sensors can complement physical load measurements and ensure long-period stress information of the main structural components of a wind turbine. Figure 1 summarizes the methodology applied. First, different virtual sensors are trained with a different number of input channels containing a combination of calibrated strain gauges, tower top acceleration, and Supervisory Control and Data Acquisition (SCADA) (see Table 1). The estimated virtual loads are compared with the measurements to evaluate the difference between the virtual and real loads.Time series representatives of three Design Load Cases (DLC) recommended by the IEC 61400-1 [8] are identified within the wind turbine measurement database and are used to train the virtual models before they are deployed: Normal power production (DLC 1.2), Normal start-up (DLC 3.1), and Normal shutdown (DLC 4.1). The selection of such operations is justified by their effect on fatigue accumulation and optimized operational strategy [9]. In the current study, these three operations account for more than 99 % of the yearly tower bottom damage consumption. Finally, a scenario of the real strain gauges failing and the virtual sensors taking over as a gap-filling method is evaluated in terms of an uninterrupted lifetime assessment. With such a setup, the virtual sensor estimate can also be used to ensure the continuous quality of the input for damage accumulation models.

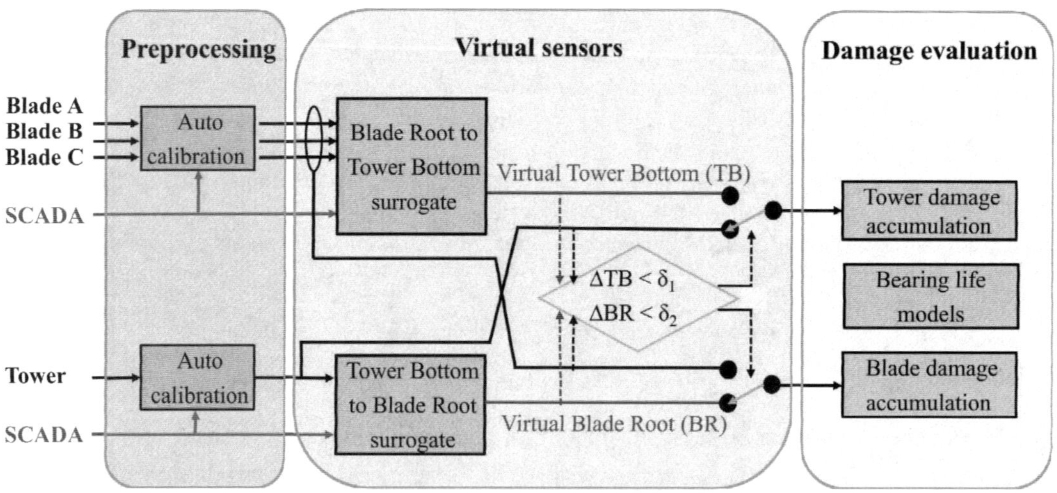

Figure 1: Schematic of the surrogate model (virtual sensors) combined within a wind turbine health monitoring framework.

2.1. Reliability of strain gauges

The approach is applied to the DTU V52 research wind turbine with an 850 kW power rating and a rotor diameter of 52 m installed at the DTU Risø campus in Denmark. The turbine is well-instrumented with high-frequency 50 Hz load and SCADA channels, as shown in Table 1.

Table 1: Description of sensor channels available in the measurement database of the DTU V52 research turbine for preprocessing and training of the virtual load sensors.

Description	Channel	Unit	Source
Controller Active power	ActPow	W	SCADA
Controller Rotor speed	ROT	rpm	SCADA
Controller Pitch position	pitch	deg	SCADA
Yaw angle	yaw	deg	SCADA*
Azimuth position	sa_ro_pos	deg	SCADA*
Tower-Top acceleration downwind	ACC3T	$[m/s^2]$	Measured
Tower Bottom Bending Moment - x-dir	MxTB	kNm	Measured
Tower Bottom Bending Moment - y-dir	MyTB	kNm	Measured
Blade A Root Bending Moment - x-dir	MxA1	kNm	Measured
Blade A Root Bending Moment - y-dir	MyA1	kNm	Measured
Blade B Root Bending Moment - x-dir	MxB1	kNm	Measured
Blade B Root Bending Moment - y-dir	MyB1	kNm	Measured
Blade C Root Bending Moment - x-dir	MxC1	kNm	Measured
Blade C Root Bending Moment - y-dir	MyC1	kNm	Measured

* Yaw angle and azimuth position are not commonly part of the SCADA data but will be classified as such in this work for the sake of simplicity of labeling the models and their inputs.

The virtual sensors include inputs such as measured blade root flapwise and tower base fore-aft bending moments, which are the result of preprocessing of the individual strain gauges combined with yaw angle (an encoder placed at the yaw ring which is not used in the training of the following models), the position of the strain gauges in the turbine, and the cross-section of the components at the measuring point. The tower bottom bending moments MxTB and MyTB are combined into the fore-aft (FA) heading, while blade strain gauges MxA1, MxB1, and MxC1 measure the main contribution to blade root flapwise bending moment but still need to be combined with the MyA1, MyB1, and MyC1 due to cross-talk effect. However, measurement campaigns may face reliability issues of the strain sensors, and long-term calibration factors of strain gauges in the tower bottom and blade root are investigated by applying automatic routines to identify and correct if zero-drift occurs (See Figure 4). Yaw sweeps [10] and Low-Speed Idling (LSI) [11] determine calibration factors when the overhanging mass of the rotor and blades are slowly yawing or rotating at low wind speeds.

2.2. Machine learning models for estimating measured bending moments time-series

Neural networks were selected as good candidates for this work as they are often well-suited for pure-measurement applications, capturing complex dependencies in high-frequency measurement data, and possibly handling noise and signal variability present in real-world measurements [4]. Three alternative neural network models have been used as benchmarks and each of these models was tuned using RandomSearch [12].

- **Feedforward Neural Network (FNN) - baseline**: Evaluates pure spatial correlation between the inputs and output (null hypothesis), as without time-lagged inputs, these networks can not capture time dependencies.
- **Lagged Feedforward Neural Network (laggedFNN)**: Receives n_{lags} inputs time-series and retrieves the same sequence from the output signal. Only a few n_{lags} are possible to keep the model trainable [4].
- **Long short-term memory (LSTM)**: Takes inputs in terms of time-series split in sequences (windows) of $w_s = 5$ to 20 seconds in length and downsampled to the wanted sampling frequency of $f_s = 50$ to 10 Hz. The LSTM is often less sensitive to noise at the cost of increased model complexity [13].

The machine learning models of this work have no inherent physics-dynamic understanding of the system. Training on more data would typically improve performance. However, larger models based on larger datasets tend to become computationally unfeasible to train. A threshold must be investigated regarding relevant and correlated inputs, whereas different model architectures are evaluated. To ensure that all models would have enough samples from the different wind turbine operation modes, a K-Means clustering [14] technique was used to extract samples from the dataset based on 10-min mean pitch angle, rotor speed, power, and operation condition itself.

2.3. Performance evaluation of damage accumulation

The Mean Absolute Error (MAE) of Damage Equivalent Loads (DELs), normalized by the reference measured DEL, shown in Equation 1 and 2, and the Normalized Root Mean Squared Error (NRMSE), shown in Equation 3, are selected to evaluate the performance of the virtual load sensors [4, 6].

$$\text{DEL} = \left(\frac{1}{N_{ref}} \sum_i n_i S_i^m \right)^{\frac{1}{m}} \tag{1}$$

$$\text{MAE} = \frac{1}{N} \sum_{i=1}^{N} \left| \frac{\text{DEL}_{\text{pred},i} - \text{DEL}_{\text{meas},i}}{\text{DEL}_{\text{meas},i}} \right| \times 100 \tag{2}$$

$$\text{NRMSE} = \frac{1}{\sigma_y} \sqrt{\frac{\sum_{i=1}^{N} (Y_{\text{pred},i} - Y_{\text{meas},i})^2}{N}} \tag{3}$$

DEL monitors the fatigue consumption by rainflow counting a time-series, according to [15], and accumulates its damage over time. N_{ref} represents the reference cycle count, n_i the number of cycles counted, S_i the stress amplitude of a cycle, and m denotes the Wöhler exponent. It was assumed m equal to 4, averaging the bi-linear curve from [16], for the tower virtual load sensor and equal to 10, from [17], for the blade sensor. In Equation 3, Y_i represents the bending moment at each timestamp, σ_y is the standard deviation of the 10 min window, and N is the number of data points. The NRMSE is the only metric sensitive to the mean load level, which guarantees the virtual sensors can be used on bearing life models or blade fatigue damage, accounting for the mean stress effect. This metric is qualitative in this study as operation modes with different oscillation amplitudes and levels are combined to evaluate the models' performance.

3. Results

3.1. Long-term calibration factors of strain gauges

Figure 2 shows the results of the automatic calibrations. The amplitude of the tower bottom strain gauges is the constant bending moment (kNm) due to the overhanging mass of the Rotor-Nacelle Assembly (RNA), and the amplitude at the blade root shows the constant bending moment due to the weight of the blade in periods of wind speed below the cut-in of the turbine. Tower bottom strain gauges display no zero-drift over the years, whereas blade root strain gauges on blade A MxA1 show a step change in 2020, and blade C MyC1 and blade B MyB1 have a constant drift. Failing to apply such calibration factors led to a systematic error in the damage accumulation models.

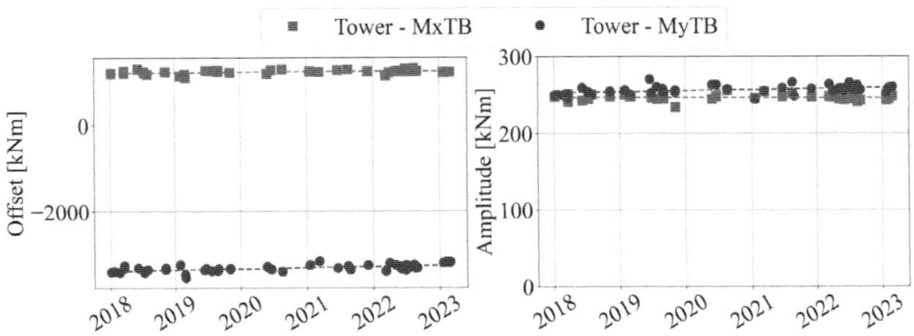

(a) Tower bottom strain gauges

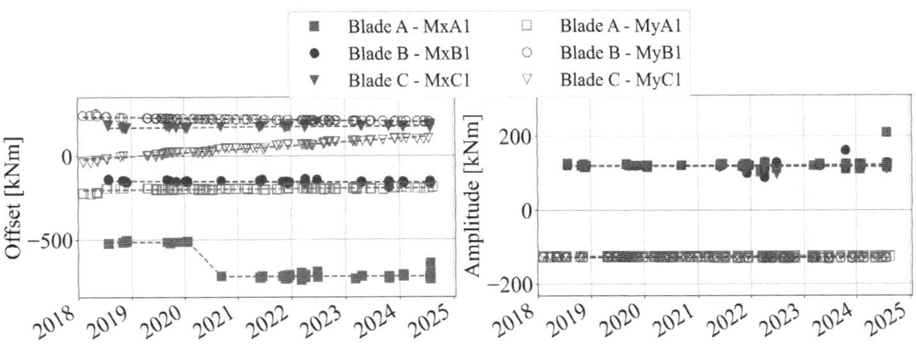

(b) Blade root strain gauges

Figure 2: The DTU V52 research turbine tower bottom (a) and blade root (b) strain gauge calibration factors identified from yaw sweep and Low-Speed Idling (LSI) operations. The offset trend lines highlight the zero drift, while an amplitude spread could point to a sensor malfunctioning.

3.2. Deploying models in different operation conditions

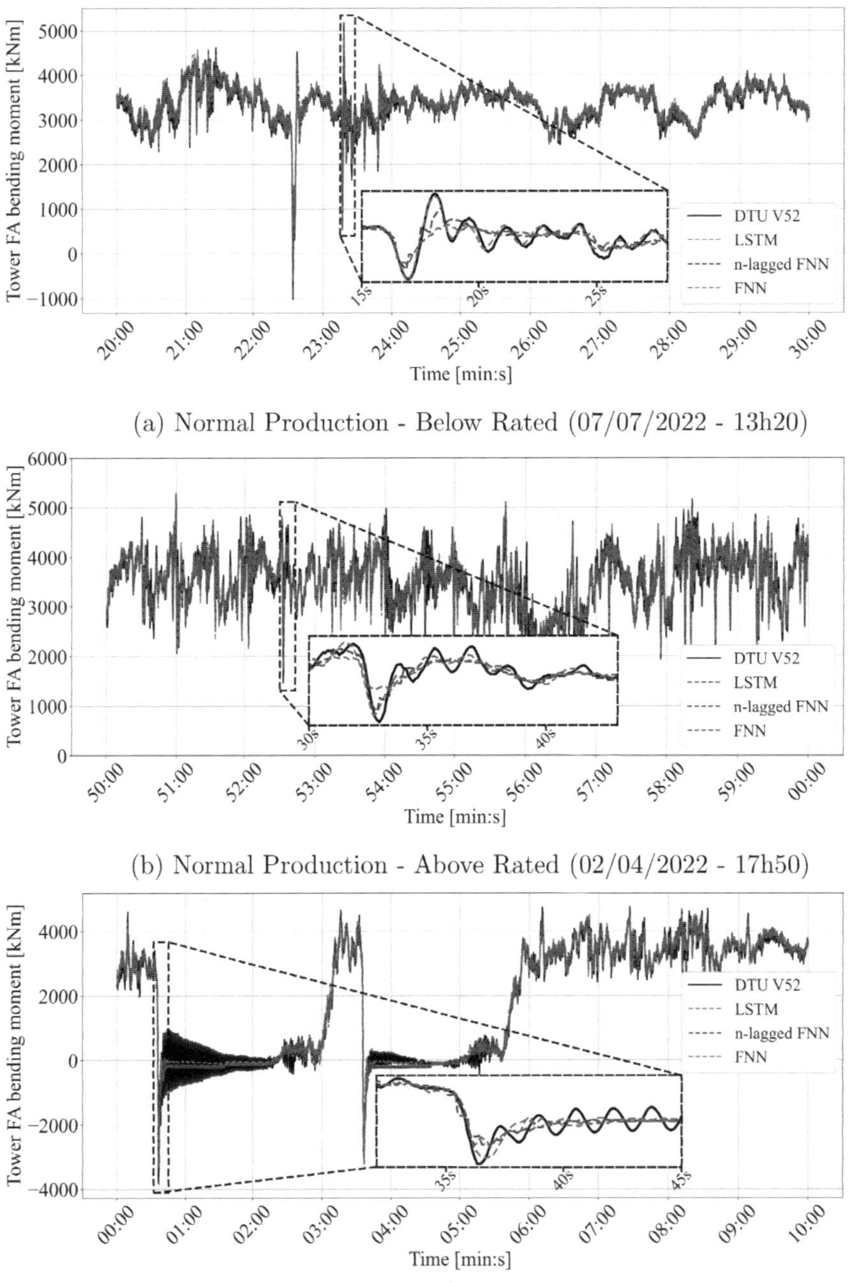

(a) Normal Production - Below Rated (07/07/2022 - 13h20)

(b) Normal Production - Above Rated (02/04/2022 - 17h50)

(c) Start-up and Shutdown (06/07/2022 - 16h00)

Figure 3: Tower bottom fore-aft (FA) bending moment predictions of the virtual sensor models (LSTM, n-laggedFNN, and FNN) compared to the real measured load signal in the operations conditions: (a) Normal Production - Below Rated, (b) Normal Production - Above Rated, and (c) Start-up and Shutdown. Insets: Short time window large oscillations. The virtual sensor models were trained with SCADA, tower top acceleration, and all three blade root flapwise bending moments (refer to Table 1).

Journal of Physics: Conference Series **3025** (2025) 012011 doi:10.1088/1742-6596/3025/1/012011

The virtual sensor models for tower bottom load estimation were trained on 160 hours of blade root bending moments and SCADA data down-sampled to a 10 Hz sampling frequency, while the blade root virtual load sensors were trained on 30 hours of tower root bending moments and SCADA data with 50 Hz. A thorough sensitivity analysis justified such a subset, whereas the performance evaluation is done on 10 hours of test samples. Three 10-min tower virtual sensor predictions are shown in Figure 3.

3.3. Training performance versus features selected

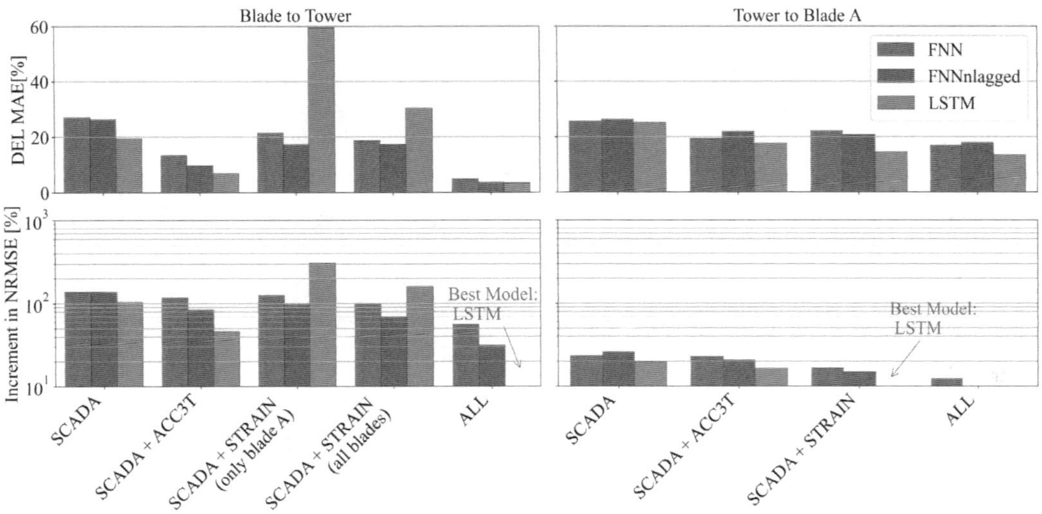

Figure 4: Performance metrics of virtual sensor models trained on different types of input sensors. The left column represents the tower virtual load performance, and the right column represents the blade root virtual load. The increment in NRMSE is a comparison to the best-performing virtual load sensor (normalized to 100% NRMSE).

Figure 4 shows the Mean Absolute Error (MAE) of Damage Equivalent Loads (DELs) of the different virtual sensor models when trained with subsets of the measurement and SCADA data or the full data set. The bars at the left show the error obtained using only the SCADA data, the SCADA data and the tower-top accelerometers(ACC3T), the SCADA data and the blade A strain gauge, the SCADA and the strain gauges of all the blades, and finally the full set of strain gauges, tower-top acceleration, and the SCADA data. It is observed that the combination of strain, acceleration, and SCADA data can benefit the learning of the LSTM model, while the addition of noisy inputs challenges the other models. The LSTM model achieved errors of 3.4% on DEL MAE for tower bottom load estimation when all inputs are used, while the second best performing model (SCADA + ACC3T) has an NRMSE 45% larger. This metric is still only qualitative because the simple rainflow counting method does not include the influence of the mean load on the damage. A more quantitative investigation of the virtual sensors can be evaluated by including the effect of the mean load using a Goodman diagram or by investigating the lifetime of the main and pitch bearings, where cycle counting of the absolute bearing load is needed [18].

On the other hand, the blade load estimations, in line with expectations, showed poorer results, as the input from the tower bottom strain gauges has partially filtered out higher frequency response components, and the blade root flapwise loads are mainly turbulence-driven, which is not included within the framework. The performance could be improved by further input selection, calibration, and tuning of hyperparameters.

3.4. Example of gap-filling technique deployment

One example of the virtual load being fed into the "Damage evaluation" framework (Figure 1) is carried out in Figures 5 and 6. The LSTM model with all inputs is trained with 160 hours of representative data in 2022 and deployed for consecutive 6 months in 2023, where the real strain gauges are assumed to fail to test the proposed approach. Instead of evaluating the Damage Equivalent Load (DEL) of the tower bottom virtual sensor, the virtual output is rainflow counted, and actual tower damage at the lowest circumferential weld is evaluated based on the two-slope SN curve from DNVGL-RP-C203 [16] category "D" (Steel S355 - butt weld) in air. The long-term lifetime is assessed using linear damage accumulation based on the Palmgren-Miner rule, following a similar approach detailed in [10]. Figure 5 shows the tower bottom damage accumulation from 2022-2024 as obtained from the measured tower bottom strain gauge as the reference-measured curve. In case the tower bottom strain gauge is lost from March to August 2023, there is a gap in the data, and the damage accumulation continues from August 2023 as the strain gauge is assumed to be repaired or replaced. The resulting damage accumulation at the end of 2024 will only reach 75 % of the measured accumulated damage because the March-August 2023 period is missing. Besides the underestimation of the accumulated damage, there is also an associated uncertainty if an extreme event happened in the period of missing strain signal. The alternative of switching to the virtual tower bottom load sensor from March 2023 is shown in the case of the LSTM surrogate model trained on all available inputs "Good virtual load gap-filling" in Figure 5). It is observed that the virtual sensor replicates the real signal reasonably well in the period of the missing strain gauge, and from August 2023, the measured accumulated damage is shown. The accumulated damage obtained from the virtual tower bottom load sensor is about 5 % below the measured reference accumulated damage showing the advantage of the virtual sensor compared to the case of the missing signal. The accumulated damage obtained from a "poor" virtual load sensor is added to Figure 5 by evaluating a virtual load sensor trained with blade root strain gauges showing a zero-drift issue as shown in Figure 2 as input and no automatic calibration as shown in Figure 1. The accumulated virtual damage shows an overestimation of the accumulated damage by the end of 2023 of 13 % compared to the damage obtained from the measurements in Figure 5. This illustrates that the automatic strain gauge calibration is important to

IOP Publishing
doi:10.1088/1742-6596/3025/1/012011

make the virtual sensor gap-filling damage accumulation technique accurate, but it is still better than the evaluation with 6 months of missing tower bottom strain sensors.

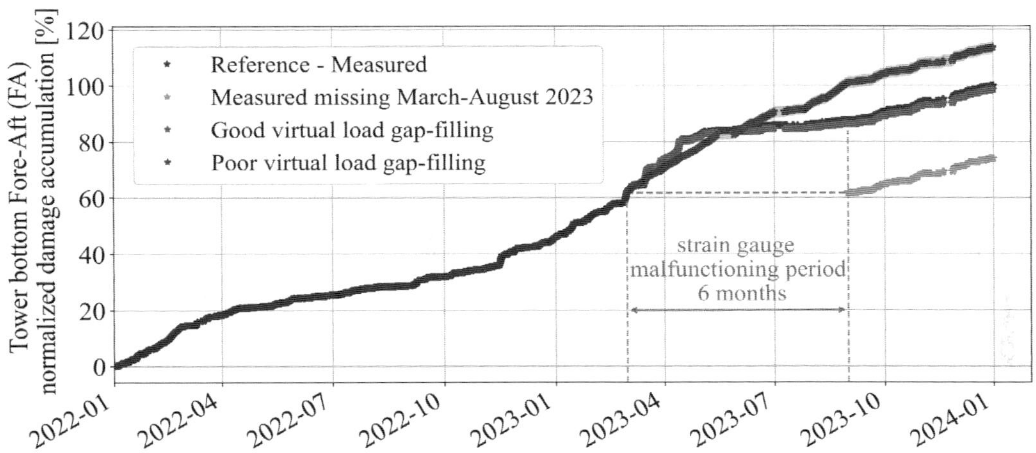

Figure 5: Deployment of the gap-filling tower bottom damage accumulation technique, where the tower bottom strain gauge is assumed to fail in March 2023 and to be repaired in August 2023. The "Good virtual load gap-filling" curve shows the damage accumulation obtained from the virtual tower bottom load sensor in the period of March-August 2023 using the best-performing surrogate model LSTM with all available inputs.

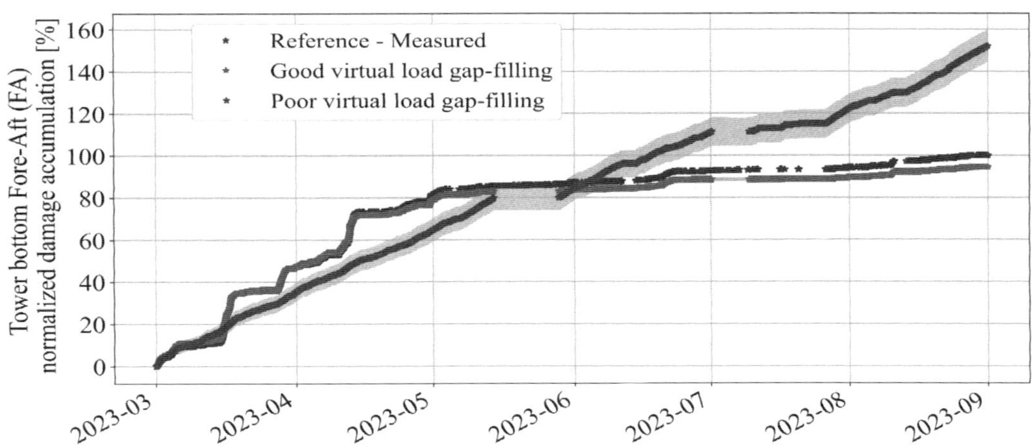

Figure 6: Focus on the replacement of the tower bottom strain signal by the virtual load sensors within the period where the tower bottom strain gauges are assumed to fail from March to August 2023. The transparent filled areas (light blue and green) represent the 97.5% confidence interval of the virtual load estimations.

4. Discussion

The tower bottom DEL MAE of the LSTM model's decrease from around 7.0% without using blade strain gauges (SCADA + ACC3T) to 3.4% with strain gauge models (ALL) looks small in Figure 4 compared to the effort of maintaining the setup, but such uncertainty when converted to a tower fatigue lifetime accumulation would be scaled due to the SN curve exponent (m) to levels around 21% and 10% (m=3) to 35 and 17% (m=5). In addition, models without strain sensors as input showed a bias in underpredicting the DEL estimation. Further analysis should be pursued to evaluate whether such an underprediction is even more pronounced for the specific cases of start-up/shutdown or curtailment operations, as the accelerometers are normally tuned for specific frequency ranges and might fail to incorporate slower oscillations or change in the overall load level. Figure 6 shows the 6-month period in which the virtual load sensor is imagined activated due to the absence of the real strain gauge. When the virtual load sensor is fed with uncalibrated blade root strain sensors, the absolute damage accumulation overestimates the measured damage by 52 % (blue), and the model uncertainty increases to 10 % (light blue). At the same time, the estimated accumulated fatigue damage relying on properly calibrated input data (Good virtual gap-filling) is under-predicted by only 5 % with a narrow confidence interval in the order of 1 % (light green area) due to low model uncertainty. It should be noted that the deployment of the virtual load sensors took only 5 hours, using the equivalent of a standard desktop-level computational capacity, to generate 6 months of 10 Hz time-series estimations.

5. Conclusions

Overall, the results show that the high-frequency virtual load sensors that use the LSTM model deliver the best performance when multiple measurements are used as input. The inclusion of a strain gauge signal has improved the model performance to an average tower bottom DEL estimation of 3.4% error while considering the different operation modes. However, such improvements depend on applying long-term calibration routines to correct for strain gauge drift. As a next step, one could evaluate a main bearing lifetime model based on the presented virtual load sensors and their sensitivity to different models' inputs and performances.

Acknowledgments

This work has received funding from the European Union's Horizon 2020 research and innovation program under grant agreement No 791875 of the ReaLCoE project. DTU (Technical University of Denmark) made the DTU V52 research wind turbine measurements available. A great thanks to Mathieu Nicolas Xin Yu Pelle, Søren Oemann Lind, Hazal Ozcan, and Steen Arne Sørensen for clarifying the DTU V52 measurement setup and Moritz Johann Gräfe and Vasilis Pettas for discussing virtual sensors.

References

[1] DNV GL AS. DNV-ST-0262 "Lifetime extension of wind turbines". Edition Amended November 2021

[2] Barlas T, Göçmen T and Riva R, "Development of a machine learning model for wind turbine fatigue and ultimate loads based on static loads", in Journal of Physics: Conf. Ser. 2767 052009, 2024, doi 10.1088/1742-6596/2767/5/052009

[3] Stetco A, Dinmohammadi F, Zhao X, Robu V, Flynn D, Barnes M, Keane J, Nenadic G, "Machine learning methods for wind turbine condition monitoring: A review," Renewable Energy, vol. 133, pp. 620-635, 2019, doi: 10.1016/j.renene.2018.10.047.

[4] Dimitrov N, Göçmen T, "Virtual sensors for wind turbines with machine learning-based time series models". Wind Energy. 2022; 25(9): 1626-1645. doi:10.1002/we.2762

[5] Kamel O, Kretschmer M, Pfeifer S et al, "Data-driven virtual sensor for online loads estimation of drivetrain of wind turbines", Forsch Ingenieurwes 87, 31–38 (2023). https://doi.org/10.1007/s10010-023-00615-4

[6] Grafe MJ, Pettas V, Dimitrov N, Cheng PW. "Machine learning based virtual load sensors for mooring lines using motion and lidar measurements", Wind Energy Science Discussions, 2024, 1–29. https://doi.org/10.5194/wes-2024-25

[7] Azzam B, Schelenz R, Rosche, B et al, "Development of a wind turbine gearbox virtual load sensor using multibody simulation and artificial neural networks", Forsch Ingenieurwes 85, 241–250 (2021). https://doi.org/10.1007/s10010-021-00460-3

[8] International Electrotechnical Commission "IEC 61400-1: Wind energy generation systems – Part 1: Design requirements, Edition 4" 2019

[9] Requate, N., Meyer, T., and Hofmann, R.: From wind conditions to operational strategy: optimal planning of wind turbine damage progression over its lifetime, Wind Energ. Sci., 8, 1727–1753, https://doi.org/10.5194/wes-8-1727-2023, 2023.

[10] Faria B, Abrahamsen, A, Sadeghi N, Dimitrov N and Kolios A, "Inclusion of low-frequency cycles on tower fatigue lifetime assessment through relevant environmental and operational conditions" in Journal of Physics Conf. Ser. 2767 042021, 2024, doi: 10.1088/1742-6596/2767/4/042021

[11] Pacheco J, Pimenta F, Guimarães S, Castro G, Cunha Á, Matos JC, and Magalhães F, "Experimental evaluation of fatigue in wind turbine blades with wake effects" in Engineering Structures, vol. 300, 117140, 2024, doi: 10.1016/j.engstruct.2023.117140.

[12] O'Malley, T., Bursztein, E., Long, J., Chollet, F., Jin, H., Invernizzi, L., and others, "Keras Tuner," 2019. [Online]. Available: https://github.com/keras-team/keras-tuner

[13] Bengio Y, Simard P and Frasconi P, "Learning long-term dependencies with gradient descent is difficult," in IEEE Transactions on Neural Networks, vol. 5, no. 2, pp. 157-166, March 1994, doi: 10.1109/72.279181.

[14] Pedregosa, F., Varoquaux, G., Gramfort, A., Michel, V., Thirion, B., Grisel, O., Blondel, M., Prettenhofer, P., Weiss, R., Dubourg, V., Vanderplas, J., Passos, A., Cournapeau, D., Brucher, M., Perrot, M., and Duchesnay, E., "Scikit-learn: Machine Learning in Python," Journal of Machine Learning Research, vol. 12, pp. 2825–2830, 2011. [Online].

[15] ASTM, "ASTM E1049-85 Standard Practices for Cycle Counting in Fatigue Analysis," 1985.

[16] DNV GL AS., "DNVGL-RP-C203 Fatigue Design of Offshore Steel Structures" - Recommended Practice. Edition April 2016.

[17] DNV GL AS., "DNV-ST-0376 Rotor blades for wind turbines". Edition April 2024.

[18] Kenworthy, J. , Hart, E., Stirling, J., Stock, A., Keller, J., Guo, Y., Brasseur, J., and Evans, R., "Wind turbine main bearing rating lives as determined by IEC 61400-1 and ISO 281: A critical review and exploratory case study". Wind Energy 27, 179-197, 2024, doi.org/10.1002/we.2883.

Journal of Physics: Conference Series **3025** (2025) 012012 doi:10.1088/1742-6596/3025/1/012012

Sensitivity of Drivetrain Condition Monitoring Signals to Wind Turbine Blade Damage

Simon J Watson[1*], Sumit K Pal[1], Donatella Zappalá[1] and Amir R Nejad[2]

[1] Faculty of Aerospace Engineering, TUD, Kluyverweg 1, 2629HS, Delft, NL
[2] IMT, NTNU, Jonsvannsveien 82, 7050 Trondheim, Norway

*E-mail: S.J.Watson@tudelft.nl

Abstract. This paper studies the sensitivity of drivetrain condition monitoring system (CMS) signals to blade damage, exploring how these signals, particularly vibration, can serve as a potential tool for detection and tracking damage progression. This is achieved using a decoupled simulation approach, combining an aeroelastic solver with a drivetrain model. First, aeroelastic simulations are performed in OpenFAST, where the low-speed shaft (LSS) forces, moments, and tower top position vector are extracted and transferred to the drivetrain model. The drivetrain is modelled using the multi-body simulation environment SIMPACK. Blade damage is introduced in OpenFAST by reducing stiffness in the flap-wise or edgewise direction. The reference DTU-10MW onshore wind turbine is used as a test case. First, the impact of blade damage on LSS shear forces is analysed. Then the drivetrain response is assessed using virtual velocity sensors placed at the main bearing, rear bearing and gearbox housing. Results indicate that damage occurring in the blade mid-span region shows higher sensitivity compared to tip and root locations. A positive correlation is observed between LSS shear force and bearings side-side velocity, with higher forces leading to increased vibration. Additionally, the trend suggests that higher stiffness reduction results in higher velocity, indicating damage progression.

1. Introduction

Wind turbine blades represent a substantial portion of the overall cost of a wind turbine and exhibit a notable failure rate in comparison to other components [1]. Blade failures can reduce turbine yield or, in extreme cases, stop turbine operation and require replacement, which is a time-consuming and costly operation involving heavy lift equipment. Blade monitoring is mainly carried out by visual inspection, either from the ground, using drones, or by physically climbing the blade. Such methods can detect external problems, but not the internal defects. In this study, a preliminary sensitivity analysis of how drivetrain condition monitoring signals respond to blade damage is presented to assess their potential for damage identification. One of the key advantages of using drivetrain CMS signals is that they are commonly available in commercial CMS installed on operational wind turbines, eliminating the need for additional dedicated blade sensors. Leveraging these existing systems for blade damage detection could offer a cost-effective and scalable alternative to more invasive or expensive monitoring solutions, with the potential to detect both internal and external blade damage.

Content from this work may be used under the terms of the Creative Commons Attribution 4.0 licence. Any further distribution of this work must maintain attribution to the author(s) and the title of the work, journal citation and DOI.
Published under licence by IOP Publishing Ltd

Fremmelev et al. (2023) [2] conducted a full scale wind turbine blade monitoring campaign for the detection of damage initiation and progression due to fatigue-driven loads during in-house testing. They demonstrated that detecting the initiation and progression of damage is possible with an actuator and a single vibration sensor positioned within 10 meters of the damage site. Recently, Varouxis et al. (2024) [3] presented an experimental study to detect fatigue damage propagation in a full scale wind turbine blade. However, deploying an array of accelerometers or strain gauges used by [3] along the entire span of operational wind turbine blades is impractical due to the high costs and additional maintenance requirements . This highlights the need for alternative, cost-effective monitoring solutions.

Popular damage detection methods for operational wind turbine blades include non-destructive testing (NDT) techniques, such as thermography, ultrasonics, acoustic emissions, and vibration analysis [4]. Zhang et al. (2022) [5] investigated the detection of wind turbine blade trailing edge cracks using aerodynamic noise in an experimental setup. Their findings suggest that tonal noise features caused by a cracks can be detected under clean or low turbulent inflow conditions at moderate angles of attack. Khoshmanesh et al. (2022) [6] used infrared thermography to study damage evolution process in the adhesive joint of a representative wind turbine blade section. They highlighted the thermography's advantage in identifying early crack formation and localisation. Although these methods are effective for detecting and localizing damage, they are often costly and time-consuming. Alternatively, leveraging existing drivetrain condition monitoring system (CMS) signals on operational wind turbines, in particular vibration, could serve as a potential practical tool for tracking blade damage progression.

This study presents a preliminary analysis aimed at quantifying the sensitivity of drivetrain CMS signals to blade structural damage. Using the reference DTU-10MW onshore wind turbine as a test case, changes in the RMS value of the first harmonic of the wind turbine rotational frequency (1P) between healthy and damaged blade conditions are analysed to identify potential damage indicators in drivetrain CMS signals. The remainder of this paper is structured as follows: the next section outlines the methodology, followed by a presentation of the results and discussion of the findings, concluding with key insights and future directions.

2. Methodology

This study employs a decoupled simulation approach to analyse the sensitivity of drivetrain condition monitoring signals to blade structural damage. The methodology incorporates:

1. *Aeroelastic Rotor Simulation* – Performed using OpenFAST to compute aeroelastic responses of the wind turbine, including main shaft loads and tower-top motion.
2. *Drivetrain Multi-Body Dynamics Simulation* – Conducted in SIMPACK, using outputs from OpenFAST as inputs to assess drivetrain dynamics and responses.
3. *Coupling* – The interaction between the rotor model in OpenFAST and the drivetrain model in SIMPACK is established to ensure consistent load transfer and dynamic behaviour.
4. *Blade damage model* – Blade structural damage is incorporated into the aeroelastic rotor simulations to assess its impact on system dynamics.
5. *Signal Processing* – The recorded time domain responses are analysed in the frequency domain to extract relevant features for condition monitoring.

2.1 OpenFAST wind turbine model

This paper utilizes OpenFAST [7] to model the coupled dynamics of a wind turbine system. The aerodynamic loads on the blade are computed using the AeroDyn 15 [8] module, which is based on Blade-Element/Momentum (BEM) theory. The structural dynamics of the tower and blades are modelled using the ElastoDyn module, which requires modal coefficients, damping properties, and elastic parameters for each degree of freedom. The ElastoDyn module allows consideration of only the first two flap wise bending modes and first edge wise bending mode. To model blade damage, the flexural rigidity (EI) in the flap-wise or edge-wise direction is modified at a specific blade location within the ElastoDyn input file. This approach is further discussed in a subsequent section. The Reference Open Source Controller (ROSCO) was used in this setup, which uses collective pitching [9]. The specifications of the DTU 10MW reference wind turbine are listed in Table 1. For this study, a steady state wind condition is considered.

Table 1. Specifications of the DTU 10MW RWT [10]

Parameter	Value
Rating	10 MW
Control	Variable speed, collective pitch, ROSCO
Rotor, Hub diameter (m)	178.3, 5.6
Hub height (m)	119.0
Cut-in, rated, and Cut-out wind speed (m/s)	4, 11.4, and 25
Cut-in, Rated rotor speed (RPM)	6, 9.6
Hub Overhang (m)	7.1
Shaft tilt angle (deg)	5.0
Pre-cone angle (deg)	-2.5
Rotor mass (kg)	227962
Nacelle mass (kg)	446036
Tower mass (kg)	628442

2.2 SIMPACK drivetrain model

Figure 1 shows the layout of the drivetrain model used in this research. A detailed study about this drivetrain model can be found in [11]. The drivetrain has a four-point support and integrates a three-stage gearbox. In the first and second planetary stages, ring gears are fixed to the gearbox housing, which is supported on the bedplate via torque arms. Input torque is applied on the planet carriers, with the sun gears serving as the output torque. In the third parallel stage, torque is transferred from the sun gear of the second planetary stage to the generator through a gear pair (Gear-Pinion) in the high-speed stage. This configuration converts high torque on the low-speed shaft to lower torque on the high-speed shaft. Gear teeth contact is modelled using the FE225 force element in SIMPACK, where the contact force is composed of stiffness, damping, and friction terms. The gears and shafts inside the gearbox are treated as rigid bodies. The specifications of the drivetrain are mentioned in the Table 2.

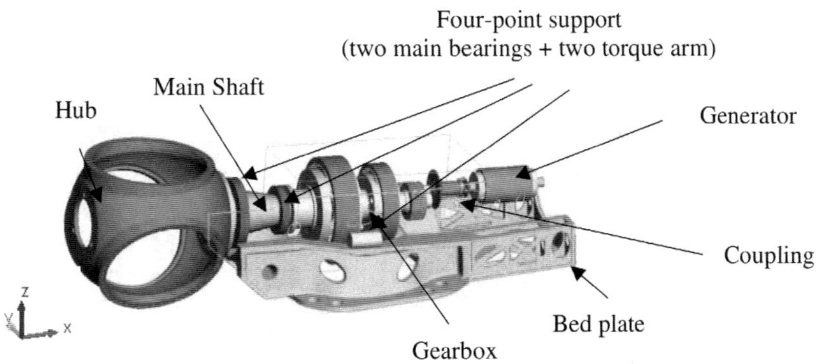

Figure 1. DTU 10MW drivetrain model [11]

Table 2. Drivetrain specifications of DTU 10MW [11]

Parameter	Value
Drivetrain type	Four-point support
Gearbox type	Two planetary + one parallel
First stage ratio	1:4.423
Second stage ratio	1:5.192
Third stage ratio	1:2.179
Total ratio	1:50.039
Rated input shaft speed (rpm)	9.6
Rated generator shaft speed (rpm)	480.4
Rated input shaft torque (kN-m)	9947.9
Rated generator shaft torque (kN-m)	198.8
Gearbox mass (tonne)	60.43
Bedplate mass (tonne)	102.39
Equivalent drive-shaft torsional-spring constant (Nm/rad)	2317025000
Equivalent drive-shaft torsional-damping constant (Nm/(rad/s))	9240560

2.3 Coupling

The coupling between the rotor model, described in Section 2.1, and the drivetrain model, described in Section 2.2, is illustrated in Figure 2. The rotor dynamics are first simulated in OpenFAST, with a simplified drivetrain setup, represented by a spring-damper system with only a torsional degree of freedom. The equivalent drive-shaft torsional spring and damping constant are given in Table 2. The tower top position ($\vec{X}$), low speed shaft (LSS) force, LSS moment ($\vec{F}, \vec{M}$), and generator speed are used as input to the SIMPACK drivetrain setup. A proportional-integral (PI) controller is used to couple the input generator speed (ω_{Gen}) from OpenFAST and the resulting generator speed in SIMPACK due to applied loads and tower top motion. This is to maintain the same dynamic condition between the two independent models.

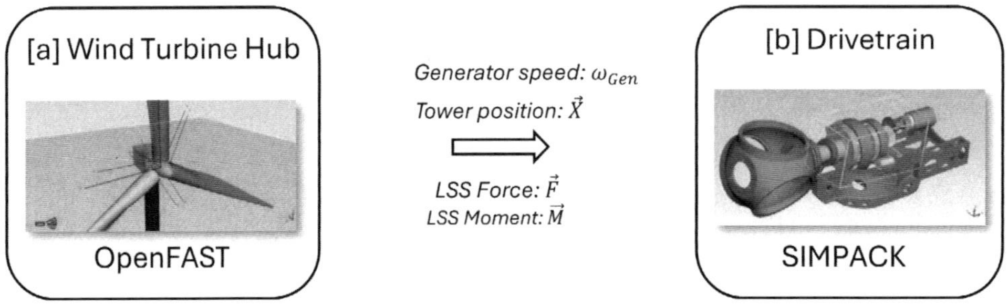

Figure 2. Flow of information between the two models in the decoupled approach.

2.4 Blade damage model

Khoshmanesh et al. [6] conducted a fatigue tension test on a spar-to-shear web assembly representative of a wind turbine blade. During the initial damage phase, marked by the appearance of transverse cracks, the stiffness decreased by less than 2%, whereas the damping increased by about 18%. During the crack saturation phase, the average stiffness reduction was 4.7%. The test specimen experienced a total average stiffness reduction of 8.6% before final failure. While this study focused on only a part of the wind turbine (WT) blade cross-section, and its findings may not directly apply to the more complex structure of a full blade, it provides valuable insights into the relationship between blade damage and stiffness reduction.

In this paper, to study the effect of blade damage on the drivetrain system, the blade stiffness in either the flap-wise or edge-wise directions is reduced in a range between 0.1 and 2% from its healthy value. Since structural damping is inherently related to stiffness, it must also be adjusted. The relationship between the modal damping ratio (ζ) and stiffness proportional damping (β) is expressed as follows [12] :

$$\beta_j = {}^{\zeta_j}/_{\pi\omega_j} \quad (1)$$

where, j is the jth mode ($j\epsilon\{1,2,3\}$), corresponding to the first two flap-wise bending modes and first edge-wise bending mode; β is the stiffness proportional damping (Rayleigh damping) co-efficient, ζ_j is the modal damping ratio for the jth mode, and ω_j is the natural frequency of the jth mode given by

$$\omega_j = (1/2\pi)\sqrt{K_{jj}/M_{jj}} \quad (2)$$

with K_{jj} and M_{jj} represent the stiffness and mass terms for the jth mode, respectively. Using Eq. (1), the effect of local stiffness changes can be incorporated into the modal damping ratio, as follows:

Step 1: Derive initial β_j

Use Eq. (1) to calculate intial β_j using the original modal damping (ζ_j) and natural frequency (ω_j) associated with the healthy blade stiffness distribution (EI)

Step 2: Modify stiffness distribution $E'I'$ and compute new natural frequency (ω_j')

Apply stiffness reduction at the selected blade location and use a Finite Element (FE) solver, e.g. BModes [13], to obtain the modified natural frequency (ω_j')

Step 3: Calculate New Modal Damping Ratio (ζ_j')

Use Equation (1) to calculate the new modal damping ζ_j' with the updated natural frequency (ω_j') from step 2, while keeping β_j constant.

Step 4: Run Aeroelastic Simulations

Incorporate modified damping ratio (ζ_j') for each mode ($j\epsilon\{1,2,3\}$) and stiffness distribution ($E'I'$), into OpenFAST simulations.

Following the above procedure ensures that the contribution of local stiffness changes to the global modal damping is accurately captured, as shown in Figure 3. The blade damage conditions investigated include various local stiffness reduction levels, with changes in flap or edge-wise stiffness of -0.1%, -0.5%, -1.0%, and -2.0%. Each case considers a 1.0 m damage length at nine locations along the blade, ranging from 5 m to 85 m from the root, with a 10 m interval between locations. This sums up to a total of 504 simulations. For each damage case, a separate damping ratio is calculated based on the steps outlined above. Figure 4 shows the change in modal damping ratio of the 1st and 2nd flap-wise modes due to a stiffness reduction of 0.1 to 2% at 45 m from the blade root. It can be seen from Figure 4 that the modal damping ratio (ζ) reduces with stiffness reduction, which is a consequence of the proportional damping model.

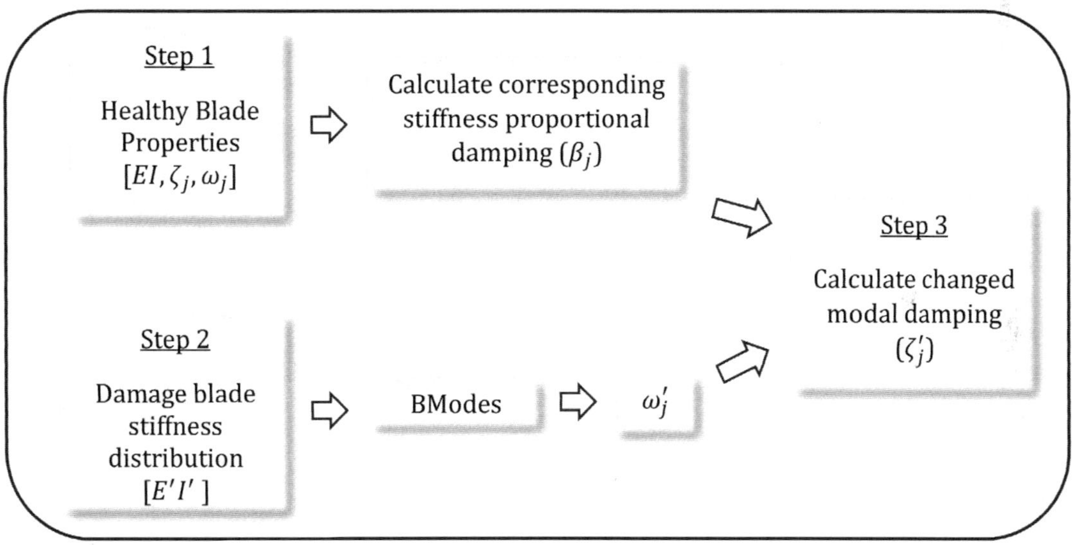

Figure 3. Procedure for including stiffness reduction into the modal damping ratio calculation

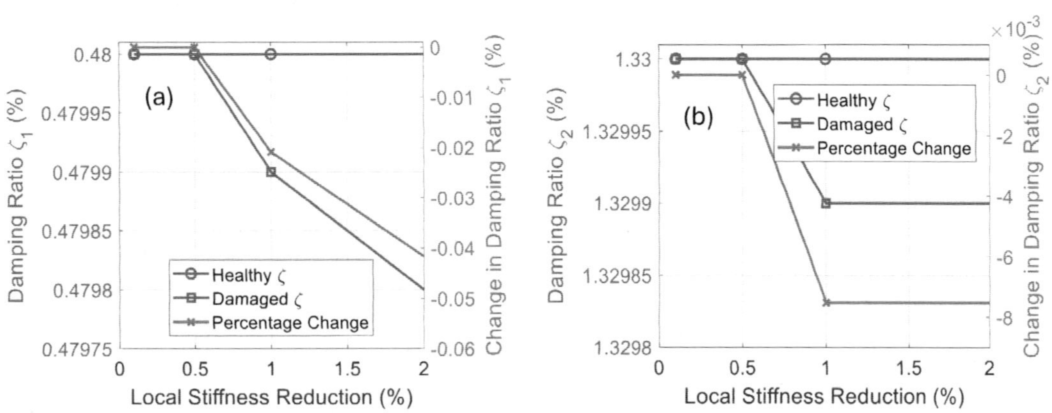

Figure 4. Modal damping ratio change with flap-wise stiffness reduction from 0.1 to 2% at 45 m location from blade root (a) 1st flap mode (b) 2nd flap mode.

2.5 Signal processing

All the blade damage cases listed in section 2.4 were simulated for 3000 s with a signal sampling frequency (f_s) of 400 Hz, discarding the initial 1500 s to avoid initial transient effects. The remaining signals were divided into five segments with 50% overlap and each window was anlaysed in the frequency domain. The analysis procedure involves the following steps :

1) A Discrete Fourier Transform (DFT) is applied to the windowed signal (x_n). The DFT is defined as,

$$X_k = \sum_{n=1}^{N} x_n e^{-2\pi i \frac{k}{N} n} \tag{3}$$

Where $k \in [0, N-1]$ is the frequency index.

2) To mitigate energy leakage, the Hanning window function (w_n) is applied to the sampled signal x_n to obtain the PSD:

$$S_{xx}(k) = \frac{2X_k \bar{X}_k}{W f_s} \tag{4}$$

Where $W = \sum_{n=1}^{N} w_n$ and $\bar{X}_k$ is the complex conjugate of X_k.

3) The Root Mean Square (RMS) of the 1P frequency with a $\pm 0.02\ Hz$ bandwidth is calculated as:

$$RMS = \sqrt{\sum_k S_{xx}(k) . \frac{f_s}{N}} \tag{5}$$

Taking the RMS value within a certain bandwidth ensures that fluctuations in 1P-related energy are captured robustly, providing a more stable indication of damage progression compared to tracking only the peak amplitude.

3. Results and discussion

The proposed sensitivity analysis is divided into two parts. The first sub-section focuses on the system dynamic response from the rotor side, followed by the analysis of the drivetrain system responses in the second sub-section.

3.1 Rotor dynamics

Figure 5 shows the absolute differences in the 1P RMS of LSS shear force in the side-side direction (F_{ys}), between healthy and damaged conditions across different blade spans and wind speeds. Figures 5 (a) and (b) refer to flap-wise and edge-wise stiffness reduction cases, respectively. LSS F_{ys} is the input force to the drivetrain setup and is measured relative to the fixed reference frame. Figure 5 includes a total of nine bar groups, corresponding to nine locations from the blade root to the tip. Each bar group represents seven wind speeds, from 8 to 25 m/s, listed in increasing order. The magnitudes corresponding to different damage levels (0.1-2% stiffness change) are stacked on top of each other for each wind speed within a given blade location. A comparison between Figure 5 (a) and (b) reveals that the 1P harmonic of LSS F_{ys} exhibits the highest sensitivity to flap-wise stiffness reduction at around the blade mid-span (~45 m) and at the rated wind speed of 11.4 m/s for same amount of stiffness change. This could be due to the mid-span of the blade being more susceptible to maximum bending stresses.

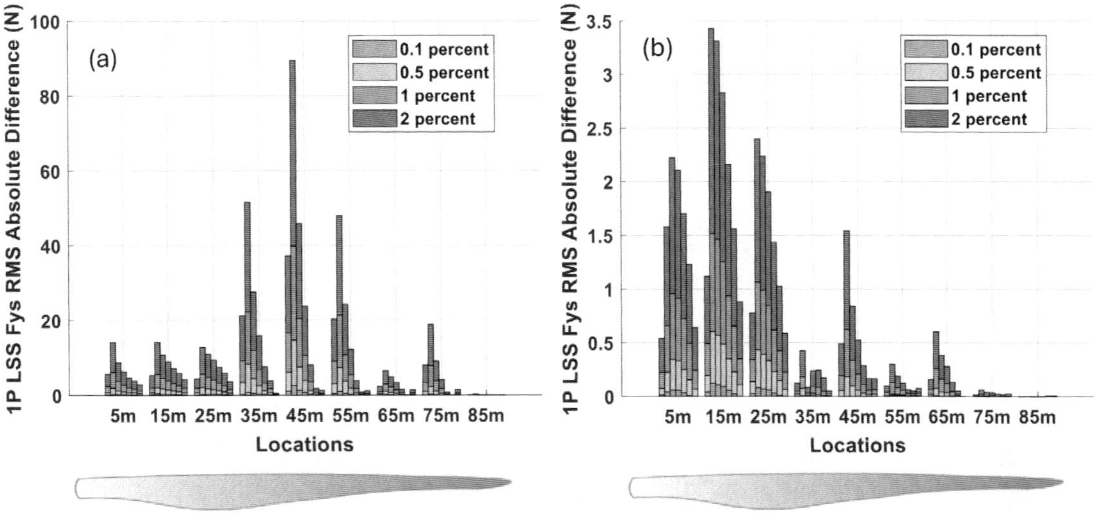

Figure 5. Difference in 1P LSS F_{ys} RMS between healthy and damage conditions due to change in blade (a) flap-wise (b) edge-wise stiffness at different span from root for the wind speeds 8, 11.4, 13, 15, 18, 21, 25 m/s.

While a 2% stiffness change contributes to around 50 N change in the 1P RMS of LSS F_{ys} in the flap-wise direction at ~45 m and rated wind conditions (Figure 5 (a)), it only contributes to a 1.9 N increase in the edge-wise direction at ~ 15m and rated wind conditions, (Figure 5 (b)). From Figure 5 (a), it is also evident that a 0.1% reduction in flap-wise stiffness alters the 1P RMS by only 2.6 N at ~45 m and rated wind conditions.

From both subplots, it can be concluded that blade damage associated with flap-wise or edge-wise stiffness reduction near the blade tip contributes significantly less to the change of 1P RMS. Therefore, in this case, in the absence of additional information, it may be challenging to distinguish damage at the blade tip based only on the value of 1P RMS. A similar conclusion applies to damage near the blade root (Figure 5 (a-b)).

3.2 Drivetrain response analysis

In Figure 6, the 1P RMS values are plotted for the gearbox housing side-side velocity, upwind main bearing (INP_A: Input A) and rear bearing (INP_B: Input B) side-side velocity, all due to flap-wise stiffness changes at 45 m from the blade root. It can be observed in all the subplots that the damage progression from 0.1% to 2% is distinctly visible around rated wind speed (11.4, 13, and 15 m/s). The damage progression trends near the cut-in (8 m/s) and cut-out wind speeds (21 and 25 m/s) are less pronounced. This may be attributed to the fact that the 1P change in LSS F_{ys} at those wind speeds shows a significantly lower contribution compared to the rated wind speed (Figure 5 (a)). It is also noticeable that out of all these subplots housing velocity seems more sensitive to damage progression.

In Figure 7, the correlations between the LSS F_{ys} and upwind main bearing and rear bearing side-side 1P RMS velocities are presented for flap-wise stiffness reduction at 45 m from the blade root. The figure shows how the local damage in a wind turbine blade is translated into system responses observed in different drivetrain components.

Journal of Physics: Conference Series **3025** (2025) 012012 doi:10.1088/1742-6596/3025/1/012012

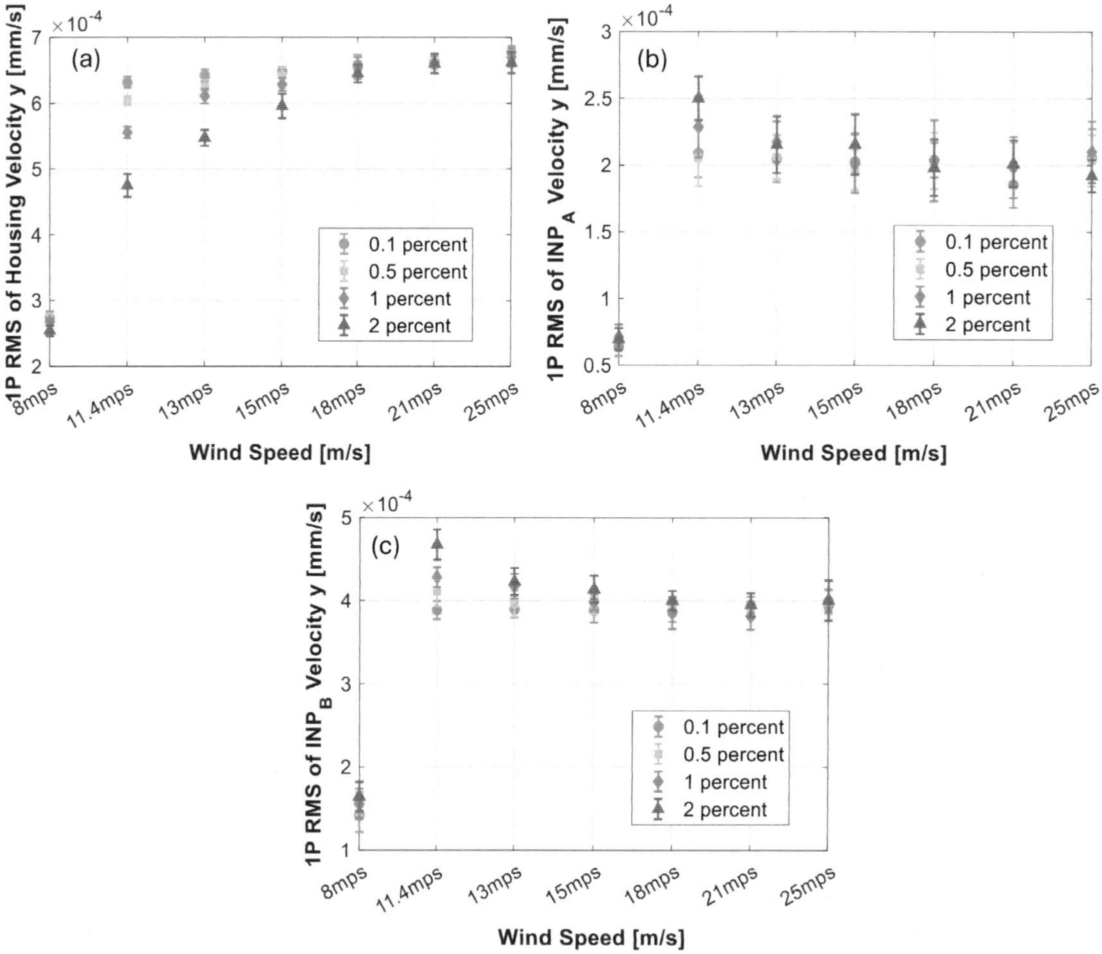

Figure 6. 1P RMS values for (a) gearbox housing side-side velocity, (b) upwind front bearing (INP_A), (c) upwind rear bearing (INP_B) side-side velocity at different wind speeds due to blade flap-wise stiffness changes at 45m from blade root.

From Figure 7, it can be observed that at most wind speeds (especially around rated), the drivetrain responses increase slightly with LSS F_{ys}, showing clear damage progression due to flap-wise stiffness reduction (0.1 to 2%) at the 45 m blade span. While the induced damage causes a change in the shear force and 1P RMS velocities, the magnitude of this change is relatively small and would likely be difficult to detect on a real operating turbine [14].

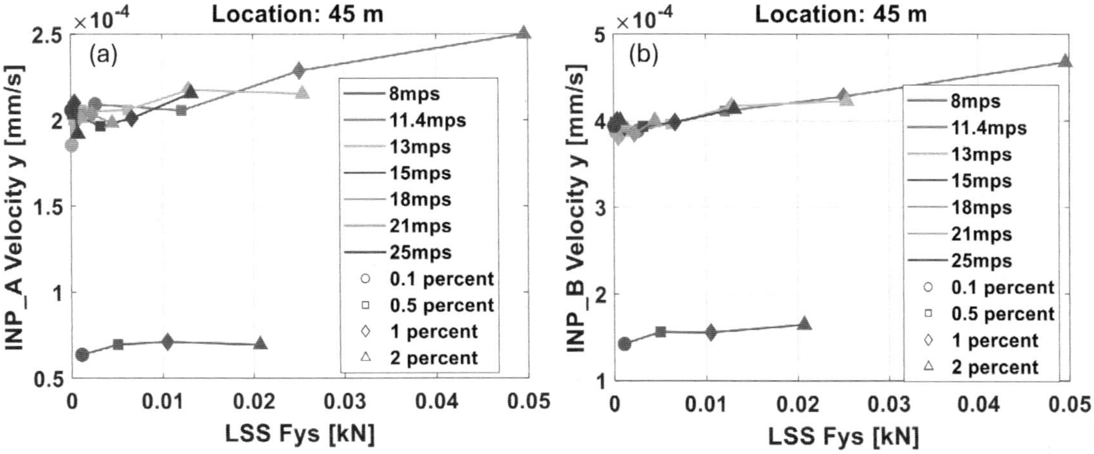

Figure 7. Correlations between the LSS F_{ys} and (a) upwind main bearing (INP_A) side-side velocity (b) rear bearing (INP_B) side-side 1P RMS velocities, due to different stages of the flap-wise stiffness reduction at 45 m from blade root.

4. Conclusions

In this paper, the effect of local stiffness reduction in wind turbine blades on drivetrain CMS signals was studied for the DTU 10MW reference wind turbine. A decoupled approach, combining an aeroelastic solver and an independent drivetrain model, was adopted to achieve the proposed goal. Only changes in the 1P RMS values caused by the introduction of the blade damage were considered in this study, serving as an initial step in the investigation. First, the impact of blade local stiffness change on the rotor dynamics was investigated by analysing the 1P peak amplitudes of the LSS side-side shear force. The results showed that, compared to the blade tip and root regions, the mid-span region is the most sensitive to the same amount of damage. The maximum change in the 1P RMS values occurred near the rated wind speed (11.4, 13, 15 m/s).

Thereafter, the simulated LSS shear forces, moments, and tower top position vector were transferred to the independent drivetrain setup in the SIMPACK multibody-dynamics platform. The 1P RMS of gearbox housing, main and rear bearing side-side velocities for both healthy and damaged conditions were analysed. Similar to the LSS side-side shear force, the drivetrain responses exhibited clear damage progression at the rated wind speed. However, for the wind speeds close to the cut -in and cut-out regions, the damage progression trend was unclear. Considering the magnitude of the associated error bars, a longer simulation time might be required for more clarity. Finally, the correlation between the LSS side-side shear force and the front and rear bearing side-side velocities at the blade mid-span (~45 m location) demonstrated a clear progression of damage as the stiffness reduction increased from 0.1% to 2.0%. Future studies should investigate the impact of higher local blade stiffness reduction and the inclusion of turbulence on drivetrain CMS signal sensitivity. Such work will be used to determine at what level of stiffness change (and thus damage) the changes to system response become significant and could be used in a practical condition monitoring system.

References

[1] D. Zappalá and P. J. Tavner, 'Wind Turbine Reliability - Maintenance Strategies', in *Comprehensive Renewable Energy*, Elsevier, 2022, pp. 353–370. doi: 10.1016/B978-0-12-819727-1.00154-0.

[2] M. A. Fremmelev, P. Ladpli, E. Orlowitz, N. Dervilis, M. McGugan, and K. Branner, 'A full-scale wind turbine blade monitoring campaign: detection of damage initiation and progression using medium-frequency active vibrations', *Structural Health Monitoring*, vol. 22, no. 6, pp. 4171–4193, Nov. 2023, doi: 10.1177/14759217231163471.

[3] T. Varouxis, V. K. Dertimanis, I. Abdallah, E. Chatzi, V. Pakrashi, and A. Malekjafarian, 'Deep-Learning-based damage detection of an experimentally tested full-scale wind turbine blade.', *eJNDT*, vol. 29, no. 7, Jul. 2024, doi: 10.58286/29852.

[4] K. Kong, K. Dyer, C. Payne, I. Hamerton, and P. M. Weaver, 'Progress and Trends in Damage Detection Methods, Maintenance, and Data-driven Monitoring of Wind Turbine Blades – A Review', *Renewable Energy Focus*, vol. 44, pp. 390–412, Mar. 2023, doi: 10.1016/j.ref.2022.08.005.

[5] Y. Zhang, F. Avallone, and S. Watson, 'Wind turbine blade trailing edge crack detection based on airfoil aerodynamic noise: An experimental study', *Applied Acoustics*, vol. 191, p. 108668, Mar. 2022, doi: 10.1016/j.apacoust.2022.108668.

[6] S. Khoshmanesh, S. J. Watson, and D. Zarouchas, 'The effect of the fatigue damage accumulation process on the damping and stiffness properties of adhesively bonded composite structures', *Composite Structures*, vol. 287, p. 115328, May 2022, doi: 10.1016/j.compstruct.2022.115328.

[7] National Renewable Energy Laboratory, 'OpenFAST Documentation - Release v2..3.0', Oct. 2020.

[8] J. Jonkman, 'AeroDyn v15 User's Guide and Theory Manual'. [Online]. Available: https://www.nrel.gov/wind/nwtc/assets/pdfs/aerodyn-manual.pdf

[9] N. J. Abbas, D. S. Zalkind, L. Pao, and A. Wright, 'A reference open-source controller for fixed and floating offshore wind turbines', *Wind Energ. Sci.*, vol. 7, no. 1, pp. 53–73, Jan. 2022, doi: 10.5194/wes-7-53-2022.

[10] C. Bak *et al.*, 'Light Rotor: The 10-MW reference wind turbine', Jan. 2012.

[11] S. Wang, A. R. Nejad, and T. Moan, 'On Initial Design and Modelling of a 10 MW Medium Speed Drivetrain for Offshore Wind Turbines', *J. Phys.: Conf. Ser.*, vol. 1356, no. 1, p. 012024, Oct. 2019, doi: 10.1088/1742-6596/1356/1/012024.

[12] M. Liu and D. G. Corman, 'Formulation of Rayleigh damping and its extensions', *Computers & Structures*, vol. 57, no. 2, pp. 277–285, 1995.

[13] NREL, 'BModes'. [Online]. Available: https://www.nrel.gov/wind/nwtc/bmodes.html

[14] ISO 10816-21, 'Mechanical vibration - Evaluation of machine vibration by measurements on non-rotating parts - Part 21: Horizontal axis wind turbines with gearbox'. 2015.

Dynamic Reliability Estimation of Wind Turbine Gearboxes Using Real-Time Operational Data

Georgo Angelis, Laurens Krudde

Sensing360, Utrecht, Netherlands

E-mail: georgo.angelis@sensing360.com, laurens.krudde@sensing360.com

Abstract.

Wind turbine gearbox often fail to reach their expected lifetime of 20 years. This causes great maintenance costs and unexpected downtime. Similarly, gearboxes that do reach their expected lifetime are often replaced while they might be able to stay in operation for longer, missing out on revenue. There is a need for a reliability estimation based on the operating conditions such that premature failure and extended lifetime risks can be assessed. This work proposes a method to dynamically compute the gearbox's reliability based on real-time measurements of the operating conditions. This is done by introducing the concept of "used capacity", which denotes the accumulated damage on the gearbox during operation. It generalizes the traditional Weibull model by allowing for varying operating conditions that cause damage to accumulate similar to Miner's rule. This results in a more accurate reliability estimate than commonly used static estimates. Furthermore, other key performance indicators such as the mission risk and the remaining useable life can be derived from the model. The approach was tested using synthetic data for a Siemens Gamesa G97 2MW Turbine as part of a research project together with National Renewable Energy Laboratory (NREL) and Siemens Gamesa. Furthermore, the approach was applied to real wind turbine data from a Penmanshiel wind park. The results demonstrate realistic reliability indicators for the gearbox and its components. Using this result, one can improve maintenance strategies, optimize resource allocation, and assess risk more accurately. Moreover, the approach is adaptable to other types of rotating machinery, offering broad applications beyond wind turbines.

1. Introduction

Wind energy has become an important component in the global transition to sustainable energy. As wind turbines are increasing in size and capacity, it is becoming more important to ensure the reliability of critical components as failures and downtime become more expensive. From all components in a wind turbine, the gearbox has been shown to be the component that experiences the greatest number of failures [1]. Wind turbine gearboxes frequently fall short of their expected 20-year lifespan and few make it past the 10-year mark [2, 3]. Wind turbine gearbox failures are one of the main reasons for significant increase in operation and maintenance (O&M) costs and wind turbine downtime

While there is a wide variety in failure prognostic models, reliability assessments are commonly performed using the L_{10} life formula and the Weibull distribution due to their simplicity and ease of use [4]. The reliability calculation is done using expected average operating conditions such as torque and rotational speed. Based on the expected operating conditions and the

Content from this work may be used under the terms of the Creative Commons Attribution 4.0 licence. Any further distribution of this work must maintain attribution to the author(s) and the title of the work, journal citation and DOI.
Published under licence by IOP Publishing Ltd

time of operation, a reliability estimate can be calculated. However, once a wind turbine is in operation, the operating conditions naturally vary due to wind conditions, grid interaction, and environmental factors. This can lead to deviations from the initial assumed reliability distribution, resulting in an unreliable reliability estimate.

This paper proposes a novel dynamic reliability assessment by integrating real-time operational data in a stochastic model. This model introduces a 'used capacity' metric, extending Weibull-based reliability estimation to account for variable operating conditions. By keeping track of the used capacity, an accurate estimate of the reliability can be calculated. Based on the used capacity formulation, additional key performance indicators can be calculated, such as the mission risk and the remaining useful life. The proposed method is applied to real operational data of wind turbines to compare it to traditional methods and to show the application to a wind park. This work aims to provide improved reliability metrics that can be used for predictive maintenance and lifecycle management and ultimately reduce O&M costs for wind turbine operators.

2. Background

In this section, we recall the common reliability calculation methods; the L_{10} formula, the Weibull distribution and Miner's rule.

2.1. L_{10} formula

The basic rating life, L_{10}, is the million of revolutions that the bearing can be expected to endure with 90% probability, where failure is defined as the appearance of subsurface material fatigue as per ISO 15243 [5]. In the case of gears, tooth pitting and bending fatigue are considered as per ISO 6336 [6]. The basic rating life, from Lundberg and Palmgren in 1924 as cited by [7], is given by

$$L_{10} = \left(\frac{C_p}{F}\right)^p \tag{1}$$

with C_p the dynamic load capacity, F the load and p the life exponent. The parameters C_p and p are often known from literature (i.e. given by bearing or gear manufacturers) or can be determined analytically or experimentally. The L_{10} formula is an initial estimate for the expected life but can not provide an adapted estimate during operating.

2.2. Weibull distribution

More generally, it has been found that fatigue life follows a Weibull distribution, as first developed by Weibull in 1939 [7]. Which can be written using L_{10} :

$$\mathrm{P}(L > l) = S \quad \text{where} \quad \log(1/S) = AL^k \tag{2}$$

where S denotes the survival probability. Substituting $S = 0.9$ for L_{10} allows us to solve for A and rewrite the Weibull equation:

$$S = e^{(L/\lambda)^k} \quad \text{where} \quad \lambda = \frac{L_{10}}{\sqrt[k]{-\log(0.9)}} \tag{3}$$

The Weibull distribution gives more insight in the failures we expect over the course of an operation. However, the Weibull distribution is only dependent on the number of revolutions and is defined for a fixed load. Varying operating conditions cannot be incorporated in this formulation.

2.3. Miner's rule
Miner's rule is a simple method to incorporate varying operating conditions and calculate the accumulated damage. If L_i is the number of revolutions under load F_i and L_{10}^i is the L_{10} at load F_i, then the total damage under n stress levels is

$$D = \sum_{i=1}^{n} \frac{L_i}{L_{10}^i} \tag{4}$$

According to Miner's rule, failure occurs when the total accumulated damage D is equal to or exceeds 1. This is clearly a conservative estimate, since only 10% is expected to fail at that point, but large margins are common in engineering design to guarantee successful operation. While Miner's rule has been successful in engineering practice, it has been advised to be extended using probabilistic approaches [8, 9].

2.4. Gearbox modelling
The previous reliability models are suited to model a single component. In order to model the reliability of a gearbox as a whole, the components can be modelled separately and combined to derive reliability model for a gearbox [10, 11]. It is common to consider the reliability of the system, i.e. the gearbox, as the product of the reliabilities of the components. However, the reliabilities of the components can also be modelled as being dependent based on the mechanics of the gearbox [12]. Another approach to estimate the remaining useful life of planetary gearboxes is to use a state space model based on the Gamma process [13].

2.5. Conclusion
While these models provide useful reliability estimates, they assume static operating conditions or a deterministic failure threshold. In reality, wind turbine gearboxes experience fluctuating loads and failure is a stochastic occurrence. The following section introduces a probabilistic approach that adapts reliability estimation to real-time operating data.

3. Methodology
This section proposes the reliability calculation based on the used capacity variable. First, the used capacity and reliability calculation is done for a single gearbox component; a bearing. Then, the used capacity notion is extended to a gearbox, being a system of certain components. Lastly, the additional prognostic metrics mission risk and remaining useful life are defined.

3.1. Used capacity metric
In order to incorporate varying operating conditions, we define a new random variable which we call the capacity and is denoted by C. The capacity combines the operating conditions load and speed in a single variable, similarly to Miner's rule. We define capacity C as

$$C = A^{1/k} L \tag{5}$$

meaning that the bearing can last L revolutions under operating condition A. To incorporate varying operating conditions, we let capacity increment as

$$c^* = \int_0^{L_f} A_f(L)^{1/k} dL \tag{6}$$

as illustrated in Figure 1. Here, the simplifying assumption from Miner's rule that the order of applied loads has no effect is used.

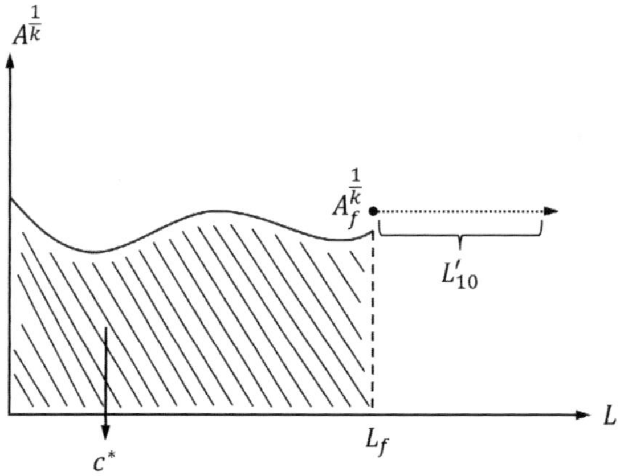

Figure 1. Used capacity accumulation.

3.2. Reliability

The way that the used capacity accumulates is similar to Miner's rule. However, we treat the capacity C as a random variable. The Weibull distribution from Equation 2 can be written in terms of capacity as

$$\mathrm{P}(C > c) = S \quad \text{where} \quad \log(1/S) = c^k \tag{7}$$

and the reliability of the bearing can be determined based on the used capacity. Important is that the distribution now only depends on the capacity C and not on A and l separately. This allows us to incorporate varying operating conditions.

3.3. Gearbox as a system of components

In the previous sections, we took the example of a single bearing. Now we will generalize to a system of components. For example, a wind turbine gearbox containing various bearings and gears. Suppose we have n components that define a system such that if 1 of the components fails this implies that also the system fails. In this case the reliability of the system

$$S_s = \prod_{i=1:n} S_i \tag{8}$$

Now the used capacity of the system C_s can be computed as

$$C_s = \Big(\sum_{i=1:n} C_i^{k_i}\Big)^{1/k_s} \tag{9}$$

As such, we can define the used capacity of a wind turbine gearbox and thus calculate its reliability.

3.4. Additional prognostics

The reliability follows directly from the Weibull distribution, but additional metrics can be computed. If the gearbox has been in operation and has exhibited a certain used capacity, then the remaining capacity has a conditional (left truncated) Weibull probability. Formally, the probability of remaining capacity c' with confidence level α after having accounted for evidenced used capacity $\tilde{c}$ is described as follows.

$$P(C \leq \tilde{c} + c' | C > \tilde{c}) = 1 - \alpha \tag{10}$$

This implies that only knowledge of the used capacity $\tilde{c}$ alters the likelihood of the remaining capacity c'. Two interesting metrics can be formulated: the mission risk and the remaining useful life. Both of which can be determined based on purely the used capacity.

3.4.1. Mission risk Given some used capacity $\tilde{c}$ and a mission that requires future capacity c'. What is the probability α that the bearing can exhibit c' more capacity? We can fix for some capacity c' (the mission) and solve for the probability α. This gives an estimate for the risk of keeping a wind turbine in operation for a certain mission duration.

3.4.2. Remaining useful life Given some used capacity $\tilde{c}$ and confidence level α. How much capacity c' can the wind turbine exhibit with confidence α? We can fix a confidence level α and solve for remaining useful capacity c'. The most interesting case is the remaining life with 90% reliability conditioned on the fact that the bearing has already exhibited a certain capacity. Since L_{10} is often denoted as the useful life of new components, the conditional L_{10} determines the remaining useful life of used components.

Both Remaining Useful Life (RUL) and Mission Risk statements depend on the known used capacity to date and require a prediction into the future, i.e. the capacity needed for the mission or the future operating conditions to estimate RUL in terms of revolutions instead of capacity. The prediction can be based on the expected operating conditions, but can also be done using exhibited used capacity from the past. As such, only the used capacity and its history is necessary to determine all metrics.

4. Data and case study

Two different datasets, consisting of torque and speed measurements on the wind turbine rotor, were used to illustrate the proposed method. First, a dataset for a single wind turbine over 20 years was used to compare the proposed method with traditional methods. Second, real measurement data from wind turbines from a 4.5-year-old wind park was used to illustrate its application to a wind park.

4.1. Single gearbox: synthetic NREL data

A dataset over the course of wind turbine's typical design life, e.g. 20 years, is appropriate to compare the proposed method with traditional methods. Due to the lack of available data over such a period, a synthetic time-series of torque and speed was created based on binned operating conditions. The binned operating conditions were derived from real-time measurements of the operating condition on a Siemens Gamesa G97 2MW Turbine as part of a research project together with National Renewable Energy Laboratory (NREL) and Siemens Gamesa, of which these results are a continuation [14]. While the synthetic time-series are similar to the binned data, the synthetic data had slightly higher operating conditions.

4.2. Wind park: real-world Penmanshiel data

A dataset of multiple wind turbines is needed to illustrate the different reliability estimates among wind turbines in a wind park as a result of the different operating conditions that they experience. This creates insights that can be used on a wind park management level. As such, the method is applied to open SCADA data of 14 Senvion MM82 2.05 MW wind turbines over 4.5 years from Penmanshiel [15]. Figure 2 shows the relative location of the wind turbines.

5. Results

We present the results for a single gearbox to compare the proposed method with traditional methods and the results for real-world data of a wind park.

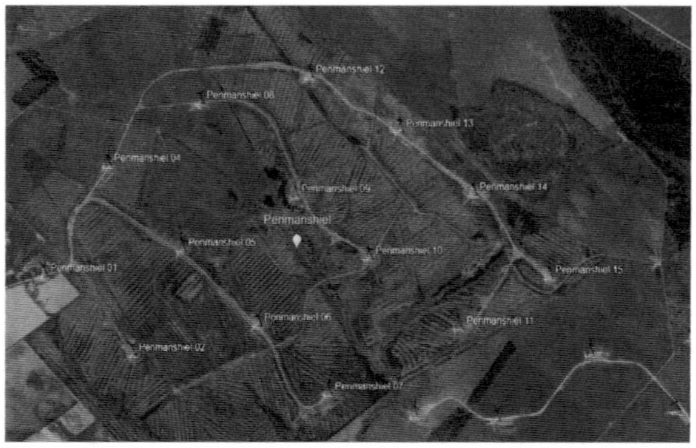

Figure 2. Penmanshiel wind turbine locations on Google Earth [16].

5.1. Single gearbox

Figure 3 shows the difference between the traditional L_{10}, Miner's rule and the proposed conditional L_{10}. Since L_{10} is only based on expected operating conditions and not on measurements, it does not change over time. Miner's rule states that failure can be expected after L_{10} is reached. The proposed method is able to calculate the remaining useful life, the life with over 90% conditioned on the accumulated used capacity. It can be seen that, if the gearbox has survived the initial L_{10}, it may still have significant remaining useful life, depending on the accumulated used capacity. This demonstrates the proposed method in providing a more accurate estimation of remaining useful life beyond the initial reliability expectations.

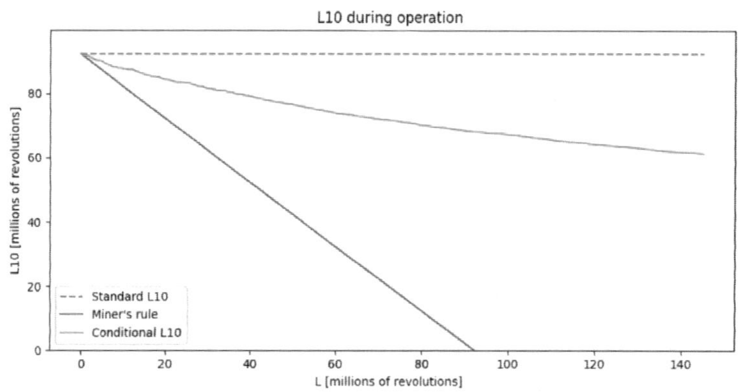

Figure 3. The remaining useful life L_{10} according to Miner's rule and the conditional L_{10} from the proposed method based on synthetic data of a wind turbine gearbox. It shows that a gearbox may have remaining useful life after surviving the initial L_{10}.

Figure 4 shows the Weibull reliability estimates based on expected operating conditions and based on the synthetic operating conditions for a single gearbox. It can be seen that the reliability estimate based on expected operating conditions perfectly crosses the L_{10} line at 90% reliability, as the L_{10} is also determined based on expected operating conditions. The proposed method incorporates the experienced operating conditions, and it can be seen that the estimates decrease faster or slower based on these experienced operating conditions. Furthermore, it can be seen that reliability estimate based on the used capacity decreases faster than the initial Weibull estimate, because the synthetic operating conditions were higher than the expected operating conditions.

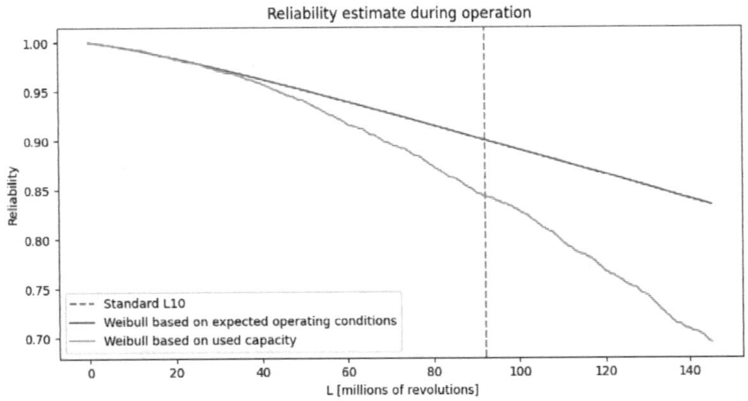

Figure 4. Weibull reliability estimation using the expected operating conditions and the used capacity baed on synthetic data of a wind turbine gearbox. It shows that the reliability estimate based on used capacity takes into account the higher experienced operating conditions.

5.2. Wind park

Figure 5 illustrates the reliability trends across the gearboxes of the 14 turbines in the Penmanshiel wind park. It can be seen that similar patterns are visible for all turbines, which is what one would expect since the turbines are experiencing similar wind conditions. However, the reliabilities start to diverge over time. Comparing this to the location of the turbines in Figure 2, it is clear that the turbines at the edge of the wind park are experiencing higher operating conditions, as they are more exposed to wind inflow. In contrast, the turbines in the centre are catching reduced wind, leading to lower operating conditions and thus slower decrease in reliability.

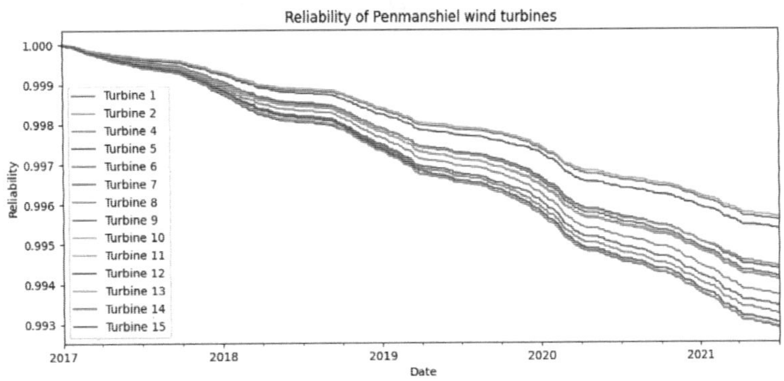

Figure 5. Reliability over time for wind turbines in Penmanshiel wind park, showing how edge turbines degrade faster due to higher loads.

Figure 6 shows the generated energy and the failure probability (1 - reliability) of the turbines after 4.5 years. As expected, it can be seen that the turbines that have generated more energy and have lower reliability due to increased accumulated stress on the gearboxes. The relationship between energy production and reliability follows an exponential trend, and it can be seen that turbines with lower cumulative energy output exhibit disproportionately higher reliability. This is a direct consequence of the life exponent in the L_{10} formula.

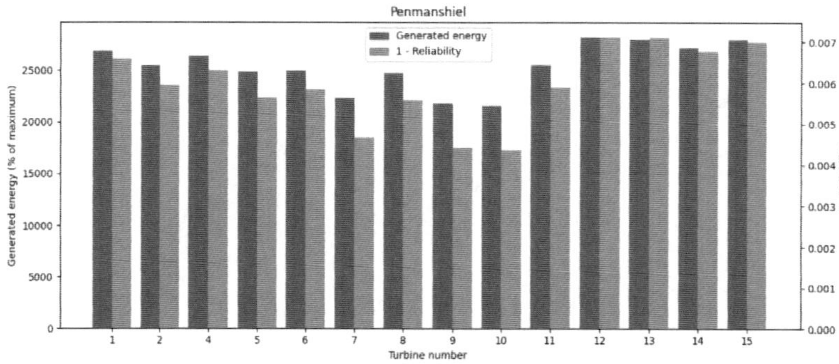

Figure 6. The generated energy and the reliability of the Penmanshiel wind turbines. It shows that turbines with lower generated energy have a larger reliability.

These variations in reliability suggest that a wind park operator can optimize the distribution of reliability across the fleet by strategically managing turbine exposure to wind. For instance, reducing the workload on turbines with higher accumulated used capacity while allowing other turbines to operate under increased loads can help balance failure risk and maximize overall uptime and energy production.

6. Discussion

This method forms the basis for a new reliability estimate, but some assumptions and limitations should be mentioned. First, this method estimates wind turbine gearbox reliability based on fatigue failure due to wear on the material. Bearing failure is defined as subsurface fatigue appearance and gear failure is defined by root pitting or bending fatigue, as per ISO standards [6, 5]. It does not account for early failures or other failure modes such as misalignment, lubrication issues, or material defects. Also, failure of one component is considered as failure of the whole gearbox and therefore operation with a partially faulty gearbox is not considered. Faulty components would increase the stress on other components and the gearbox as a whole, which complicates a reliability model, but inspiration can be taken from [12]. Second, the reliability estimate relies on measured torque and speed data, making the accuracy dependent on data quality. Uncertainty in measurements induces uncertainty in the reliability estimate. This uncertainty could be modelled, but it is advised to ensure accurate measurement data. Lastly, it should also be noted that the current approach only considers torque on the gearbox carrier, capturing torque-induced loads but ignoring non-torque loads, i.e. disturbance forces, that also contribute to fatigue. Future work should incorporate these additional forces acting on the gearbox, if measurable.

7. Conclusion

This study presents a novel method to improve the reliability estimation of wind turbine gearboxes by utilizing real-time measurements of the operating conditions. By introducing the concept of "used capacity", the proposed approach generalizes traditional Weibull distribution models, making them adaptable to varying conditions that a gearbox experiences over time. This method provides accurate predictions of key performance indicators such as reliability, mission risk and remaining useful life. This can be the basis for better maintenance strategies and extended operational lifetimes for wind turbines. The application of synthetic data demonstrates the method's effectiveness, laying a foundation for future implementation in real-world scenarios. Future work could extend this approach to other rotating machinery applications, such as offshore turbine drivetrains.

References

[1] Lu B, Li Y, Wu X and Yang Z 2009 *2009 IEEE Power Electronics and Machines in Wind Applications* pp 1–7 URL https://ieeexplore.ieee.org/abstract/document/5208325
[2] Froese M 2018 Why wind-turbine gearboxes fail to hit the 20-year mark https://www.windpowerengineering.com/wind-turbine-gearboxes-fail-hit-20-year-mark/ [Online]
[3] Pall 2023 The consequence of gearbox failures in wind turbines https://www.pall.com/en/oil-gas/blog/wind-turbine-gearbox-failures.html [Online]
[4] Zhang W, Vatn J and Rasheed A 2022 *Journal of Physics: Conference Series* **2362** 012043 ISSN 1742-6588, 1742-6596 URL https://iopscience.iop.org/article/10.1088/1742-6596/2362/1/012043
[5] 2017 Rolling bearings — damage and failures — terms, characteristics and causes Standard International Organization for Standardization Geneva, CH
[6] 2019 Calculation of load capacity of spur and helical gears Standard International Organization for Standardization Geneva, CH
[7] Zaretsky E V 2016 Rolling Bearing Life Prediction, Theory, and Application Tech. rep.
[8] Saunders S C 1970 A Review of Miner's Rule and Subsequent Generalizations for Calculating Expected Fatigue Life: Tech. rep. Defense Technical Information Center Fort Belvoir, VA URL http://www.dtic.mil/docs/citations/AD0717283
[9] Saunders S 1974 The Theory Relating the Woehler Equation to Cumulative Damage in the Distribution Fatigue Life Tech. rep. National Technical Information Service Washington State University
[10] Savage M 1982 RELIABILITY MODEL FOR PLANETARY GEAR TRAINS Tech. rep.
[11] Dinner H 2021 How Reliable is a Reliability Calculation? Tech. rep.
[12] Gao P, Xie L and Hu W 2018 *Shock and Vibration* **2018** 9106404 ISSN 1070-9622, 1875-9203 URL https://onlinelibrary.wiley.com/doi/10.1155/2018/9106404
[13] Ni X, Zhao J, Chen J and Li H 2015 . *ISSN* **5**
[14] Angelis G and Santiago U G 2024 Field demonstration of Fiber-Optical Sensing for Wind Turbine Gearbox Monitoring & Reliability
[15] Effenberger N and Ludwig N 2022 *Wind Energy* **25** 1659–1683 ISSN 1099-1824 _eprint: https://onlinelibrary.wiley.com/doi/pdf/10.1002/we.2766 URL https://onlinelibrary.wiley.com/doi/abs/10.1002/we.2766
[16] Google Earth URL https://earth.google.com/web/

Data imputation for SCADA data using Graph Neural Networks

Florian Hammer, Sarah Barber
University of Applied Sciences Rapperswil, Oberseestrasse 10, 8640 Rapperswil, Switzerland
E-mail: `florian.hammer@ost.ch`

Abstract. Missing data in wind turbine SCADA systems can arise due to sensor failures, software issues, or maintenance, impacting data mining and analysis tasks. This work investigates the use of Graph Neural Networks (GNNs) for imputing missing wind speed data by leveraging global and local spatial relationships between turbines. The goal is to improve data quality and completeness for downstream tasks such as energy loss estimations and performance analysis. A GNN model was developed using SCADA data from the Kelmarsh wind farm, incorporating wind speed components and turbine nacelle orientations as node features. The model was trained to predict missing wind speeds using information from neighbouring turbines. Its performance was compared against three baseline methods, namely a simple mean model, a k-nearest neighbors (KNN) imputer, and a Generative Adversarial Imputation Network (GAIN). Results show that the GNN model improves wind speed prediction accuracy, measured by the root mean square error, by around 20% and downstream power predictions, evaluated using the mean absolute percentage error, by around 3% compared to the GAIN, which was the second best model. However, we also found that more accurate wind speed and power predictions do not necessarily result in more accurate energy loss estimations and Weibull distribution fitting, likely due to over-estimations and under-estimations cancelling each other out.

1. Introduction

In the era of digitalisation, an abundance of data is available that data-driven approaches, especially deep learning methods, can leverage. In the recent years those approaches have been employed more and more in the wind energy sector [1].

The reliability of these methods, however, depends on the quality as well as the quantity of the data. In the case of wind turbine generators (WTGs), data comes from the Supervisory Control and Data Acquisition (SCADA) system, containing variables such as wind speed, wind direction, power output and more. However, faulty and missing data arise due to sensor errors, software issues, icing, dirt, or even human intervention. This becomes especially problematic when long periods of data are missing or faulty, and this is not sufficiently documented, as can be the case for wind speed measurements on WTGs where upgrades were installed [2].

Missing data can be dealt with by either deletion or imputation, with the former discarding all data where a missing value is observed and the latter trying to find appropriate values to fill those gaps [3]. In the context of machine learning, this task can be formulated as a prediction problem [4]. In the field of wind energy, data-driven methods for data imputation such as k-nearest neighbours, random forests [5], generative adversial nets [6, 7, 8], adaptive neuro-fuzzy inference system [9, 10], Gaussian process regression [11], long short-term memory based

Content from this work may be used under the terms of the Creative Commons Attribution 4.0 licence. Any further distribution of this work must maintain attribution to the author(s) and the title of the work, journal citation and DOI.
Published under licence by IOP Publishing Ltd

Journal of Physics: Conference Series **3025** (2025) 012014 doi:10.1088/1742-6596/3025/1/012014

recurrent neural network [12], spatio-temporal correlation and feature correlation [13] as well as deep auto-encoders [14] have been employed.

Recent advances in data imputation strategies include the use of Graph Neural Networks (GNNs) [3, 4], although these have not yet been applied to wind turbine data to our knowledge. GNNs are part of graph representation learning, which encodes structural information of graph structured data into features for machine learning tasks. A graph is a data structure that allows representation of a wide variety of different and complex systems that consist of different objects and their paired connections [15]. The advantage of GNNs compared to other common machine learning architectures such as Support Vector Machines, Random Forest Trees and more is that local and global spatial relationships between WTGs can be easily captured [16]. In recent years they have shown to taking into account and predicting the influencing patterns between turbines [17, 18, 19, 20, 21, 22]. Furthermore, GNNs have the ability to generalise well and be transferred to different wind farm layouts, as recently demonstrated by [23].

This work aims to employ a GNN to leverage local and global spatial relationships between WTGs for imputing missing wind speed data and estimating energy loss during WTG downtime based on the imputed wind speeds.

2. Methodology

2.1. Dataset

For this work, the open-source dataset of the Kelmarsh wind farm in the UK was chosen [24], due to its high quality and availability, as well as our existing collaboration with the data providers. The dataset includes 10-minute SCADA and events data, with over 300 different channels, of six 2.05 MW Senvion MM92 WTGs for the period from 2016 to 2022. Figure 1 shows the layout of the wind farm.

Figure 1: Kelmarsh Wind Farm, UK, comprising six 2.05 MW Senvion MM92 wind turbines. Each turbine has 10-minute averaged measurement records spanning 2016 to 2022.

We focused on three years worth of data and a small subset of available channels to showcase the feasibility of the proposed method. For the years from 2020 - 2022 there are a total of 938 645 measurement points; 156 440 per WTG. To build the data imputation model, the wind directions, the nacelle directions as well as the wind speed components u and v from all six WTGs were used.

Figure 2 shows the power curve for WTG "Kelmarsh 5" (*KWF 5*) before and after the data cleaning step. Firstly, real missing values were dropped from the dataset. This is important as we introduce artificial missing data points in a later step, in order to assess the method. Secondly, to filter out periods of WTG downtime, power values that are below 1% of the rated power as well as wind speeds below the WTGs' cut-in wind speed of $3.5\,m/s$ were removed. After that, the data was grouped into $0.5\,m/s$ wind speed bins, and all points in each respective bin that deviated by more than two standard deviations from the bin's power values were removed. Lastly, all points were dropped that were not available at the same timestamp for all six WTGs. After these filtering steps, 45% – 422 400 measurement points – of the original dataset were left; 70 400 samples per WTG. For the model training, data from the years 2020 and 2021 was used,

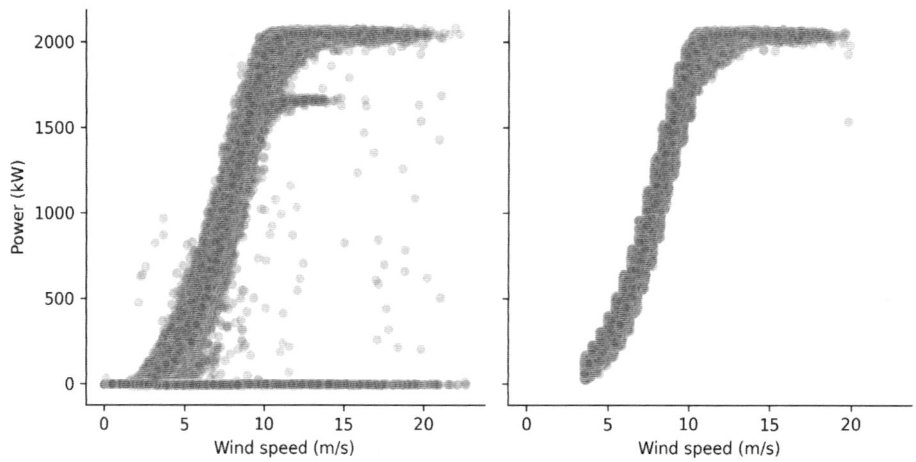

Figure 2: The power curve of WTG "Kelmarsh 5" before (left) and after (right) the data cleaning step.

consisting of 305 490 samples, around 70% of the total dataset. This set was further split into a training and validation set, with a ratio of 4:1. For the final testing step, data from year 2022 was set aside and served to evaluate the models on completely unseen data.

2.2. Graph Neural Network

The main goal of our model was to predict the missing wind speeds, i.e. data imputation, for a given turbine based on the wind speeds of its neighbouring turbines. Additionally, we also looked at a downstream task, which required accurate wind speed predictions. When a WTG experiences downtime due to maintenance or other causes, it is of interest to estimate the potential energy loss for a given time range. In order to do so, the wind speed predictions were used in combination with a power curve model to estimate the energy yield.

GNNs encode structural information of graph-structured data into features for machine learning tasks. A graph, $\mathcal{G} = (\mathcal{V}, \mathcal{E})$, consists of a set of nodes, $\mathcal{V}$, and edges, $\mathcal{E}$, as illustrated in figure 3. When representing a wind farm by a graph, the nodes are the individual WTGs, which are described by the available SCADA measurements. The nodes are connected by edges based on user defined criteria. Within a wind farm these criteria might be the relative distance and angle between WTGs. The closer the WTGs are to each other, the stronger the influence on the aerodynamic performance on the respective WTGs. This influence can be represented in the edges by adding, for example, weight factors, physical equations, overlap measures based on

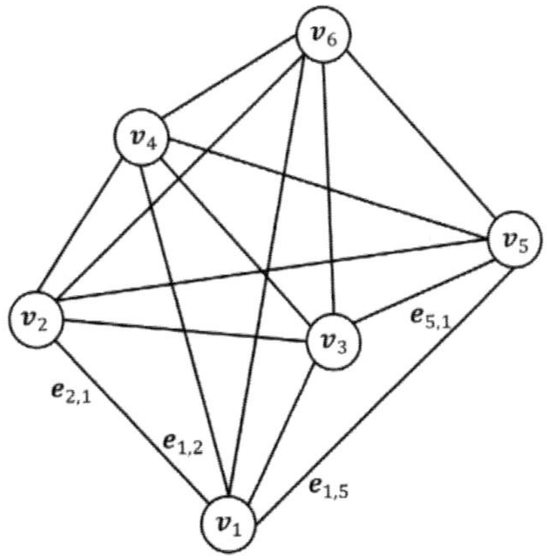

Figure 3: A graph comprising nodes in $\mathcal{V}$ and edges in $\mathcal{E}$. Source: [25]

wind direction, distances and relative angles. The GNN tries to "learn" the strength of these connections between WTGs based on measurement data.

As mentioned earlier, compared to other approaches, GNNs readily capture local and global spatial relationships based on the constructed graph of the wind farm.

In this work, the PyTorch Geometric [26] library was used to build the GNN. It is based on the popular PyTorch library, an open-source machine learning framework, a very common choice in academia containing the latest models and methods developed by researchers. Furthermore, a graph attention approach was chosen [27], which was shown to allow for some degree of interpretation of the complex model in terms of WTG wake interactions [17].

The chosen model architecture is shown in figure 4. Here, the input graph, $\mathcal{G}_{in}$, is passed to our GNN block (green), which was stacked $n = 4$ times. The GNN block itself is composed of a graph attention layer [27], with four attention heads, followed by a leaky ReLU activation function, a skip connection and an edge dropout layer. The resulting latent features are passed through a final linear layer that outputs the absolute wind speeds for each WTG.

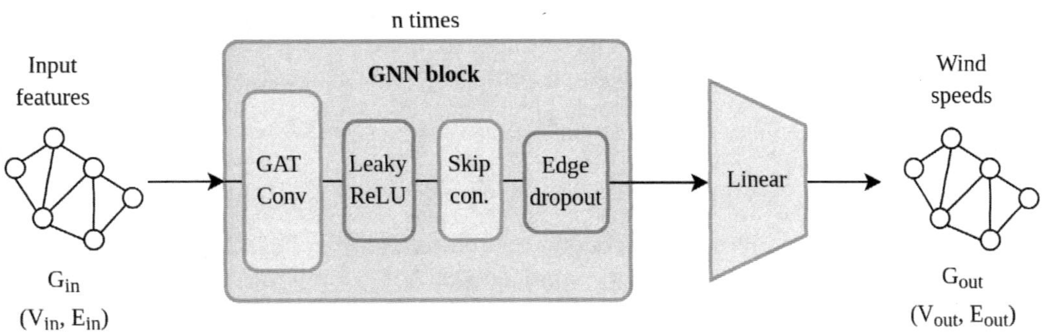

Figure 4: The chosen GNN architecture. The GNN block comprises a graph attention layer with four attention heads, followed by a leaky ReLU activation, a skip connection, and an edge dropout layer. A final linear layer outputs the absolute wind speeds for each WTG.

Let $\mathbf{D} \in \mathbb{R}^{n_t \times n \times m}$ be our wind farm dataset with the number of WTGs, n_t, the number of samples, n, per WTG and the number of features, m. Additionally, let $\mathbf{C} \in \mathbb{R}^{n_t \times 2}$ be a dataset containing the geospatial coordinates of each individual WTG. Based on these two datasets the graphs, $\mathcal{G}_{in} = (\mathcal{V}_{in}, \mathcal{E}_{in})$, were constructed. The edge features, E_{in}, were chosen to be the x and y components of the relative distance between neighbouring turbines, which can be calculated with **C**. The node features, $\mathcal{V}_{in}$, contain the u and v components of the wind speed vectors as well as the nacelle orientation components parallel to u and v. This forms a fully connected graph that captures the global and local relationships in the data [4].

For the imputation task, *KWF 5* was randomly selected to be the WTG with missing data, and its u and v components were set to zero in the training dataset. The goal for the GNN was then to learn to predict the absolute wind speeds of *KWF 5* based on the information of the neighbouring WTGs. Prior to training, the input features were scaled by using `scikit-learn`'s `MinMaxScaler`. The model was trained for 800 epochs with a learning rate of 4×10^{-5} and a batch size of 128. The GNN model has a total of 600 833 learnable parameters. The training was performed on a Quadro RTX 4000 and took around 72 minutes.

In order to compare the performance of our GNN model, three additional models with increasing complexity were considered. Firstly, a simple averaging approach was tested, in which the wind speed measurements of all WTGs but *KWF 5* were averaged at each timestamp. The averaged wind speeds were then taken to be the imputed wind speeds for *KWF 5*. This model serves as the lower bound in terms of model performance. Secondly, `scikit-learn`'s k-nearest neighbour imputer, `KNNImputer`, a non-parametric and local similarity-based approach, was trained with `n_neighbors` = 2 and uniform weights. It is important to note that `n_neighbors` does not relate to actual neighbouring WTGs, but to the smallest euclidean distance between pairs of samples. Lastly, a generative adversarial imputation net (GAIN [6]) was trained. The GAIN model consists of a generator, which predicts the missing data, and a discriminator, which assesses both the real and generated values to determine which are observed and which are imputed. During training, the generator iteratively improves its predictions until the discriminator can no longer distinguish between real and imputed values.

Compared to the GNN model, the chosen comparison models only rely on the information given in dataset **D** and not $\mathcal{G}_{in}$. The simple averaging model does not require any training and gives instant results. As opposed to our model, where the u and v components of the wind speed vector were simply set to zero for *KWF 5*, the KNN imputer and the GAIN model require data to be flagged as missing. For this, random 20% of the u and v wind speed components of *KWF 5* in the training set was flagged. GAIN is the most complex model of the three comparison approaches and contains 2 300 learnable parameters, significantly fewer than our GNN model, and takes around 20 minutes for 800 epochs at a learning rate of 1×10^{-4}. While the increased complexity of our GNN model could, in itself, contribute to better performance, the comparison remains valuable as it highlights the benefits of leveraging graph-structured data alongside the dataset information.

3. Results and discussion

In order to evaluate the performance of all four approaches, data from year 2022, $\mathbf{D}_{\text{test}}$, was used. For this dataset, all u and v wind speed components of *KWF 5* were either set to zero – for the GNN model – or flagged as missing – for the three comparison models. The performances were evaluated based on wind speed prediction accuracy and a downstream task. In the downstream task, the imputed wind speeds were first used to predict power output, which was then used to estimate energy losses during WTG downtime.

The root mean squared error

$$\text{RMSE}(y, \hat{y}) = \sqrt{\sum_{i=1}^{n} \frac{(\hat{y}_i - y_i)^2}{n}} \tag{1}$$

as well as the mean absolute percentage error

$$\text{MAPE}(y, \hat{y}) = \frac{1}{n}\sum_{i=1}^{n} \frac{|\hat{y}_i - y_i|}{\max(\epsilon, |y_i|)} \tag{2}$$

were used for the performance evaluation, with $\hat{y}_i$ denoting a predicted value and y_i denoting a measurement. The RMSE scores for the training, validation and testing datasets are shown in table 1. The first thing to notice is the close convergence of training and validation scores for all but the "mean" model, suggesting good generalisation capabilities and indicating neither significant overfitting nor underfitting. The "mean" model is clearly not a good choice. The scores for the test dataset show the performances of the models on completely unseen data. The GNN model performs best, followed by the GAIN. Surprisingly, the performance of the KNN imputer significantly drops for the test dataset, indicating that it is not suitable for continuously missing data as encountered in the test set.

Table 1: The RMSE scores of all four models for the training, validation, and testing datasets. Boldface indicates the best score.

	mean	KNN	GAIN	GNN
Train	1.02056	0.618132	0.556517	**0.410646**
Validation	0.785994	0.618115	0.553141	**0.424536**
Test	0.803096	0.855552	0.570268	**0.451556**

Figure 5 shows a more qualitative comparison between the actual measured wind speeds and the predicted wind speeds of WTG *KWF 5* for the test dataset. As can be seen, all models are able to predict a reasonable spread of wind speeds for low and high wind speed regimes, with the GNN model performing markedly best. This is also reflected in the R^2 correlation values. All three comparison models, however, struggle – to different degrees – with lower wind speed predictions, as can be observed in the boxplot in figure 6.

The boxplot also reveals that all models, except for the KNN imputer, tend to overestimate rather than underestimate wind speeds, as evident by the points above the 75th percentile in nearly all wind speed bins. In contrast, the KNN imputer overestimates and underestimates wind speed values in roughly equal measure within the medium wind speed range.

The predicted wind speed Weibull distributions are shown in figure 7. The black dashed line denotes the wind speed distribution of the measured wind speeds, with scale factor $A = 3.64$ and shape factor $k = 1.60$. An important point to remember is that all distributions were shifted by $3.5\,m/s$, as all wind speeds below this value were removed (see figure 2). The KNN imputer achieved the best fit to the original distribution. Both the mean model and the KNN imputer reasonably match the scale and shape factors, whereas the GNN model noticeably overestimates the Weibull coefficients. The GAIN model leads to a significant overestimation of these coefficients. Additionally, all models except the KNN imputer shift the mode to a higher value. Moreover, the mean model increases the density around the mode, while the GAIN model decreases it. This shift will have a crucial impact on energy yield estimations, which will be investigated later.

To assess the impact of imputed wind speeds on a downstream task, specifically power loss estimation, we developed a simple power curve model using the `XGBoost` library, which has been shown to perform well for this task in previous work [28, 29, 21]. The power curve model simply takes the wind speed at the *KWF 5* WTG and predicts its power output.

The MAPE between the predicted and measured power outputs for the test dataset are shown in figure 8. Here "Measurement" refers to the error for the power predictions based on

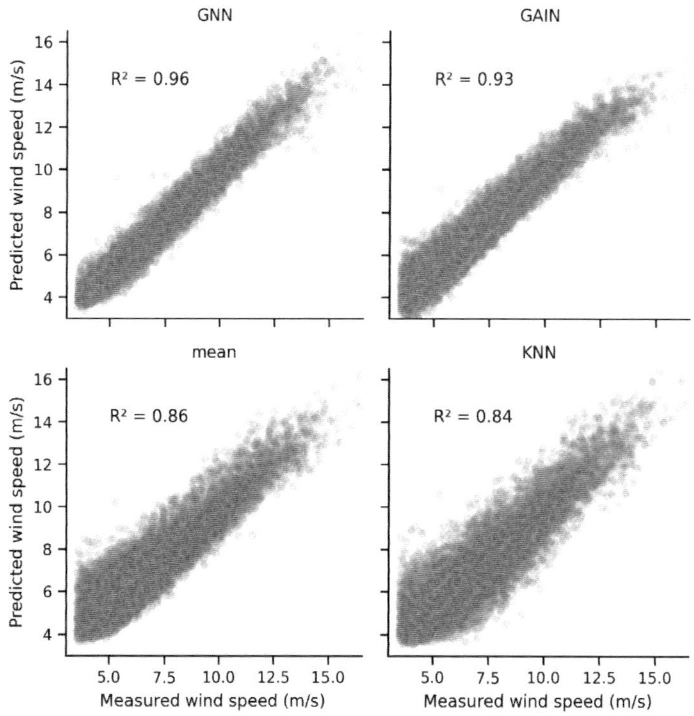

Figure 5: Measured wind speed versus the predicted wind speed for WTG *KWF 5* on the test dataset using all four models.

the measured wind speeds and serves as a reference. When using the imputed wind speeds of the GNN model, a 6% increase in MAPE compared to the measured wind speeds can be observed. For the GAIN model the error increases additionally by roughly 3%. Both the "mean" and the KNN imputer models result in a 20% error increase with respect to the baseline.

Figure 9 shows the RMSE values of the predicted versus the measured power for different wind speed bins. The KNN imputer performs poorly for all wind speed bins. For the smallest wind speeds, the "mean" model is on par or even slightly better than the GNN and GAIN models. Overall, our GNN model results in the lowest RMSE values throughout all bins compared to the baseline, where the measured wind speeds were used for the power predictions. Interestingly, the highest RMSE values occur around the wind speeds corresponding to the inflection point of the power curve.

The final goal of this comparison is to evaluate the model performances in terms of energy loss estimations due to WTG downtimes. Figure 10 shows the differences in the estimated energy production minus the measured produced energy for the complete time span of the test dataset. The energy yield was calculated by simply summing up the predicted 10-minute power values. Again, "Measurement" serves as the baseline, where the measured wind speeds were used with the power curve model. Interestingly, the KNN imputer model performs considerably better than all other models. Although the GNN model previously showed the lowest errors, it overestimates the energy production by approximately 45 MWh compared to the baseline, in this case. These results make sense in light of the Weibull distributions (see figure 7). For all models but the KNN imputer the mode increased compared to the original wind speed distributions. Hence, the higher frequency of higher wind speeds led to the higher predicted energy yields, especially in case of the "mean" and the GAIN model.

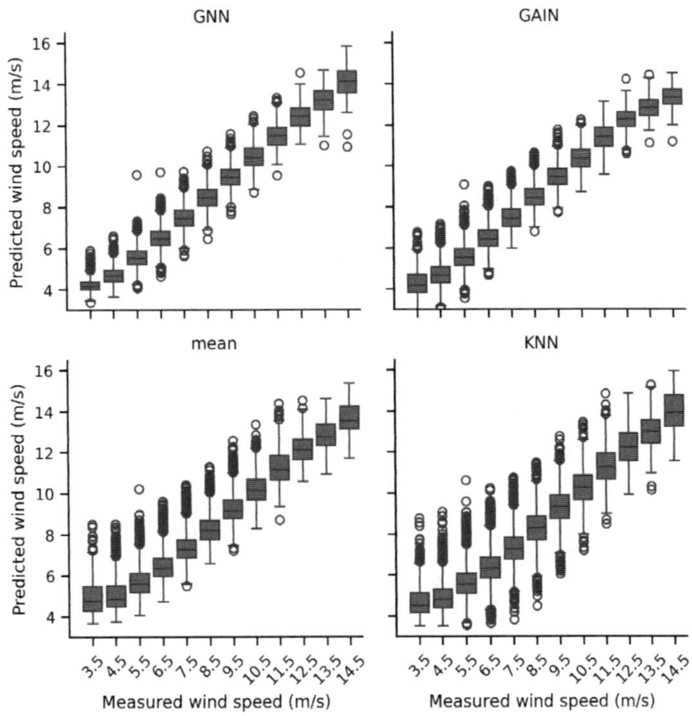

Figure 6: Distributions of the measured versus the predicted wind speed across different wind speed bins for WTG *KWF 5* on the test dataset. The boxes represent the 25th, 50th (median), and 75th percentiles.

The better performance of the KNN imputer in both Weibull distribution fitting and energy yield estimations is most likely due to negation effects, where over-estimations are canceled out by under-estimations. This is further supported by the wide spread of wind speed predictions in the case of the KNN imputer (see figure 5 and 6). Therefore, one must be cautious when estimating the total energy loss over longer periods of time and will be further explored in future work.

Overall, the GNN model achieves the best performance in both wind speed and power predictions based on the imputed wind speeds. As mentioned earlier, the GNN model is significantly more complex than the other models, and a better performance was to be expected. Nevertheless, this comparison remains valuable as it demonstrates how wind farm data can be utilised for both a data imputation task and a downstream task by structuring the dataset as a graph. Another advantage of using a GNN is its ability to generalise and transfer to other wind farms [23] and will be investigated in the context of data imputation in the future.

4. Conclusions

This work aimed to test and evaluate a GNN model for the imputation of missing wind speed data in wind farm SCADA systems. Accurate wind speed predictions are crucial for applications such as estimating power losses during WTG downtimes.

To evaluate our approach, we developed a GNN model that leverages spatial relationships between WTGs and compared it against three baseline methods, namely a simple mean model, a KNN imputer, and a GAIN model. The models were assessed based on their ability to predict wind speeds and their impact on a downstream energy loss estimation task. For this, we selected

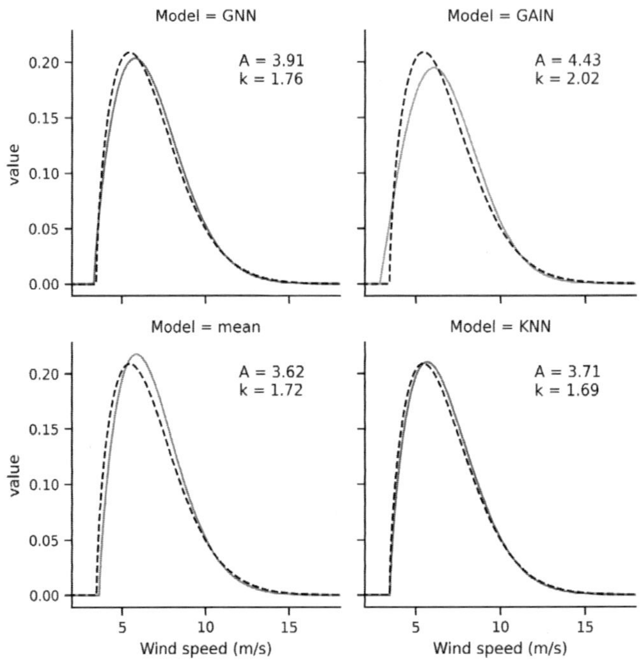

Figure 7: Weibull distribution fits of the predicted wind speeds on the test dataset, including scale factor, A, and shape factor, k.

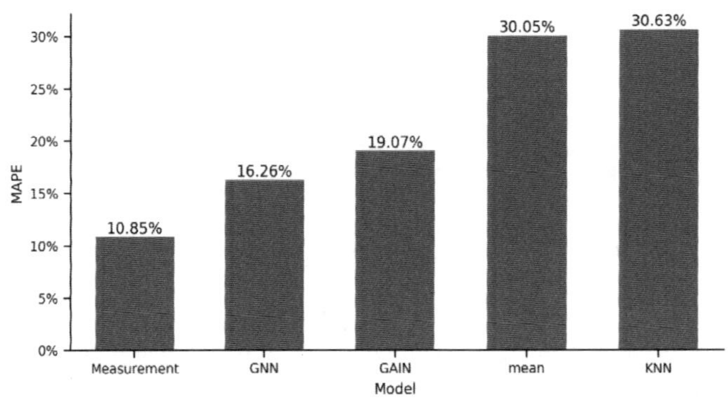

Figure 8: MAPE of the predicted power output for WTG *KWF 5* on the test dataset. "Measurement" refers to the error in power predictions based on measured wind speeds.

the open-source dataset of the Kelmarsh wind farm in the UK [24], which includes 10-minute SCADA and events data from six 2.05 MW Senvion MM92 WTGs spanning the period from 2016 to 2022.

The results showed that the GNN model achieved the highest accuracy in both wind speed imputation and power predictions, followed by the GAIN. The "mean" model and the KNN imputer performed poorly, especially for the test dataset, where continuously missing data was encountered. Surprisingly, for the total energy loss estimation as well as the Weibull distribution fitting the KNN imputer achieved the most accurate result, which was most likely due to

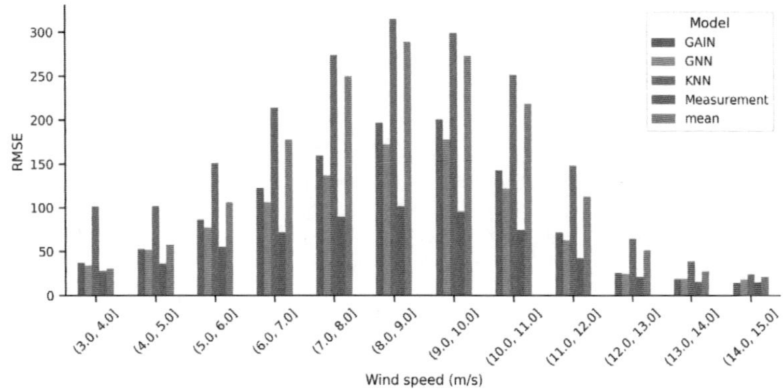

Figure 9: RMSE of the predicted power output across wind speed bins for WTG *KWF 5* on the test dataset. "Measurement" refers to the error in power predictions based on measured wind speeds.

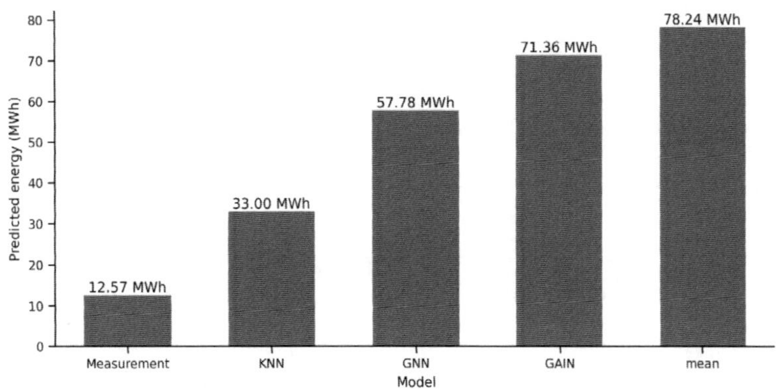

Figure 10: Differences between estimated and measured energy production over the full duration of the test dataset.

cancellation effects, where over-estimations cancel out under-estimations.

While the GNN model's greater complexity naturally contributes to its better performance, the comparison remains valuable as it illustrates how wind farm data can be effectively used for both data imputation and downstream tasks by representing the dataset as a graph.

The issues of Weibull distribution fitting and misleading energy loss estimates over longer periods due to negation effects will be investigated in future work. Additionally, in ongoing work we investigate transfer learning techniques for GNNs for wind farm data imputation.

References

[1] Clifton A, Barber S, Bray A, Enevoldsen P, Fields J, Sempreviva A M, Williams L, Quick J, Purdue M, Totaro P and Ding Y 2023 *Wind Energy Science* **8** 947–974 ISSN 2366-7451 URL `http://dx.doi.org/10.5194/wes-8-947-2023`

[2] Astolfi D, Castellani F and Terzi L 2018 *Energies* **11** 1300 ISSN 1996-1073 number: 5 URL `http://www.mdpi.com/1996-1073/11/5/1300`

[3] You J, Ma X, Ding D Y, Kochenderfer M and Leskovec J 2020 Handling Missing Data with Graph Representation Learning arXiv:2010.16418 [cs, stat] URL `http://arxiv.org/abs/2010.16418`

[4] Spinelli I, Scardapane S and Uncini A 2020 *Neural Networks* **129** 249–260 ISSN 08936080 [Online; accessed 2024-02-23] URL `https://linkinghub.elsevier.com/retrieve/pii/S0893608020302185`

[5] Becker R and Thrän D 2017 *Applied Energy* **208** 252–262 ISSN 03062619 [Online; accessed 2024-02-28] URL `https://linkinghub.elsevier.com/retrieve/pii/S0306261917314587`

[6] Yoon J, Jordon J and van der Schaar M 2018 GAIN: Missing Data Imputation using Generative Adversarial Nets (*Preprint* 1806.02920)

[7] Qu F, Liu J, Hong X and Zhang Y 2018 *Data Imputation of Wind Turbine Using Generative Adversarial Nets with Deep Learning Models* vol 11301 (Cham: Springer International Publishing) pp 152–161 ISBN 978-3-030-04166-3 978-3-030-04167-0 series Title: Lecture Notes in Computer Science URL `http://link.springer.com/10.1007/978-3-030-04167-0`$_1$4

[8] Qu F, Liu J, Ma Y, Zang D and Fu M 2020 *Mechanical Systems and Signal Processing* **139** 106610 ISSN 08883270 [Online; accessed 2024-02-23] URL `https://linkinghub.elsevier.com/retrieve/pii/S0888327019308313`

[9] Yang Z, Liu Y and Li C 2011 *Renewable Energy* **36** 993–998 ISSN 09601481 [Online; accessed 2024-02-28] URL `https://linkinghub.elsevier.com/retrieve/pii/S0960148110004040`

[10] Morshedizadeh M, Kordestani M, Carriveau R, Ting D S K and Saif M 2017 *Energy* **138** 394–404 ISSN 03605442 [Online; accessed 2024-02-28] URL `https://linkinghub.elsevier.com/retrieve/pii/S0360544217312094`

[11] Liu T, Wei H and Zhang K 2018 *Applied Soft Computing* **71** 905–916 ISSN 15684946 [Online; accessed 2024-02-23] URL `https://linkinghub.elsevier.com/retrieve/pii/S1568494618304150`

[12] Li T, Tang J, Jiang F, Xu X, Li C, Bai J and Ding T 2019 Fill Missing Data for Wind Farms Using Long Short-Term Memory Based Recurrent Neural Network *2019 IEEE 3rd International Electrical and Energy Conference (CIEEC)* (Beijing, China: IEEE) pp 705–709 ISBN 978-1-72811-675-4 [Online; accessed 2024-02-28] URL `https://ieeexplore.ieee.org/document/9077310/`

[13] Sun C, Chen Y and Cheng C 2021 *Energy* **229** 120777 ISSN 03605442 [Online; accessed 2024-02-28] URL `https://linkinghub.elsevier.com/retrieve/pii/S0360544221010252`

[14] Liu X and Zhang Z 2021 *IEEE Sensors Journal* **21** 10933–10945 ISSN 1530-437X, 1558-1748, 2379-9153 [Online; accessed 2024-02-23] URL `https://ieeexplore.ieee.org/document/9360630/`

[15] Hamilton W L *Synthesis Lectures on Artificial Intelligence and Machine Learning* **14** 1–159

[16] Wu Z, Pan S, Chen F, Long G, Zhang C and Yu P S 2019 Publisher: arXiv Version Number: 4 URL `https://arxiv.org/abs/1901.00596`

[17] Bentsen L O, Dilp Warakagoda N, Stenbro R and Engelstad P 2022 *Journal of Physics: Conference Series* **2265** 022035 ISSN 1742-6588, 1742-6596 number: 2 URL `https://iopscience.iop.org/article/10.1088/1742-6596/2265/2/022035`

[18] Park J and Park J 2019 *Energy* **187** 115883 ISSN 03605442 [Online; accessed 2024-02-28] URL `https://linkinghub.elsevier.com/retrieve/pii/S0360544219315555`

[19] Yu M, Zhang Z, Li X, Yu J, Gao J, Liu Z, You B, Zheng X and Yu R 2020 *Future Generation Computer Systems* **113** 145–157 ISSN 0167739X [Online; accessed 2024-02-28] URL `https://linkinghub.elsevier.com/retrieve/pii/S0167739X19330626`

[20] Bleeg J 2020 *Journal of Physics: Conference Series* **1618** 062054 ISSN 1742-6588, 1742-6596 [Online; accessed 2024-02-28] URL `https://iopscience.iop.org/article/10.1088/1742-6596/1618/6/062054`

[21] Hammer F, Helbig N, Losinger T and Barber S 2023 *Journal of Physics: Conference Series* **2505** 012047 URL `https://dx.doi.org/10.1088/1742-6596/2505/1/012047`

[22] Stańczyk T and Mehrkanoon S 2021 Publisher: arXiv Version Number: 1 URL `https://arxiv.org/abs/2101.10041`

[23] Duthé G, De N Santos F, Abdallah I, Weijtjens W, Devriendt C and Chatzi E 2024 *Data-Centric Engineering* **5** e29 ISSN 2632-6736

[24] Plumley C 2022 Kelmarsh wind farm data [Online; accessed 2024-03-01] URL `https://zenodo.org/record/8252025`

[25] Ghaderi A, Sanandaji B M and Ghaderi F 2017 *CoRR* **abs/1707.08110** URL `http://arxiv.org/abs/1707.08110`

[26] Fey M and Lenssen J E 2019 Fast Graph Representation Learning with PyTorch Geometric URL `https://arxiv.org/abs/1903.02428`

[27] Brody S, Alon U and Yahav E 2021 How Attentive are Graph Attention Networks? URL `https://arxiv.org/abs/2105.14491`

[28] Barber S, Hammer F and Tica A 2022 *ASCE-ASME J Risk and Uncert in Engrg Sys Part B Mech Engrg* ISSN 2332-9017, 2332-9025

[29] Hammer F and Barber S 2022 *Journal of Physics: Conference Series* **2151** 012006 ISSN 1742-6588, 1742-6596

Journal of Physics: Conference Series **3025** (2025) 012015 doi:10.1088/1742-6596/3025/1/012015

Push power where you can and prevent damage where you need

J Pacheco[1*], M Berezyak[1], J Martins[1], N Teixeira[1], C Pereira[1], R Figueiredo[1] and C Oliveira[1]

[1] Fibersail, Oporto, Portugal

*E-mail: joao.pacheco@fibersail.com

Abstract. The sustainable development of the wind energy industry faces several challenges, one of which is the lack of flexibility in adapting to increasing power system complexity. From the wind farm operator's perspective, flexibility means the ability to adjust energy generation patterns to different constraints. However, in current industry practice, curtailment strategies are primarily driven by regulatory restrictions and market-based economic factors. The effects of these strategies on turbine loading, lifetime, and their associated costs or benefits are mostly overlooked. Currently, operators lack tools to evaluate curtailment strategies while also considering their potentially conflicting effects on turbine component lifespan. In this sense, a condition monitoring system (CMS) based on an advanced shape sensing solution for wind turbine blades is being developed by Fibersail. This work demonstrates the capability of Fibersail's CMS to deliver critical data to assess the impact of curtailment strategies on turbine blades in terms of fatigue loads and deflection. By continuously tracking blade shape and loads throughout turbine's lifetime, operators can make more informed operational decisions. This enables a transition to active asset management, where operators can strategically choose between "Power Mode" - maximizing energy production or "Lifetime Mode" - minimizing component wear and extending turbine lifespan.

1. Introduction

The wind power industry in Europe has experienced rapid growth during the last two decades, reaching a total installed capacity of 272 GW by the end of 2023 [1]. However, the sustainable development of wind energy industry faces several challenges, one of which is the lack of flexibility in adapting to increasing power system complexity [2].

From the wind farm operator's perspective, flexibility refers to the ability to adjust energy generation patterns based on various constraints. There are two key scenarios where flexibility tools can be highly beneficial for owners and operators. First, during periods of high renewable energy production and low electricity demand, market prices can drop, sometimes even turn negative. This often drives to the curtailment of wind farms. Additionally, grid operator may require wind farms to temporarily curtail their turbines due to grid overload [3]. Second, there is growing industry interest in quantifying the consumed and remaining useful lifetime of wind assets, particularly with a view to potential lifetime extension. Wind farm owners often encounter critical operating conditions that increases structural wear and damage. For example, highly

Content from this work may be used under the terms of the Creative Commons Attribution 4.0 licence. Any further distribution of this work must maintain attribution to the author(s) and the title of the work, journal citation and DOI.
Published under licence by IOP Publishing Ltd

turbulent wind sectors can significantly increase fatigue loads on wind turbine components, even during periods of low power production.

However, in current industry practice, curtailment strategies are often based solely on regulatory restrictions and market-driven economic factors. The effects of these strategies on turbine loading, lifetime, and their consequential costs or benefits are mostly overlooked. In fact, with lifetime extension becoming an increasingly important topic for wind farm operators [4], it is crucial to assess the impact of curtailment on asset longevity. Currently, operators lack tools to evaluate curtailment strategies while also considering their potentially conflicting effects on turbine components lifespan.

Recent studies have explored the effects of curtailment on wind turbine components. Robbelein et al. [5] assessed the impact of curtailment on the monopile support structures of offshore wind turbines. Their findings reveal that some curtailment scenarios can lead to an increase in damage equivalent loads (DELs), mainly due to a higher frequency of shutdowns and start-ups. This study shows that curtailment strategies may either shorten or extend the remaining useful lifetime of monopiles, depending on the selected curtailment intervals. Another study, [6] Natarajan et al. investigated curtailment through derating and its positive effects on the loads and lifetime of a Vestas V-52 onshore wind turbine, showing that DELs of the tower and blade root edge were reduced. While curtailment regimes can offer benefits, they can also lead to a potentially critical operational conditions that affect fatigue life, such as increased standstill time and additional transitional events [5].

Despite these findings, there are no fully implemented commercial solutions that can respond to operator's needs. To address this gap, a Condition Monitoring System (CMS) based on an advanced shape sensing solution for wind turbine blades is being developed by Fibersail. This innovative technology involves the installation of fiber-optic sensors within turbine blades to continuously monitor and measure their real-time shape, deformation, and structural health during operation. By collecting various parameters such as blade curvature, twist, and deflection, this technology provides valuable data for inferring aerodynamic performance and monitoring blade health throughout its lifespan.

This work intents to demonstrate the capability of Fibersail's CMS in delivering critical data/information to assess the impact of different curtailment strategies on wind turbine blades, particularly in terms of fatigue load and blade deflection.

The paper is structured as follows. Section 2 provides a brief description of the experimental campaign and introduces the Fibersail CMS. It also presents notes on the implemented curtailment strategies and data processing methods. Section 3 is dedicated to assessing and discussing the effects of curtailment on blade behaviour. The main conclusions and potential future developments are outlined in Section 4.

The main motivation for this study lies in the upward interest of the industry towards the use of monitoring data for structural assessment during operation, understanding the impact of curtailment scenarios on lifetime extension and the need for data-driven support in asset management decisions. This works aims to contribute to the development of sustainable curtailment strategies, enabling the flexible and efficient use of wind energy.

2. Description of the case study

This work uses results from an extensive ongoing experimental campaign, which started in May 2023 and involves the instrumentation of the rotor of a large wind turbine, designed for offshore deployment.

2.1 Shape-sensing of wind turbine blades

The tested wind turbine was instrumented with a CMS based on an advanced shape-sensing solution for wind turbine blades, developed by Fibersail. The versatility of this fiber-optic based sensor allows for the design of customized instrumentation solutions tailored to specific measurement objectives. The system is easy to transport and install onsite and can be retrofitted into existing wind turbines, independent of OEM systems.

In this case, the CMS consists of the Fibersail Shape Sensor (FSSS) and the Fibersail Load Sensor (FSLS). Simply put, the FSSS consists of four parallel arrays of distributed FBGs (Fiber Bragg grating) embedded in a composite beam and installed longitudinally from the root to the tip inside the blade. As schematically represented in Figure 1, the FSSS was attached to the shear-web, with one sensor per blade.

Additionally, four FSLS were installed at the blade root, aligned with the blade's main directions (see Figure 1). FSLS are labelled according to their position on the blade as follows: *PS* (pressure side), *LE* (leading edge), *TE* (trailing edge) and *SS* (suction side). Each FSLS also incorporates an optical temperature sensor for thermal compensation.
The monitoring data collected from the blades is complemented by SCADA data, which characterizes the environmental and operational conditions on-site. Calibration of the system is performed during slow roll conditions.

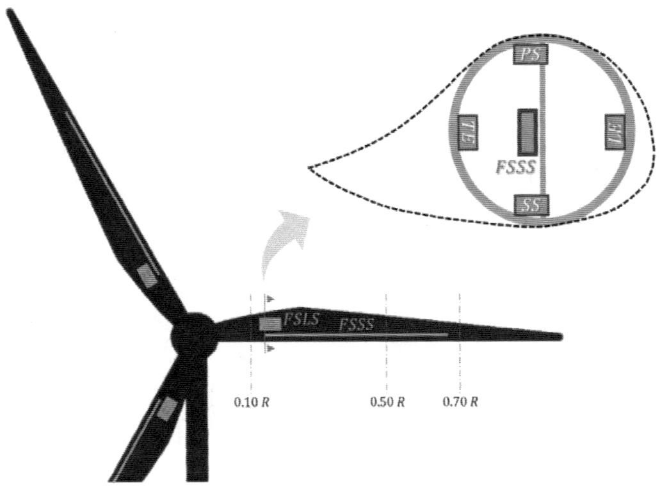

Figure 1. Schematic representation of the instrumentation layout adopted in the experimental campaign (R: total blade length). FSSS in green and FSLS in orange.

2.2 Load and deflection assessment of wind turbine blades

The effects of curtailment are assessed using experimental data measured by Fibersail CMS. Blade loads and deflections, measured at 20 Hz, are calculated throughout the monitoring period.

The SCADA data and Fibersail CMS data were divided into 10-minute time series and its respective mean values were calculated. To link the 10-minute time series with specific wind turbine operating conditions, CMS parameters were clustered into different load cases, such as normal power production, Start/Stop events, parked or idling states, and shutdowns. The clustering method uses environmental parameters (wind speed and turbulence intensity) and operational parameters (rotor speed, power output and pitch angle) provided by the SCADA system, recorded at 1 Hz. In this work, only the load scenarios classified as power production (10-

min during which the turbine generated power) are considered, as they represent the majority of operating time. The remaining loading categories, due to their specific characteristics, will be analyzed in future studies.

2.3 Curtailment of wind turbine

The primary objective of this work is to study the impact of curtailment regimes on blade loads and lifetime. In view of this, two operational scenarios are defined:normal turbine operation and a curtailment scenario where power output is limited to 35% of rated capacity. The experimental power curves obtained for both operational scenarios are represented in Figure 2.

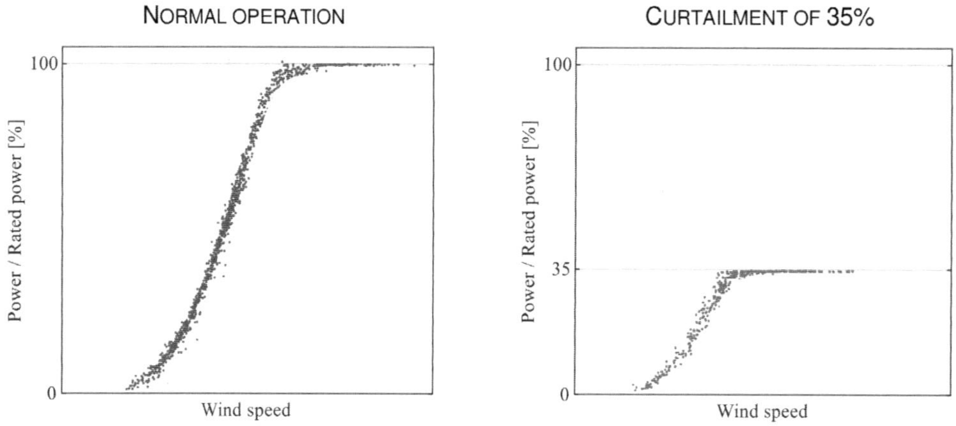

Figure 2. Comparison of the experimental power curves for the two operating scenarios (10-min mean values from SCADA data). All values are normalized.

It is important to note that the curtailment defined in this work does not imply stopping the wind turbine production, but rather limiting the power generated. As shown in Figure 3, by adjusting the pitch controller it is possible to limit the power output. When curtailment is activated, the pitch angle increases and, subsequently the lift force decreases. As a result, both rotor speed and power output remains constant despite increasing wind speed.

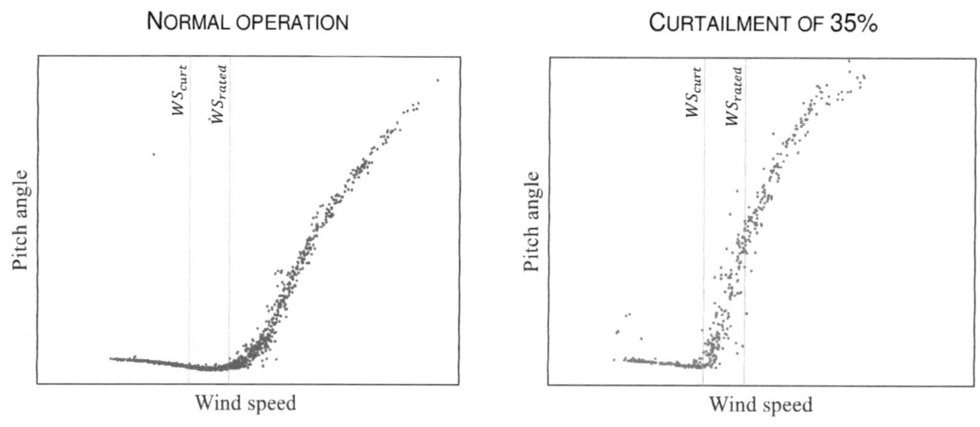

Figure 3. Comparison of the behaviour of the blade pitch controller in the two operating conditions addressed, where WS_{rated} is the rated wind speed and WS_{curt} is the wind speed at which curtailment is activated (10-min mean values from SCADA data).

3. Monitoring the curtailment effects on blade behaviour

Figure 4 presents a measured time series during curtailment activation. Focusing on environmental and operational parameters, when wind speed increases and exceeds the defined threshold of WS_{curt}, the turbine control system is activated, and the pitch angle value increases to reduce lift force on the rotor. Consequently, rotor speed and power generation remain stable, without exceeding the predefined limits.

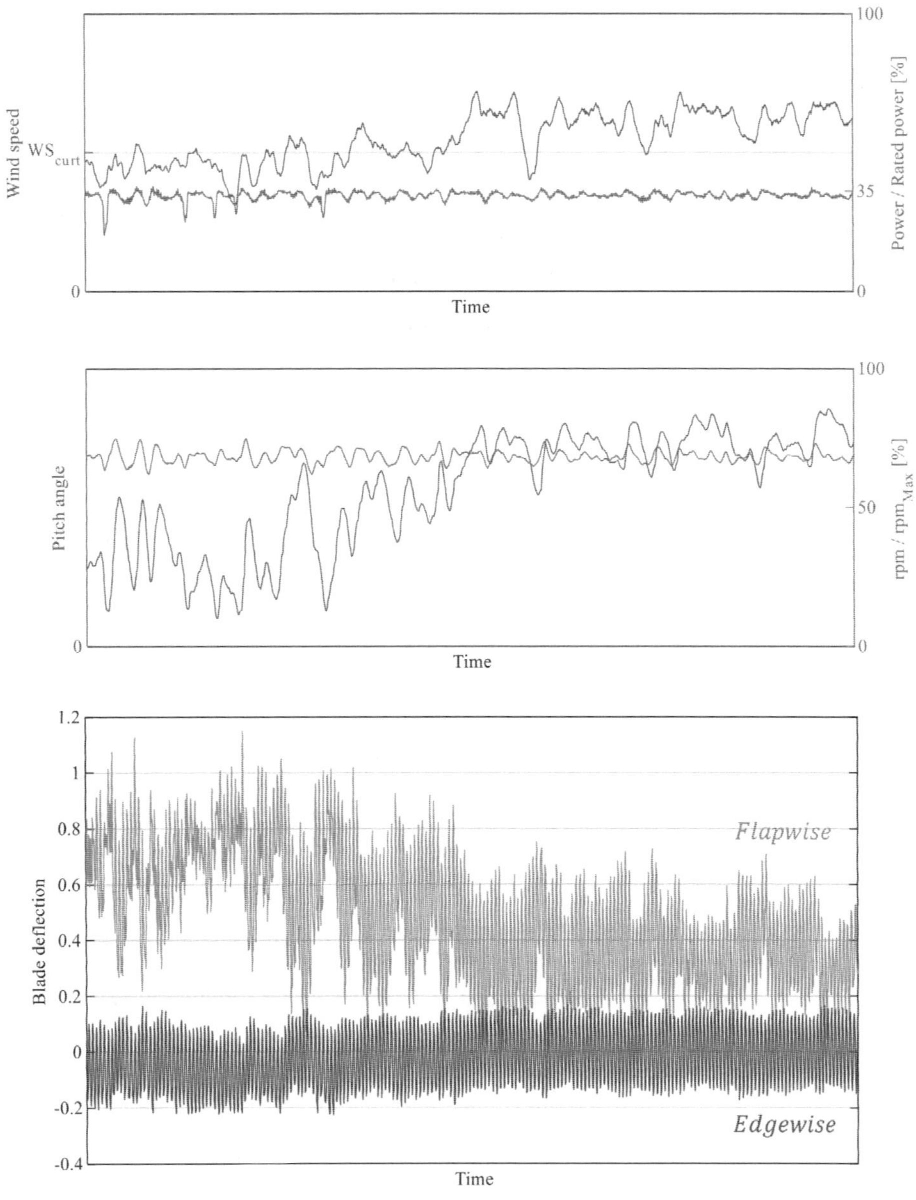

Figure 4. Time series during wind turbine curtailment. Operational and environmental conditions: wind speed, power output, pitch angle and rotor speed. Flapwise and edgewise blade deflection measured at blade radius $R = 0.5$. All values are normalized. (The time window is 30 min).

Looking now at the blade behaviour, it can be seen in Figure 4, as soon as curtailment is activated, the increase in pitch angle leads to a decrease in flapwise blade deflection (green line). On the other hand, there is a slight increase in blade deflection in the edgewise direction (red line). In fact, edgewise forces are directly related to rotor torque.

A similar analysis is also performed for blade root loads, with flapwise and edgewise bending moments represented in Figure 5. The results obtained in flapwise are similar to Figure 4, since the reduction in drag force caused by the increase in pitch angle leads to less deflection and load in this direction. As for the bending moments, no significant reduction is observed in the edgewise direction. In fact, it is important to emphasize that for this direction the gravitational loads are predominant compared to aerodynamic loads, in contrast to the flapwise direction.

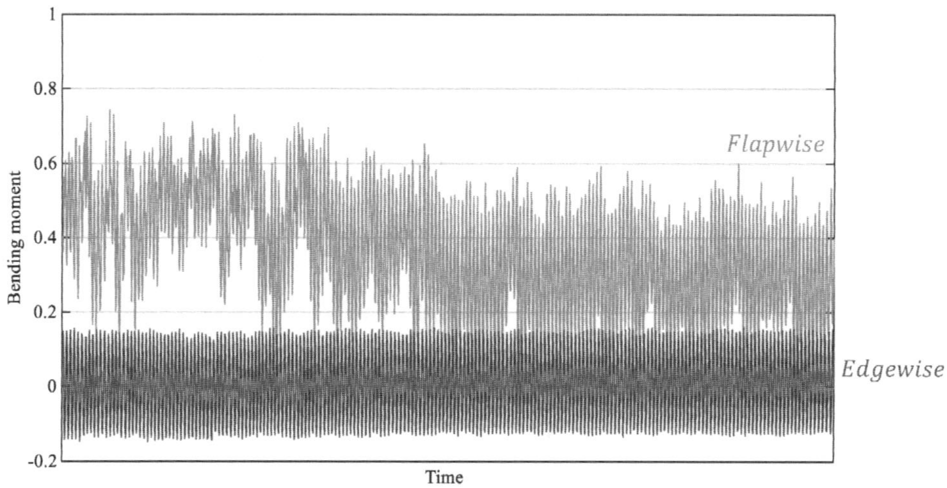

Figure 5. Time series during wind turbine curtailment. Flapwise and edgewise bending moments at blade root. All values are normalized.

The presented results make it easy to identify distinct blade behaviour patterns when curtailment is activated, particularly in terms of loads and deflection. Based on these findings, it is possible to characterize the blade's behaviour in terms of deflection and loads over the analysed period. Considering the turbine operating under power production, Figure 6 represents how blade deflection varies as a function of wind speed and rotor speed.

Initially, blade deflection increases with wind speed until it reaches the turbine´s rated speed. At this point, the pitch control mechanism comes into action with the intention of limiting the aerodynamic loads on the blades. Consequently, flapwise blade deflection decreases as wind speed continues to increase.

Journal of Physics: Conference Series **3025** (2025) 012015 doi:10.1088/1742-6596/3025/1/012015

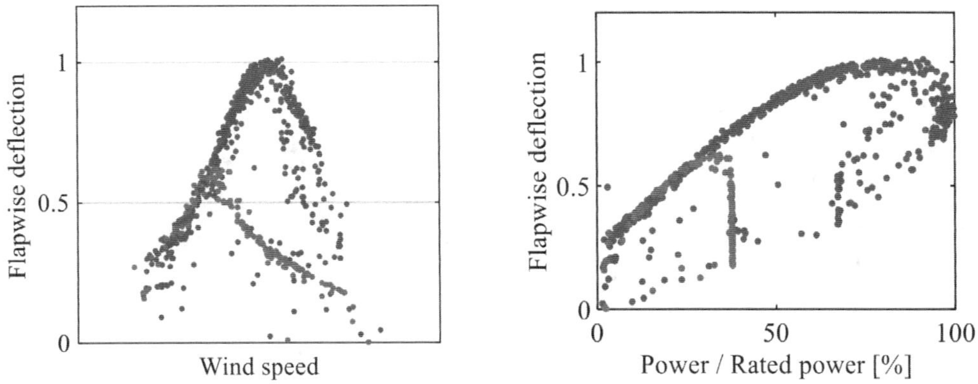

Figure 6. Curtailment strategy effect on blades behaviour. Flapwise blade deflection at radius R = 0.5 as function of wind speed (left) and power output (right), considering power production conditions (normal operation: blue; curtailment 35%: orange). All values are normalized.

Similar conclusions can be drawn for the flapwise and edgewise bending moments at blade root, represented in Figure 7.

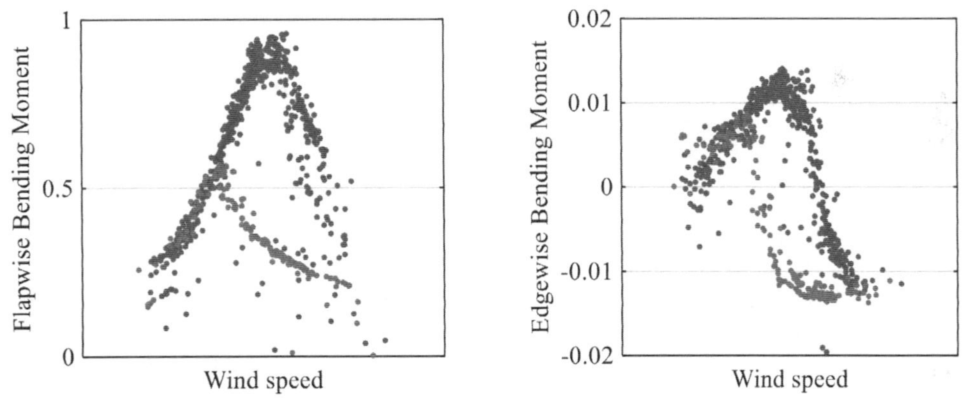

Figure 7. Curtailment strategy effect on blades behaviour. Flapwise (left) and edgewise (right) bending moments at blade root as function of wind speed, considering power production conditions (normal operation: blue; curtailment 35%: orange). All values are normalized.

The Fibersail CMS accurately characterizes variations in deflection and load for different turbine operating conditions, closely correlating with wind speed in both main blade directions. Moreover, the results obtained are in accordance with expectations. Notably, the high precision in deflection and load tracking allows to clearly identify the two implemented control strategies. The results are highly promising, serving as essential inputs for calculating fatigue damage at the blade root and effectively assess and compare the impact of curtailment on load lifetime.

By precisely tracking blade shape and loads throughout wind turbine's lifetime, operators can make informed operational decisions. This enables a shift towards active asset management strategies, allowing operators to choose between "Power Mode" - maximizing energy production or "Lifetime Mode" - minimizing component wear and extending turbine lifespan.

4. Conclusions and Future Works

Wind farm operators need to optimize curtailment strategies to account for their effects on blade loads, wear, and lifetime. This paper presents an approach to obtain valuable insights that enable owners and operators to make informed decisions.

It utilizes measurements from Fibersail CMS as well as high-resolution SCADA data (1 Hz) from a large offshore wind turbine. Blade deflection and loads are calculated for normal operation, disregarding, at this stage, transient events (shutdown, start-stop, and idling). The suggested methodology is appealing due to its simplicity in implementation.

As shown, blades benefit from curtailment due to the reduction of blade loads and deflection compared to normal operation. Despite an increase in edgewise deflection, the total bending moment decreased in curtailment situations as aerodynamic forces acting on the rotor were reduced.

The Fibersail CMS system addressed in this work emerges as a key component of predictive and preventive maintenance strategies. By precisely tracking blade shape changes during wind turbine operation, operators can visualize/track the aerodynamic performance of the blades in real time, optimize the production, detect anomalies early, and adjust maintenance interventions to minimise losses. The main topics addressed in this work are:

- Industry interest towards the use of monitoring data for structural assessments of the assets during operation.
- The importance of understanding long-term curtailment effects, particularly in the context of lifetime extension.
- Opportunity to provide data-driven support in asset management decisions towards sustainable curtailment strategies supporting the flexible use of wind energy assets.

Finally, it is important to note that these results depend strongly on turbine type as well as on environmental conditions and thus cannot be generalized.

In future work, the results presented in this work will be instrumental to assess the impact on the blade's fatigue life when additional event load cycles are introduced by curtailment regimes. The impact of additional event cycles under various curtailment regimes on the fatigue life of the blades cannot be neglected and should be quantified with data-driven lifetime assessment methods using real-world load measurements.

Additionally, the effects of transient events during curtailment strategies will be analysed in future work. Assessing curtailment impacts on wake-exposed wind turbines will also be relevant, with a view to optimizing this type of strategies for wind farm operation.

Future research should also investigate the costs-benefit trade-off of curtailment induced loads and wear and integrate these factors into curtailment optimization strategies. For example, a parameter such as 'lifetime cost per curtailment duration' could account for loss of future production due to the reduced lifespan of support structures.

The methodology developed in this work aims to safeguard the flexible use of wind energy assets in the short-term while ensuring long-term reliability. With the ambition of lifetime extension, curtailment strategies may be recommended as a proactive approach.

Journal of Physics: Conference Series **3025** (2025) 012015 doi:10.1088/1742-6596/3025/1/012015

Acknowledgements

The authors acknowledge the financial support from COMPETE 2030, Portugal 2030 and the European Union, through the R&D project "Fibersail 2.0 - Advanced Condition Monitoring Solutions for Wind Turbines", with reference COMPETE2030-FEDER-00579500, in which this work was developed.

References

[1] Wind Europe. (2024). Wind energy in Europe 2023. Statistics and the outlook for 2024-2030.

[2] Wind Europe & Hitachi Energy. (2024) Maximising the power of wind through grid flexibility.

[3] Linnemann, C., Echternacht, D., Breuer, C., & Moser, A. (2011). Modeling optimal redispatch for the European Transmission grid. In 2011 IEEE Trondheim PowerTech (pp. 1-8). IEEE.

[4] Ziegler, L., Gonzalez, E., Rubert, T., Smolka, U., & Melero, J. J. (2018). Lifetime extension of onshore wind turbines: A review covering Germany, Spain, Denmark, and the UK. Renewable and Sustainable Energy Reviews, 82, 1261-1271.

[5] Robbelein, K., Daems, P. J., Verstraeten, T., Noppe, N., Weijtjens, W., Helsen, J., & Devriendt, C. (2023). Effect of curtailment scenarios on the loads and lifetime of offshore wind turbine generator support structures. In Journal of Physics: Conference Series (Vol. 2507, No. 1).

[6] Natarajan, A., & Pedersen, T. F. (2018). Remaining life assessment of offshore wind turbines subject to curtailment. In ISOPE International Ocean and Polar Engineering Conference.

Journal of Physics: Conference Series **3025** (2025) 012016 doi:10.1088/1742-6596/3025/1/012016

Performance Analysis of Offshore Wind O&M Activities based on Vessel and Helicopter Movement Data

David Baumgärtner[1], **Jonas Kaczenski**[1], **Jürgen Mackeprang**[2], **Thomas Pracejus**[3]

[1]Fraunhofer Institute for Wind Energy Systems IWES, Bremerhaven, Germany, [2]TRACTEBEL DOC OFFSHORE GmbH, Bremerhaven, Germany, [3]Siemens Gamesa Renewable Energy, Hamburg, Germany

E-mail: david.baumgaertner@iwes.fraunhofer.de

Abstract.
Offshore wind energy is emerging as a crucial part of the global energy transition, yet the complexity and cost of operations and maintenance (O&M) continue to challenge the sector. This study introduces a novel analytical framework that leverages vessel and helicopter movement data—captured through AIS and ADS-B signals—and integrates hourly ERA5 weather data to comprehensively evaluate O&M performance at offshore wind farms (OWFs). By automatically identifying key operational events such as turbine transfers, port departures, and idling periods, the methodology computes a suite of key performance indicators that capture both the effectiveness (i.e., workload capacity under varying weather conditions) and the efficiency (i.e., transit, idle, and transfer times) of diverse logistical concepts. Analysis across multiple OWFs, including a detailed case study for the Butendiek wind farm, reveals that while efficiency scores remain relatively consistent, effectiveness varies significantly—particularly in response to weather constraints and fleet composition. The transition from a solely vessel-based strategy to a mixed fleet incorporating service operation vessels, crew transfer vessels, and helicopters resulted in improved workload capacity and enhanced overall performance, despite minor trade-offs in operational efficiency. These findings demonstrate the potential of integrating real-time movement and weather data to optimize O&M strategies, thereby reducing downtime and operational costs of offshore wind farms.

1 Introduction

Offshore wind energy has rapidly evolved from a novel renewable technology into a cornerstone of the global energy transition. With ambitious expansion targets, such as the European goal of 450 GW by 2050 [1], the offshore wind market is growing not only in scale but also in operational complexity. As offshore wind farms (OWFs) increase in both number and capacity, the challenges associated with the Operations and Maintenance (O&M) phase become increasingly pronounced. Studies have highlighted that O&M costs can represent up to 30% of the total cost of offshore wind projects [2, 3, 4], underscoring the need for more efficient logistical solutions.

The complexity of O&M logistics is further compounded by the harsh and variable weather conditions prevalent in offshore environments. Traditional strategies, which have relied solely on vessels for service and repair operations, are being re-evaluated in favor of more diversified approaches that include the use of helicopters [5]. This diversification is critical for minimizing downtime and ensuring that maintenance activities can be carried out even under challenging conditions.

Content from this work may be used under the terms of the Creative Commons Attribution 4.0 licence. Any further distribution of this work must maintain attribution to the author(s) and the title of the work, journal citation and DOI.
Published under licence by IOP Publishing Ltd

Journal of Physics: Conference Series **3025** (2025) 012016 doi:10.1088/1742-6596/3025/1/012016

Although AIS data (Automatic Identification System) is primarily designed for navigational safety, allowing vessels to communicate their navigational status with one another [6], it has been the focus of various studies related to offshore wind. In the context of analyzing and evaluating O&M activities, AIS data has been used for real-time observation [7]. Additionally, a recent study applied AIS data along with a machine learning approach to assess offshore installation processes and the resulting learning curves during the deployment phase [8].

AIS data is increasingly used in OWFs to enhance the logistics and efficiency of O&M. This data enables the analysis of vessel movements near wind farms, providing valuable insights into usage patterns and resource allocation. By evaluating AIS data, operators can optimize travel times to OWFs, improve maintenance scheduling, and identify potential bottlenecks in logistics. This contributes to reducing operational costs and increasing the reliability of maintenance activities, especially under challenging weather conditions [9].

However, further research is necessary to fully understand the potential applications of AIS data and to develop advanced methodologies for its integration into O&M strategies. Moreover, as stated above, a comprehensive evaluation of O&M performance in OWFs must include both vessels and helicopters. This paper aims to bridge that gap by presenting a novel analytical framework that combines vessel and helicopter movement data with weather information. In addition to AIS data, we incorporate movement data from helicopters using Automatic Dependent Surveillance–Broadcast (ADS-B) [10], which serves a similar purpose for aircraft.

By utilizing these extensive and publicly available datasets, valuable insights into past service and repair activities can be obtained. This study explores innovative analytical methods for a joint analysis of O&M logistics and the resulting optimization opportunities by evaluating key performance indicators (KPIs) related to both the effectiveness and efficiency of logistical operations. Our approach not only provides actionable insights into the current state of O&M practices but also enables the identification of best practices across different OWFs and highlights areas where further optimization is possible.

2 Methodology

2.1 Processing of AIS and ADS-B Data

The methodology builds upon the approach presented in a previous paper [9], which is briefly summarized in the following.

The movement data of all vessels and helicopters in the German Bight is recorded by three measurement stations that were installed within the scope of the research project LogReview [11] and are operated by the Fraunhofer IWES. The stations are equipped with both AIS as well as an ADS-B antenna. For the exemplary OWF of Butendiek, the relevant AIS and ADS-B data is shown on Figures 1a and 1b, respectively. The data has been filtered for vessels and helicopters that have performed any O&M activity at the Butendiek OWF.

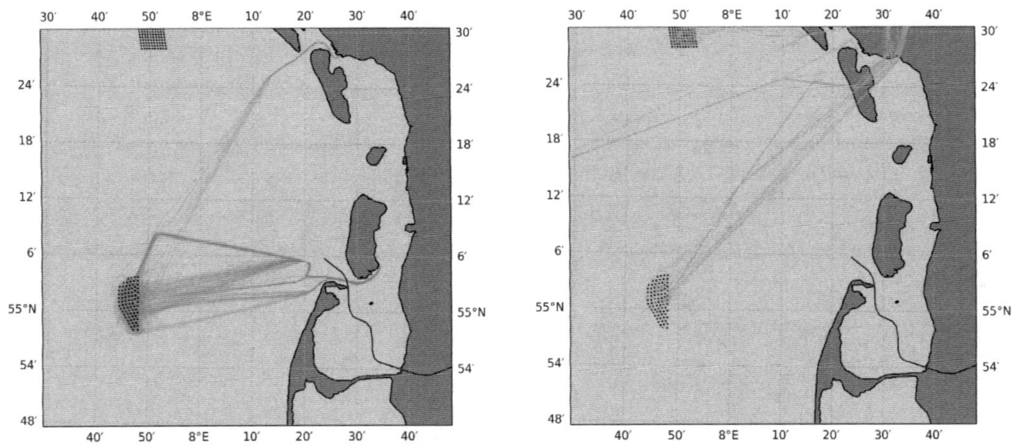

(a) AIS signals at Butendiek (b) ADS-B signals at Butendiek

Figure 1: Vessel and helicopter movement data recorded during 2024 by Fraunhofer IWES' sensor stations. The data is filtered for vessels and helicopters that operated at Butendiek during 2024.

To manage the large amount of data efficiently, static AIS data (ship identifiers and characteristics) are stored in a central database, while dynamic data (position, speed, and course) are stored per vessel. Helicopter signals are initially filtered out from the ADS-B data (which also includes other types of aircraft) and then processed in the same manner as the vessel data. Outliers are identified by comparing reported and theoretical velocities (the velocity needed to get from one data point to the next); deviations beyond set thresholds mark the data point as an outlier. Key vessel and helicopter events, such as wind farm, wind turbine, platform, and (air)port visits, are automatically identified based on proximity, time, and speed criteria. This enables detailed analysis of transit times, distances, and speeds between these events, as well as periods of idling.

For a comprehensive understanding, weather data from the ERA5 data set [12] is linked to the movement data, enabling the analysis of weather impacts on vessel and helicopter performance during O&M activities. The ERA5 weather data includes hourly values of met-ocean variables such as wind speed, wind direction, wave height, and wave period.

The methodology has been validated against operational data from our associated project partner Siemens Gamesa Renewable Energy (SGRE), who provided daily progress reports (DPRs) as well as daily flight reports (DFRs) for a certain wind farm in an electronic format. These reports include all performed transfers at the turbines/platforms by the vessels and helicopters. Figure 2 shows the number of transfers of a vessel as identified from the AIS data as well as the DPR data. Overall the AIS data matches the DPR very closely, with only minor differences on some days. Therefore, deployed logistic concepts in the different OWFs can be extracted and detailed comparisons of the combined vessel and helicopter concepts between OWFs can be performed.

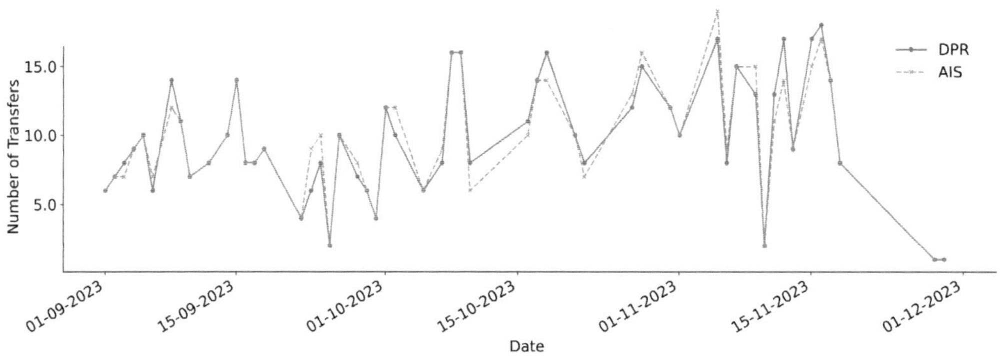

Figure 2: Comparison of the recognized turbine transfers through AIS data to the transfers listed in the DPRs

2.2 Performance Analysis

In addition to identifying logistic concepts, it is important to evaluate their performance. In the scope of this study, performance is assessed from two main perspectives:

- **Effectiveness:** The ability of the logistic concept to operate at an OWF under local weather conditions. This is evaluated by determining whether the concept can handle large workloads whenever necessary.
- **Efficiency:** The operational efficiency of vessels and helicopters at the wind farms. Reducing idle and transit times, as well as minimizing transfer times at the turbines, contributes to improved efficiency.

For each category, a set of KPIs have been developed. These KPIs were defined to enable comparisons between different OWFs and were normalized, where necessary, based on the number of turbines at each OWF. In many cases, a P95 value was determined to compute the KPI. The P95 value represents the 95th percentile of the data distribution, meaning that 95% of the observed values fall below this threshold. This approach minimizes the influence of outliers and provides a robust measure of typical performance.

The effectiveness of the logistic concept is the first metric to assess performance. This metric measures the total work capacity of the entire logistic concept while taking local weather conditions into account. The following KPIs are used to determine the effectiveness:

- **Share of Turbines Visited per Day (P95):** This KPI assesses whether there are enough vehicles (vessels and/or helicopters) and service teams to cover a significant share of turbines in a single day. For instance, if multiple failure events occur on the same day, the logistic concept must be large enough to handle them concurrently.
- **Total Work Time per Day (P95):** This indicator measures the aggregate net tool time of all work teams across the OWF. A higher total work time indicates that the logistic concept can support a larger workload when necessary.
- **Workability:** This KPI evaluates how many hours per day transfers are theoretically possible, given the maximum weather constraints encountered by all vehicles during the year. It helps assess whether the chosen setup of vessels and helicopters can operate effectively under the local weather conditions.
- **Work Time on Single Turbine per Day (P95):** This metric captures the maximum work time achievable by a single work team on an individual wind turbine. It assesses whether excessive transit times to and from the OWF are causing significant work time losses. For OWFs located far from the coast, it may be beneficial to use vessels stationed at the OWF or helicopters.

While effectiveness evaluates the overall capacity of the logistic concept under local weather conditions, efficiency focuses on how well the available resources are utilized during operations. The following KPIs are used to assess efficiency:

- **Total Operating Time per Transfer:** This KPI measures the complete operational time for each transfer, defined as the time from when a vehicle leaves the (air)port until its return. It provides insight into the overall duration of each operational cycle.
- **Average Vehicle Transit Time per Transfer:** This indicator calculates the average time that vehicles spend in transit per transfer. It reflects how quickly vehicles can move between key locations (e.g., from the port to the wind farm) and the number of transfers performed during each visit.
- **Total Vehicle Transit Time:** Expressed as a percentage, this KPI represents the proportion of transit time relative to the total operational time at the OWF. A lower percentage indicates that a greater share of time is devoted to productive tasks at the OWF rather than transit.
- **Total Vehicle Idle Time:** This KPI quantifies the percentage of time vehicles are idling, either outside the OWF or inside while moving at speeds less than 1.5 knots (excluding the time during transfers at wind turbines or offshore service stations). Reducing idle time is key to enhancing operational efficiency.
- **Total Vehicle Shift Time:** This metric measures the percentage of time that vehicles spend actively moving within the wind farm (with speeds above 1.5 knots). It captures the time spent traveling between turbines or to designated waiting spots, and helps assess the fluidity of operations.
- **Vehicle Shift Distance per Transfer:** This KPI is calculated by dividing the total distance travelled by a vehicle during the operational period by the number of transfers performed (at turbines or service stations). It provides an average distance metric per transfer, which can be useful for optimizing routing strategies.
- **Average Vehicle Transfer Time:** This metric captures the time taken for the approach phase (from 100 m inwards), the crane and gangway operations, and the departure phase (back to 100 m). Durations exceeding 60 minutes are excluded to focus on typical operational scenarios and omit major component replacements.
- **Average Vehicle Stationary Time:** This KPI measures the time that vehicles remain stationary at the wind turbine during crane and gangway operations, excluding periods longer than 60 minutes. It highlights the efficiency of on-site transfer operations.

- **Average Number of Transfers at the Same Turbines per Day:** This indicator reflects the average number of transfers at an individual wind turbine per day. The most efficient number would be two, one for dropping off the service team and one transfer for picking them up, without the need for additional transfers in between.

In the next step, thresholds were established for each KPI to translate the absolute values of each KPI into a score ranging from 1 (good) to 6 (poor). All KPIs were computed for monthly time periods as well as yearly values. Based on initial results of the analysis, the threshold values for the scores were adjusted so that each KPI, where applicable, would resemble a normal distribution.

Finally, each KPI was assigned a weighting within its category, allowing for the calculation of an overall score that reflects how effectively and efficiently the logistic concept is performing. This weighting factor allowed us to assign greater influence to certain KPIs, such as the *Workability* or the *Vehicle Shift Distance per Transfer*.

3 Distribution of Efficiency and Effectiveness

The analysis was performed on all German OWFs located in the North and Baltic Sea, as well as on the Danish OWFs. For the Danish OWFs, as well as the German OWFs in the Baltic Sea, AIS data provided by the Danish government [13] was used. The Danish wind farms close to the German border are also under the coverage of our measurement stations. All the OWFs located in the Baltic Sea are not covered by any of our antennas, resulting in no helicopter data being included in the analysis of those OWFs. We expect that this has only a limited effect on the results, as analysis of commercially available historical data from the Baltic Sea has shown that helicopters are rarely used for O&M purposes, likely due to the generally milder weather conditions compared to the North Sea.

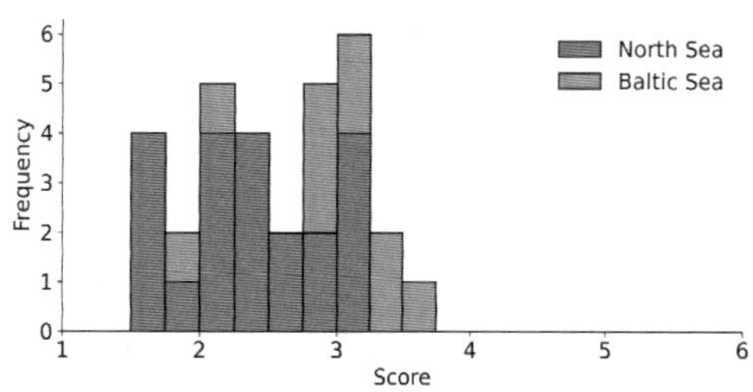

(a) Effectiveness

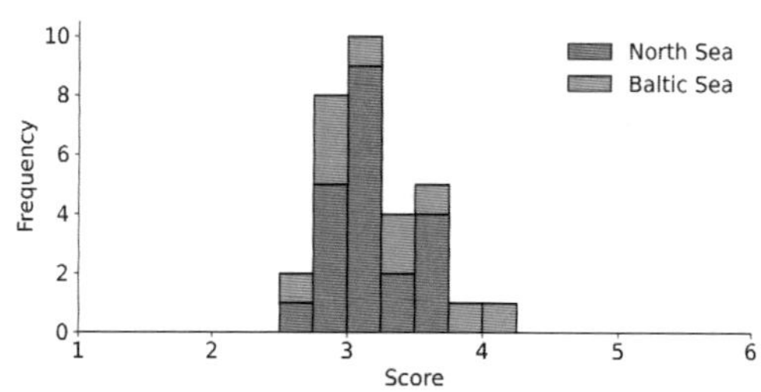

(b) Efficiency

Figure 3: Performance scores computed for all German and Danish OWFs for the year of 2024

Figures 3a and 3b show the performance scores for the computed effectiveness and efficiency for all considered OWFs. Wind farms that already use a joined logistical concept (e.g. OWFs Albatros and Hohe See) have a combined score. The data analysis indicates a relatively narrow range in efficiency scores—between 2.55 and 4.15—across OWFs, even though individual KPIs are scaled to yield scores that normally distribute from 1 to 6. This suggests that no current logistics concept excels simultaneously across all the KPIs contributing to the efficiency score. In contrast, the effectiveness scores exhibit a broader variation, ranging from 1.56 to 3.56.

Effectiveness is the primary driver for any optimization that a wind farm operator or service provider might pursue. A more effective logistics concept is better positioned to mitigate turbine downtimes, which represent some of the most significant cost drivers. The scores shown in Figure 3a and 3b are also distinguishing between the OWFs located in the North Sea and the Baltic Sea. While there is no difference in the efficiency, the OWFs in the North Sea tend to have a higher effectiveness. This comes down to the lower workability of the crew transfer vessels (CTVs) that are generally deployed in the Baltic Sea. Nevertheless, their workability is sufficient given the considerably lower chartering cost compared to a service operating vessel (SOV).

Still, it is important to optimise the efficiency of the logistic concepts, as this can directly reduce the operational cost. Further, an increased efficiency can lead to an increased effectiveness. To give an example, a SOV might be able to improve its efficiency to the extent that it becomes feasible to add an extra worker team, thereby enhancing the effectiveness of the entire logistical concept.

Similar performance scores can be computed for individual vessels and helicopters, just as for the overall logistics concepts. Figures 4a to 4f display these computed scores. Each score is calculated relative to the specific OWF where the asset operates, meaning that a single vessel or helicopter may have multiple performance scores. For our analysis, only SOVs with more than 400 turbine transfers at a given OWF are included, while the threshold is set at 200 transfers for CTVs and 40 transfers for helicopters. Using these thresholds ensures, that only the most relevant vessels and helicopters of each OWP are represented.

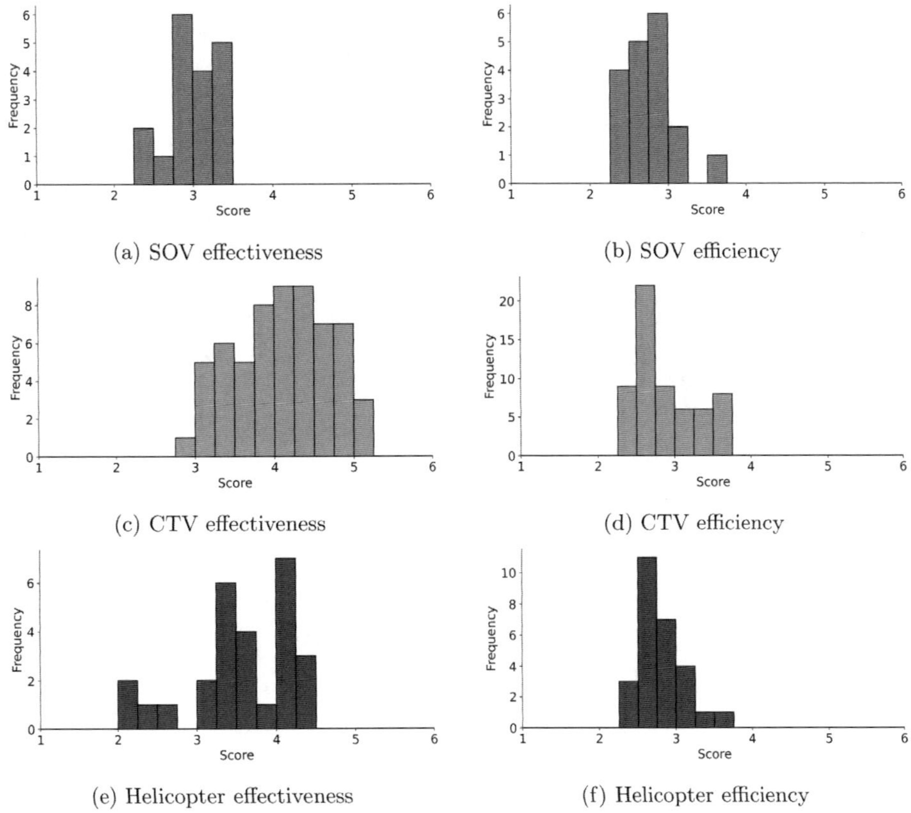

(a) SOV effectiveness (b) SOV efficiency

(c) CTV effectiveness (d) CTV efficiency

(e) Helicopter effectiveness (f) Helicopter efficiency

Figure 4: Performance scores computed for all SOVs, CTVs, and helicopters operating in German or Danish OWFs with respective turbine transfer thresholds in 2024

The data indicate that the average effectiveness of SOVs is higher than that of CTVs. This difference is likely due to several factors: SOV have a wider operational window due to higher workability and are therefore less restricted by the weather conditions, are permanently stationed at the OWF (allowing for longer daily work periods), and typically have more service teams that enable simultaneous task execution. Helicopters achieve the highest effectiveness scores, which is expected given their ability to operate even under stringent weather conditions.

Regarding efficiency, the average performance across different vehicle categories is comparable, although each option exhibits distinct advantages and disadvantages. For instance, SOVs benefit from reduced transit times because they are stationed at the OWF; however, they require longer transfer times and tend to idle more. In contrast, CTVs—despite having lower overall workability compared to SOVs—operate only when conditions permit and otherwise remain at the service port, thereby reducing idle times at the OWF. Additionally, the 'push-on' transfer method offers an efficient means of transferring technicians, although CTVs still spend more time in transit since they typically remain at the OWF for just one work shift. Helicopters, on the other hand, enjoy rapid transit, shorter on-site shift durations, and fast transfer times; however, a significant portion of their operating time is consumed by transit, as they often need to fly out twice—once to drop off and again to pick up a single technician team.

It is important to note that directly comparing a SOV to a CTV may not be entirely meaningful, as a single CTV cannot substitute for a SOV in terms of workload capacity. Instead, this analysis is most valuable when comparing the performance of multiple CTVs deployed at the same OWF or when assessing which vessel types might be best suited for a particular OWF.

4 Exemplary Analysis at Butendiek

This section demonstrates how the approach described previously can be applied to independently analyze the logistical concept deployed at the exemplary OWF Butendiek. Located 36 km off the German coast near the Danish border, Butendiek was fully commissioned in 2015. The wind farm comprises 80 turbines (SWT-3.6-120) rated at 3.6 MW each, resulting in a total capacity of 288 MW [14].

The relevant movement data of vessels and helicopters operating at the OWF is presented in the heatmaps in Figure 1. These figures reveal that both vessels and helicopters are utilized: Helicopter flights originate from Esbjerg, while vessels depart from Esbjerg and Rømø Havn.

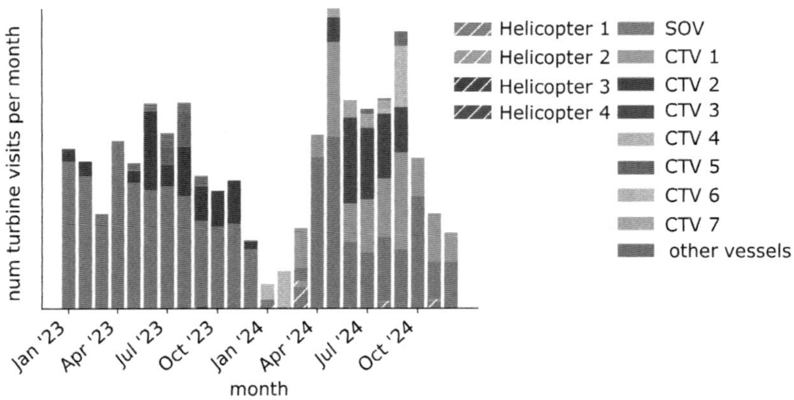

Figure 5: Turbine transfers per month at Butendiek 2023-2024 (absolute numbers are left out for confidentially reasons)

An analysis of the AIS and ADS-B data quantifies the number of transfers performed at the OWF, as shown in Figure 5. The data indicates that between 2023 and 2024 the logistical concept transitioned from being solely reliant on vessels to employing a mixed fleet that includes a SOV, CTVs, and helicopters. The impact of this revised logistical concept on both effectiveness and efficiency is discussed in the following.

The impact of these changes is reflected in the performance scores. Starting in 2023, the yearly effectiveness score improved from 2.89 to 2.11, resulting in a rise in rank from 20th to 9th out of the 31 logistical concepts analyzed. This improvement is attributed to the enhanced workload capacity of the larger fleet, which enables more turbines to be serviced concurrently and increases the total available work time at the OWF. In contrast, the efficiency score experienced a slight decline, moving from 2.70 to 2.85. Despite this small drop, the efficiency score remains among the best of all analyzed logistical concepts, with the ranking shifting only from 3rd to 4th. These results indicate that the adjustments to the logistical concept achieved the desired effect.

Figure 6 illustrates the seasonal variations of both performance scores over time. While efficiency remains relatively stable, effectiveness exhibits notable fluctuations—particularly a dip during the winter months when harsher weather conditions reduce accessibility of the OWF.

While these scores were initially computed for the entire logistic concept, the same evaluation was also performed for the individual vessels and helicopters at the OWF. This individual analysis makes it possible to identify which vehicles do not contribute positively to the KPIs of the entire fleet. Figure 7 shows the effectiveness scores for the SOV and three CTVs deployed at Butendiek. While a generally worse score for a CTV is expected compared to a SOV, the significant difference between the three CTVs of similar dimensions is notable. Identifying such underperforming vessels can lead to targeted adjustments, potentially improving the overall performance score of the logistical concept. By examining the individual KPIs that contribute to the overall effectiveness score, it became clear that the underperforming CTV did not meet the weather restrictions as well as the other two CTVs, resulting in significantly lower workability.

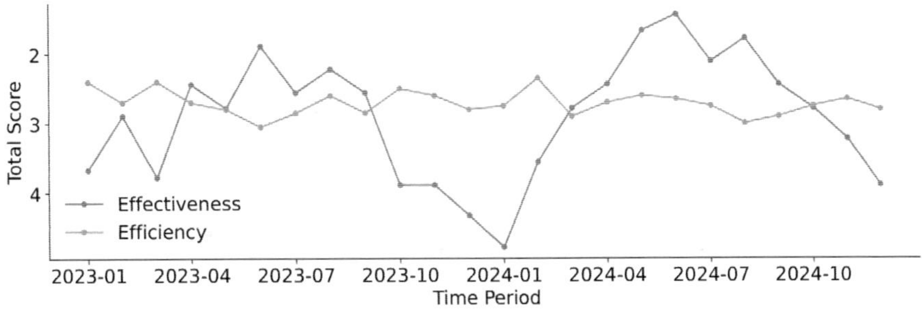

Figure 6: Seasonal variation in the effectiveness and efficiency of the entire logistic concept at Butendiek 2023-2024

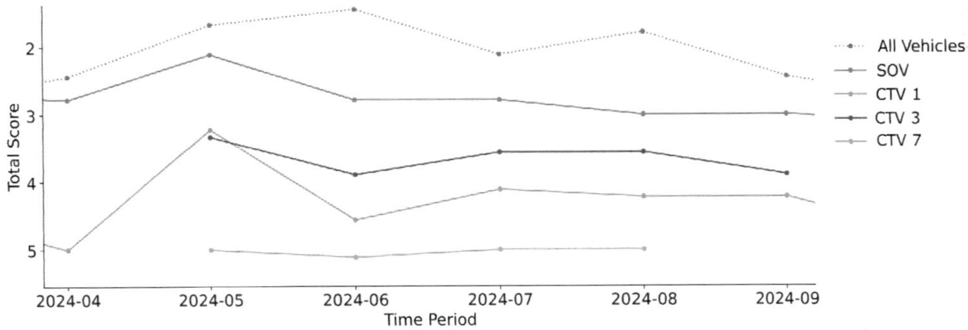

Figure 7: Effectiveness score of different vessels at Butendiek in 2024

5 Conclusions and Future Work

This study demonstrates the potential of leveraging publicly available data sources, such as AIS and ADS-B signals, to enhance the analysis and optimization of offshore wind farm logistic operations. By comparing the performance of various service concepts across multiple OWFs, the analysis not only identifies best practices but also supports strategic decision-making aimed at improving both the efficiency and effectiveness of O&M activities. Continuous assessment and benchmarking of different logistical approaches enable targeted optimizations that contribute to more sustainable and high-performing operational management.

The use of key performance indicators in this study has allowed operators to pinpoint the main drivers influencing both effectiveness and efficiency. Such insights are critical for fine-tuning logistical strategies, as they help identify underperforming vessels and helicopters and highlight areas for improvement. For example, while a larger and more diverse fleet can increase the overall work capacity (effectiveness) by servicing more turbines concurrently, it is equally important to minimize transit and idle times to control operational costs (efficiency).

Looking ahead, future research should incorporate cost estimates for all vessels and helicopters to better contextualize performance metrics relative to operational expenses. This integration would facilitate more informed decision-making by aligning performance improvements with cost-effectiveness. In addition, the development of advanced data analytics and the integration of emerging technologies could further enhance logistical strategies, ensuring that OWFs remain competitive and adaptive in the face of evolving demands and challenging weather conditions. Investigating the availability and impact of Helicopter Emergency Medical Services (HEMS) may also help identify potential bottlenecks early, thus proactively mitigating risks.

Overall, the presented analytical framework not only provides actionable insights into the current state of O&M practices but also serves as a foundation for future-oriented logistical concepts. By enabling operators to independently evaluate and optimize their logistical concepts, this approach paves the way towards improved operational excellence, enhanced sustainability, and reduced downtime across offshore wind operations.

6 Acknowledgements
This work has been carried out as part of the research project LogReview, which is funded by the German Federal Ministry for Economic Affairs and Climate Action (BMWK) under the grant number 03EE3051A [11]. The authors would like to thank the project partners for their support and cooperation.

References
[1] Kate Freeman, Ciaran Frost, Giles Hundleby, and Alun Roberts. Our energy, our future: How offshore wind will help europe go carbon-neutral. `https://maritime-spatial-planning.ec.europa.eu/practices/our-energy-our-future`, 11 2019. Accessed: 2024-02-28.

[2] BVG Associates. Guide to an offshore wind farm: Updated and extended. Technical report, BVG Associates, January 2019.

[3] B. Hu and C. Yung. Offshore Wind Access Report 2020. Technical Report TNO 2020 R11992, TNO, 2020.

[4] Christine Röckmann, Sander Lagerveld, and John Stavenuiter. Operation and Maintenance Costs of Offshore Wind Farms and Potential Multi-use Platforms in the Dutch North Sea. In Bela H. Buck and Richard Langan, editors, *Aquaculture Perspective of Multi-Use Sites in the Open Ocean: The Untapped Potential for Marine Resources in the Anthropocene*, pages 97–113. Springer International Publishing, Cham, 2017.

[5] Pascal Cotte and et al. Offshore wind: Future of logistics. `https://media-publications.bcg.com/BCG-Future-of-logistics-in-offshore-wind-March-2022.pdf`, 03 2022. Accessed: 2024-09-06.

[6] International Maritime Organization. Resolution a.1106(29) - revised guidelines for the onboard operational use of shipborne automatic identification systems (ais). Adopted on 2 December 2015 (Agenda item 10), 2015.

[7] I. Lazakis R. Martin and S. Barbouchi. Analysis of input factors to operations and maintenance of two offshore wind farm case studies; a screening process. 3rd Renewable Power Generation Conference (RPG 2014), 2014.

[8] Bo Tranberg, Kasper Koops Kratmann, and Jason Stege. Determining offshore wind installation times using machine learning and open data. *ArXiv*, 2019.

[9] David Baumgärtner and Jonas Kaczenski. Innovative benchmarking of offshore wind o&m strategies via ais data analysis. *Journal of Physics: Conference Series*, 2875(1):012033, nov 2024.

[10] United States Department of Transportation - Federal Aviation Administration. Automatic dependent surveillance - broadcast (ads-b). `https://www.faa.gov/about/office_org/headquarters_offices/avs/offices/afx/afs/afs400/afs410/ads-b`, Accessed: September 2024.

[11] Fraunhofer-Institut für Windenergiesysteme (IWES). Joint project: Logreview - use of ais data to optimise o&m logistics in offshore wind farms; subproject: Ais and ads-b data analysis for the optimisation of o&m logistics in offshore wind farms. `https://www.enargus.de/detail/?id=4014591`, Accessed: February 2024. Funded by BMWK. Reference number: 03EE3051A.

[12] European Centre for Medium-Range Weather Forecasts. Era5: Fifth generation of ecmwf atmospheric reanalyses of the global climate. `https://www.ecmwf.int/en/forecasts/datasets/reanalysis-datasets/era5`, Accessed: February 2024. ERA5 provides hourly estimates of a large number of atmospheric, land and oceanic climate variables.

[13] Danish Maritime Authority. AIS Data, Accessed: February 2024. Historical AIS data provided by the Danish Maritime Authority, supporting maritime safety and navigation.

[14] OWP Butendiek GmbH Co. KG. OFFSHORE WINDPARK BUTENDIEK. `https://www.owp-butendiek.de/`, Accessed: Februrary 2024.

A bin-wise approach to wake steering optimization under fatigue load constraints

Ibtihel Ben Gharbia[1], Nina Khvoenkova[1], Paolino Tona[2], Fabrice Guillemin[1], Marine Dupoiron[3], Emilie Quenedey[3] and Alexis Cousin[1]

[1]IFP Energies nouvelles, Institut Carnot IFPEN Transports Energie, 1-4 Av. du Bois Préau, 92852 Rueil-Malmaison, France

[2]IFP Energies nouvelles, Institut Carnot IFPEN Transports Energie, Rond-point de l'échangeur de Solaize, 69360, Solaize, France

[3]GreenWITS, Rond-point de l'échangeur de Solaize, 69360, Solaize, France

E-mail: paolino.tona@ifpen.fr

Abstract. When implementing wake steering control, AEP (Annual Energy Production) uplift comes at a cost for the wind farm: the misaligned operating mode has an impact on the mechanical loads and vibrations experienced by the turbines, so there is a trade-off between production and fatigue to be considered. AEP optimization problems under fatigue constraints can be computationally demanding, particularly when they consider cumulative fatigue and require computing multiple damage equivalent loads for various wind conditions. In this study, a simplified but tractable formulation of the AEP optimization problem under multiple fatigue constraints is proposed and solved for a small offshore wind farm. The results show that this pragmatic approach significantly improves AEP compared to the baseline scenario without wake steering, while maintaining fatigue loads within reasonable limits and distributing them more evenly among the turbines.

1 Introduction

The impact of wake effects in a wind farm, which is detrimental both to production and turbine components lifetime, can be reduced by implementing a *wind farm flow control* (WFFC) strategy (Andersson et al. 2021), where turbine control setpoints are chosen in a coordinated manner. This approach contrasts with the standard, "greedy" control strategy, which aims to maximize the individual power output of each turbine, disregarding its aerodynamic interactions with the rest of the wind farm. In this study, we focus on *wake steering control*, which consists in voluntarily misaligning the wind turbines with respect to the prevailing wind direction, by modifying the setpoints applied by the turbines' yaw control system. If the yaw setpoints sent to the turbines are correctly chosen, the turbines furthest upstream will experience a drop in production, which will be more than offset by an increase in production from the downstream turbines, leading to a net gain. However, this could come at a cost: the misaligned operating mode affects the mechanical loads and vibrations experienced by the turbines and a compromise between production and fatigue may need to be considered.

As outlined in (Houck 2022), a comprehensive review on WFFC that dedicates a large section to the effects of wake steering on loads, predicting these effects is not straightforward. Misaligning a turbine can increase some of its loads, but not in all cases. Similarly, while steering wakes away from downstream

Content from this work may be used under the terms of the Creative Commons Attribution 4.0 licence. Any further distribution of this work must maintain attribution to the author(s) and the title of the work, journal citation and DOI.
Published under licence by IOP Publishing Ltd

turbines is generally expected to be beneficial, it may have a negative impact on certain loads in cases of partial overlap.

Despite the possible negative effects on fatigue loads – and consequently on turbine lifetime – most wake steering control strategies focus only on power maximization. Typically, the computation of the power-maximizing yaw setpoints is carried out based on a low-fidelity steady-state wind farm flow model, which predicts the time-averaged effects of a control policy on the power output of the wind farm, under a predefined set of prevailing wind conditions. Yaw optimization is generally carried out offline, with its results used to populate a look-up table to be used for online control; but it can also be directly performed online based on the current operational and ambient conditions. Examples of steady-state wake simulators used in this context include FLORIS, PyWake, and FarmShadow – the tool on which this study is based.

Fatigue can be included in the optimization problem if a suitable model that relates it to wake effects from yaw misalignments is available, while ensuring computational feasibility. Surrogate models trained on aeroelastic simulations are ideal for this. In (Hulsman et al. 2020), a two-turbine setup uses surrogate models trained on LES and aeroelastic simulations to predict damage equivalent loads (DELs) -— flapwise root bending and tower bottom bending moments -— based on yaw angles and turbine spacing, without considering free-flow wind variations. These models are coupled with FLORIS (Gebraad et al. 2014), to compute wake steering control policies from the solutions of an optimization problem where the DELs are directly incorporated into the objective function and weighted against power.

Similarly, (Sood et al. 2025) introduces a weighted objective function combining normalized power gain and fatigue load increase, expressed as blade root DELs gains. The fatigue model is a look-up table computed for the DTU 10-MW reference turbine using OpenFAST. Optimal yaw setpoints for the TotalControl 32-turbine plant are determined using a Gaussian wake-merging model and applied in a quasi-static closed-loop LES simulation. Note that the study does not assess whether the cumulative fatigue caused by the optimal control setpoints, computed wind-bin-wise, remains within design limits.

In contrast, (Liew et al. 2024) optimizes yaw setpoints to maximize Annual Energy Production (AEP) while ensuring that the Lifetime Damage Equivalent Load (LDEL) of each turbine remains within constraints. A neural network-based surrogate model predicts DELs for multiple turbine components using a reduced-order rotor wind field representation, free-flow wind conditions, and yaw angles. Trained on aeroelastic simulations, it is coupled with PyWake for yaw setpoint optimization, via the derivative-free COBYLA algorithm (Ragonneau 2023). Though the surrogate fatigue model is able to compute DELs for different turbine components, only the constraint on blade flapwise DEL is considered in the application example on a fictitious 9-turbine wind farm.

In this study, a simplified but tractable formulation of the AEP optimization problem under multiple DEL constraints is proposed and solved for a small offshore wind farm, where the closest turbines are spaced about four diameters apart. Wind farm layout and wind turbine characteristics (which cannot be disclosed, for confidentiality reasons), along with typical site conditions from the Basis of Design (BoD), operational and structural health monitoring (SHM) data, provided by the operator, have been used to model the wind farm flow and the turbine fatigue as described in Section 2. Three AEP optimization problems are introduced in Section 3: one without DEL constraints, one with cumulative DEL constraints, and a simplified version with wind-bin-wise DEL constraints. Details on the solvers used are also provided. Section 4 focuses on the results for the first and third problems, since the toolchain linking the wind farm flow model and the surrogate fatigue model makes the AEP optimization problem under cumulative DEL constraints too computationally demanding for the considered case study. Section 5 concludes the paper by summarizing the key findings and outlining potential directions for future research.

2 Modeling

2.1 Wind farm flow model

Wind farm modeling is based on FarmShadow, a steady-state flow solver developed at IFP Energies nouvelles (IFPEN), where the turbines are represented as actuator disks, through their power and thrust coefficient curves, and coupled to a set of state-of-the-art analytical wake models. The wake models compute, among other effects, the velocity deficit, wake-added turbulence, wake deflection, and wake superposition. The most advanced of these models, along with their default tuning parameters, have been validated against several experimental and large-eddy simulation datasets with very good results (Blondel & Cathelain 2020, Blondel 2023). The version of FarmShadow used in this study computes a horizontal uniform wind field with vertical wind shear based on power-law exponents. Free-flow turbulence intensity is prescribed – along with the wind speed and direction – and is used as an input to all sub-models.

Figure 1 illustrates how FarmShadow, which had been initially developed for wind farm layout optimization, can be used in the context of open-loop WFFC. Its settings are chosen once and for all

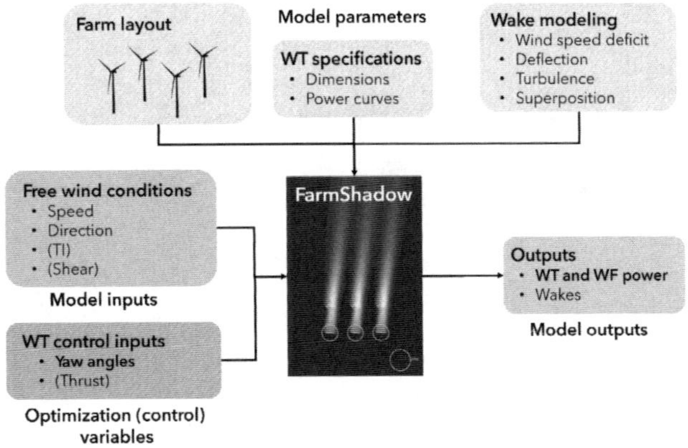

Figure 1: Use of FarmShadow for WFFC via wake steering.

Model	Choice
Velocity deficit	Super-Gaussian (Blondel et al. 2020)
Added turbulence	Ishihara (Ishihara & Qian 2018)
Deflection	Jiménez (Jiménez et al. 2010) **or** Qian (Qian & Ishihara 2018)
Superposition	Local linear sum (Zong & Porté-Agel 2020, Method C)

Table 1: Wake modeling choices.

with regards to wind farm layout, wind turbine characteristics and wake models. The WFFC optimization problem consists in finding the yaw misalignments which maximize the wind farm power production for a given set of free-wind conditions. For this study, all wind conditions except wind speed and direction are kept fixed during the optimization process. For wake modeling, the options listed in Table 1 were selected.

Two different deflection models were tested: Jiménez (Jiménez et al. 2010) and Qian (Qian & Ishihara 2018). The latter, widely used in wake-steering studies in the literature, was originally designed for a top-hat velocity deficit shape function, as in the Jensen model. As a result, it is not strictly compatible with a super-Gaussian (or Gaussian) velocity deficit shape function. However, since the only output of this deflection model is the altered wake center, it can in practice be used in conjunction with any model. This limitation has been alleviated in (Blondel et al. 2020) by slightly changing the formulation of the Jiménez model. The main reason the Jiménez model is frequently used for WFFC is its ease of parameterization, via a coefficient k_d which is directly related to the amount of deflection produced by misalignment. The value that matches best existing data is $k_d = 0.08$ (based on (Blondel et al. 2020)), which is the default value in FarmShadow. The Qian deflection model, derived using a Gaussian velocity deficit profile, considers near wake and far wake region separately and has been derived to match existing data on a large range of thrust coefficient and turbulence intensity. The Jiménez model results in greater deflection than the Qian model when using $k_d = 0.08$. Given the large uncertainty of these deflection models, we decided to conduct the yaw optimization with two models, representing an "optimistic" and a "pessimistic" scenario, as the Jiménez model is expected to yield higher AEP gains.

It would have been appropriate to also include the secondary effects of wake steering in the wind farm flow modeling, as the misalignments required to achieve a given deflection are lower when considering them. However, the implementation of secondary steering in FarmShadow has not been finalized yet.

2.2 Surrogate fatigue model

The surrogate model considered in the following for load prediction has been developed in the digital-twin study described in (Cadoret et al. 2025) following the methodology outlined in (Mc William et al. 2022).

The first step of this methodology, illustrated in the upper part of Figure 2, involves performing quasi-steady wake simulations using FarmShadow (or any similar steady-state flow solver) under varying

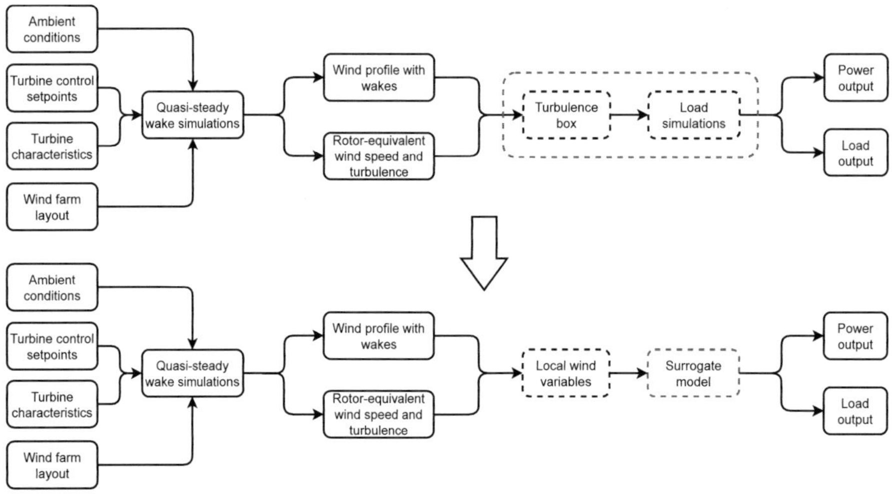

Figure 2: Wind farm wake effect parameterization using surrogates for local wind estimates.

free-flow wind conditions and for a set of yaw misalignments for each turbine. From these simulations, the vertical plane of the averaged wind speed seen by the rotor of a given wind turbine is extracted. From this vertical plane, a Gaussian function is fitted to model each mean wake effect. A turbulence box is then generated around the turbine using NREL's TurbSim. At each time step of this turbulence box, the Gaussian functions are added to integrate the wakes. Finally, a simulation is conducted with the turbine OpenFAST model, incorporating this wind field – along with the wave parameters observed in SCADA if an offshore wind farm is considered, as in our case study, – to predict fatigue DELs at locations where load measurements are available. In the second step, depicted in the lower part of Figure 2, a more efficient workflow is designed to replace the computationally demanding TurbSim and OpenFAST simulations with a set of Kriging – or Gaussian process (GP) regression (Rasmussen & Williams 2005) – metamodels trained on them. The input space of these metamodels is composed of the free-flow wind parameters and the Gaussian parameters describing the wake effects.

For the considered wind farm, the DEL channels provided for each turbine are: the blade root edge-wise bending moment; the blade root flap-wise bending moment; the tower torsional moment; the tower bottom stress at 350° with respect to the North; the tower bottom stress at 80° with respect to the North. In Section 3, they will be referred to as DEL_k, where k ranges from 1 to 5, in the order of their respective definitions.

The trained surrogate model predicts DELs that closely match the observed ones, as illustrated in Figure 3 for the tower bottom stress DELs, and can be used for digital monitoring of fatigue – the initial purpose of the methodology described in (Mc William et al. 2022) – as well as for wake steering control.

In the latter context, the inputs received by the turbine fatigue metamodels at each optimizer call to FarmShadow are:

- The reduced set of incident wake parameters representing the mean wake effect at the rotor, extracted from the simulation for the current wind bin.
- The local wind speed and turbulence intensity, U and TI, from the same simulation.
- A prescribed set of, possibly varying, sea conditions, including: the significant wave height, H_s, the peak wave period, T_p, and the wind-wave misalignment angle, β.
- The yaw misalignment angle of the turbine rotor with respect to the wind, set by the optimizer during the optimization process.

Note that the computational workflow associated with the current coupling of FarmShadow to the surrogate models is significantly more time-consuming than the call to FarmShadow itself. As discussed in the next section, this hinders the applicability of AEP optimization under multiple cumulative DEL constraints and justifies the development of an approximate, bin-wise approach.

Journal of Physics: Conference Series **3025** (2025) 012017 doi:10.1088/1742-6596/3025/1/012017

3 AEP optimization approaches

The key mathematical notations used in this section are listed in Table 2. All the optimization approaches presented in the following explore a discrete wind set, encompassing free-flow wind scenarios (or wind bins) defined in terms of possible wind speed and wind directions.

Table 2: Notations

N_t	Number of wind turbines in the wind farm.
t	Index of the wind turbine within the farm with $t \in [1 \cdots N_t])$.
w_s	Random variable representing the free-flow wind speed.
w_d	Random variable representing the free-flow wind direction.
w	Random free-flow wind variable such that $w := (w_s, w_d)$.
$\mathcal{W}$	Domain of the random variable w. $\mathcal{W} = [0, \infty[\times [-\pi, \pi]$.
γ	Vector of misalignment setpoints, $\gamma : \mathcal{W} \mapsto [-\pi, \pi]^{N_t}$.
$\mathbf{W_s}$	Discrete set of possible wind speeds.
$\mathbf{W_d}$	Discrete set of possible wind directions.
N_w	Number of elements of the discrete wind set (wind bins) $\mathbf{W_s} \times \mathbf{W_d}$ with $N_w := \text{card}(\mathbf{W_s})\text{card}(\mathbf{W_d})$.
P: $[-\pi, \pi]^{N_t} \times \mathcal{W} \mapsto \mathbb{R}$	Function whose image is the power produced by the farm for the yaw angles and wind configuration provided in argument.
F: $[-\pi, \pi]^{N_t} \times \mathcal{W} \mapsto \mathbb{R}^{N_t}$	Function whose image is the DEL for each turbine of the wind farm for the yaw angles and wind configuration provided in argument.
$(.)_i$	i^{th} coordinate of the vector in argument.
DEL	Damage Equivalent Load of a wind turbine.
$\text{DEL}_{k,t}$	Damage Equivalent Load for fatigue component k of wind turbine t.
m_k	Wöhler coefficient for component k, used to compute cumulative DEL.
AEP	Annual Energy Production of the wind farm.
$\mathbb{E}_w\varphi$	Expected value of φ with respect to the random variable w.
p_w	Probability of wind occurrence w.

3.1 *"Pure" AEP optimization*

The annual expected production of energy for a wind farm, denoted as AEP, is defined by the equation

$$\text{AEP}(\gamma) := \mathbb{E}_w \text{P}(\gamma(w), w),$$

with the power P being computed on 1 h intervals. $\mathbb{E}_w \text{P}(\gamma(w), w)$ represents the expected power production, where $\gamma(w)$ is the yaw angle policy for different wind scenarios w.

The AEP optimization problem for wake steering control can be mathematically formulated as follows:

$$\max_{\gamma} \mathbb{E}_w \text{P}(\gamma(w), w) \tag{1}$$

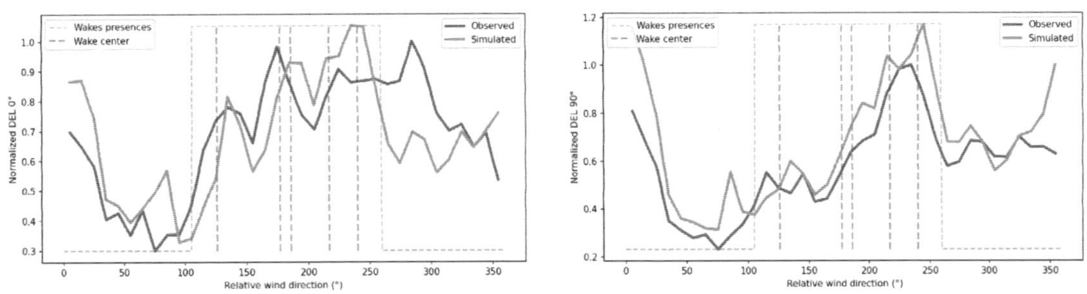

(a) At 350°. (b) At 80°.

Figure 3: Comparison between observed and simulated normalized tower bottom stress DELs.

subject to the constraint:

$$\gamma(w) \in [\gamma^-, \gamma^+]^{N_t}, \forall w. \tag{2}$$

Here, γ^- and γ^+ represent the lower and upper limits of yaw angles, respectively, and N_t denotes the number of turbines.

The optimization problem aims to compute a general optimal policy $\gamma : \mathcal{W} \to [\gamma^-, \gamma^+]^{N_t}$, which means determining the optimal yaw angles for the farm across all wind configurations.

Given that the expected value of the power production is computed by:

$$\mathbb{E}_w \mathrm{P}(\gamma(w), w) = \sum_{i=1}^{N_w} p_{w_i} \mathrm{P}(\gamma(w_i), w_i),$$

where p_{w_i} are the probabilities of different wind scenarios, the optimization problem (1)-(2) is separable in terms of $\gamma(w_i)$. The resulting independent optimization problems can be solved separately for each bin, in parallel, facilitating efficient computation.

3.2 AEP optimization under cumulative DEL constraints

From a mechanical perspective, it is important to consider the cumulative effects of loading to assess the damage sustained by a component. Damage considerations can be incorporated as constraints within the AEP optimization framework, similarly to (Liew et al. 2024). Thus, the maximization of the farm AEP can be constrained by the requirement that the cumulative optimized DELs for each turbine do no exceed the maximum nominal cumulative DELs observed among all turbines. Here, the "nominal" value refers to the DEL when the turbines are directly facing the wind, without misalignment, i.e., when the yaw angles are zero.

The cumulative DEL for fatigue component k associated with turbine t is defined as follows

$$\left(\sum_{i=1}^{N_w} p_{w_i} \mathrm{DEL}_{k,t}^{m_k}(\gamma(w_i), w_i) \right)^{1/m_k},$$

where m_k is the corresponding Wöhler coefficient. Typical values for this coefficient are 10 for the blade root and 3 or 4 for the tower. If the weights p_{w_i} are derived from the long-term wind rose observed at a wind farm site, then this cumulative DEL can be interpreted as the lifetime DEL for component k.

The optimization problem, which aims to maximize the annual energy production of a wind farm while considering "global" constraints on cumulative damage equivalent loads, can be formulated as follows:

$$\max_{\gamma} \mathbb{E}_w \mathrm{P}(\gamma(w), w) \tag{3}$$

under the following constraints

$$\left(\sum_{i=1}^{N_w} p_{w_i} \mathrm{DEL}_{k,t}^{m_k}(\gamma(w_i), w_i) \right)^{1/m_k} \leq \max_t \left(\sum_{i=1}^{N_w} p_{w_i} \mathrm{DEL}_{k,t}^{m_k}(0, w_i) \right)^{1/m_k}, \quad \forall k, \forall t, \tag{4}$$

$$\gamma(w) \in [\gamma^-, \gamma^+]^{N_t}, \forall w. \tag{5}$$

The constraint (5) ensures that the yaw angles are bounded within specified limits. The constraints (4) are defined for each component and each turbine, resulting in a total count of

$$N_t \times \text{number of DEL components},$$

Those constraints ensure that, for every component of each turbine, the cumulative DEL does not exceed that observed on the most stressed turbine.

The global constraints (4) can be relaxed and managed using a duality framework. Subsequently, a classical duality-based Uzawa algorithm can be employed to solve the optimization problem constrained by (3)-(4), as proposed in (Ben Gharbia et al. 2023b). Upon convergence, the algorithm computes the optimal yaw angles for the entire wind farm across all wind scenarios. This comprehensive approach is to be favored because the constraints, referred to as "global constraints", are based on aggregated data from all wind bins.

Unfortunately, unlike "pure" AEP optimization, the resulting problem cannot be decomposed into wind-bin independent subproblems. As a result, the method is computationally demanding, especially when multiple DELs are considered and when relatively long processing times are required to compute them. In the study described in (Ben Gharbia et al. 2023b, Ben Gharbia et al. 2023a), it was successfully applied to a 7-turbine wind farm model – corresponding to the Ablaincourt project in France – coupled with a simple look-up table providing only one DEL: the blade root flap-wise bending moment. For the case study considered here, the method proved to be too time-consuming, due to the current coupling between FARMSHADOW and the surrogate fatigue model.

3.3 AEP optimization under bin-wise fatigue constraints

To address the issue of the long computational times associated with the above-defined optimization problem, we propose an alternative approximate approach that involves solving a constrained optimization problem separately for each wind bin. The key idea is to replace the global constraint with a local constraint, which considers only the turbine DEL within the current bin. Finally, the cumulative constraints (4) are verified a posteriori across the entire wind rose.

Such a problem can be expressed for each wind bin w as follows:

$$\max_{\gamma} \mathrm{P}(\gamma(w), w) \tag{6}$$

under the following constraints

$$\mathrm{DEL}_{k,t}^{m_k}(\gamma(w), w) \leq \max_{t} \left(\mathrm{DEL}_{k,t}^{m_k}(0, w)\right), \quad \forall k, \forall t, \tag{7}$$

$$\gamma(w) \in [\gamma^-, \gamma^+]^{N_t}. \tag{8}$$

This approach is effective for rapidly obtaining results for a single wind bin; however, it does not ensure compliance with the global constraints on the cumulative DEL outlined in (4). Although the bin-wise constraints (7) prevent the DEL for each turbine from exceeding the highest DEL in the wind farm, they do not address the global constraint. To verify compliance with (4), we must compute the solution across all wind configurations and perform an a posteriori check, which reveals the inherent limitations of this less conservative method.

With regard to the number of inequality constraints (7), the count remains consistent with that of our initial fatigue-constrained optimization problem (3)–(5).

3.4 Optimization solvers

In (Ben Gharbia et al. 2023b, Ben Gharbia et al. 2023a), we had initially tried to solve both problem (1) – (2) ("pure" AEP optimization) and problem (3)–(5) (AEP optimization under cumulative DEL constraints, via the Uzawa algorithm) respectively via the Python solvers BOBYQA (Bound Optimization BY Quadratic Approximations) and COBYLA (Constrained Optimization BY Linear Approximations) (Ragonneau 2023). Those DFO solvers were favored because gradient computations were not available for the wind farm flow model, and numerical differentiation would have been computationally inefficient. Since it had been found that both solvers struggle to handle the bound constraints on the yaw angles effectively, these constraints were directly integrated into the objective function by adding a penalty term, implementing the barrier method, an interior-point technique. As a result, the problems were transformed into unconstrained problems and solved using COBYLA.

For this study, to solve both problem (1) – (2) ("pure" AEP optimization) and problem (6)–(8), we instead turned our attention to the COBYQA (Constrained Optimization BY Quadratic Approximations) solver. COBYQA, which was recently introduced in Python, represents a significant advancement in DFO techniques (Ragonneau 2023). This algorithm is particularly well-suited for handling complex optimization problems where traditional derivative-based methods are inadequate due to the absence of gradient information. COBYQA constructs successive quadratic models of the objective function using only function values, thereby eliminating the need for derivative information. It iteratively adjusts those models based on results from previous evaluations, enabling it to effectively navigate through the solution space, even with stringent boundary and inequality constraints.

Given its highly satisfactory performance, COBYQA has become our primary choice for solving AEP optimization problems for wake-steering control, regardless of whether fatigue constraints are considered. A significant difference we observed with the fatigue-constrained problem, is that COBYQA needs to be coupled with a multi-start procedure in this case, whereas it remains quite insensitive to the choice of the initial yaw angles for the "pure" AEP optimization problem. As expected, convergence is more difficult in the former case.

Notably, an approximate solution to problem (1) – (2) can be obtained via the SERIAL-REFINE method, a heuristic algorithm proposed by the NREL for fast wake-steering yaw optimization (Fleming et al. 2022), included in the FLORIS suite. It is based on the serial optimization of yaw misalignments, from upstream to downstream, using a small number of candidate angles. According to (Fleming et al. 2022), SERIAL-REFINE can find solutions that are at least as good as conventional optimization approaches, but that require much less computation time. However, comparative results in (Fleming et al. 2022) were only provided with respect to a gradient optimization method, used without multi-start, which is an unsuitable approach to wake-steering AEP optimization when no gradient information is available. In the context of this study, we have also implemented the SERIAL-REFINE algorithm for use on a FARMSHADOW wind farm model, to benchmark its performance with respect to DFO methods.

4 Application and results

The optimizations have been carried out over 4860 wind bins, obtained by discretizing the wind direction into 2° increments and the wind speed in the operating range of the turbines into 1 m/s increments. The sea conditions were fixed to their average values over the year according to SCADA data to avoid increasing the number of degrees of freedom in the optimization process to five, and consequently, the size of the resulting look-up table of optimal yaw angles. The yaw limits were set to $\gamma^- = \mathrm{deg2rad}(-20°)$ and $\gamma^+ = \mathrm{deg2rad}(20°)$.

Since the inclusion of tower DELs (expecially the torsion DEL) considerably worsens the convergence and lengthens the duration of the optimization campaigns, only the two DELs for blade root bending moments ($k = 1, 2$) were eventually included as bin-wise constraints in AEP optimization. Nevertheless, as shown in Table 3, after each optimization and upon obtaining the optimal yaw angles, the cumulative DELs are computed for all five components and the compliance with the global constraints is verified.

k	$\mathrm{DEL_k}$	Unit	$\mathrm{m_k}$	As bin-wise constraints	As global constraints
1	Blade root egde-wise bending moment	N m	10	Included in optimization	Verified a posteriori
2	Blade root flap-wise bending moment	N m			
3	Tower torsional moment	N m	4	Not included in optimization	
4	Tower bottom stress at 350°	Pa			
5	Tower bottom stress at 80°	Pa			

Table 3: DEL constraint setup.

Table 4 displays the percentage gain in annual energy production achieved through AEP optimization, both with and without local DEL constraints, using the derivative-free solver COBYQA on a FARMSHADOW simulator parameterized with the two deflection models. The increase in AEP of the optimization outputs provided by a yaw angles policy γ with respect to the nominal policy ($\gamma \equiv 0$) is given by the following relation:

$$\mathrm{AEP\ Gain}(\gamma)\,[\%] := 100\Big(\mathrm{AEP}(\gamma) - \mathrm{AEP}(0)\Big)/\mathrm{AEP}(0).$$

The results illustrate the sensitivity of AEP gains to the choice of deflection models. Notably, the inclusion of fatigue constraints only slightly diminish the AEP uplift. For "pure" AEP optimization, consistent with the conclusions of (Fleming et al. 2022), the SERIAL-REFINE algorithm closely approximates the performance of COBYQA, while requiring significantly less time to perform the optimization (about 5 min on a high-end laptop compared to more than 2 h).

	Without DEL constraints		Under bin-wise DEL constraints
Deflection model	SERIAL-REFINE	COBYQA	COBYQA
Jiménez	3.115	3.116	2.741
Qian	2.078	2.084	1.848

Table 4: AEP gains [%] for the differerent AEP optimization problems.

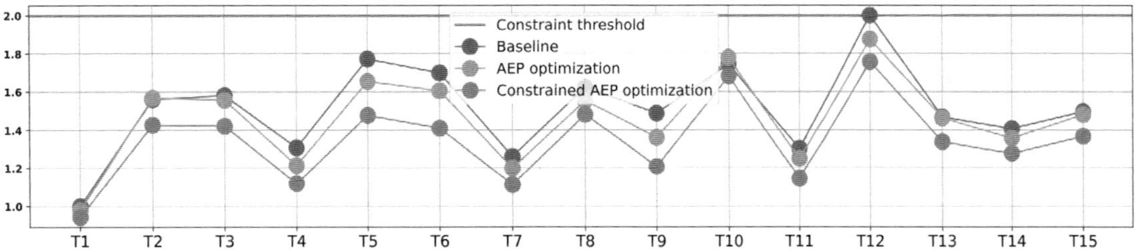

(a) For normalized blade root edge-wise moment. The relative difference w.r.t. the maximum is 1.1%.

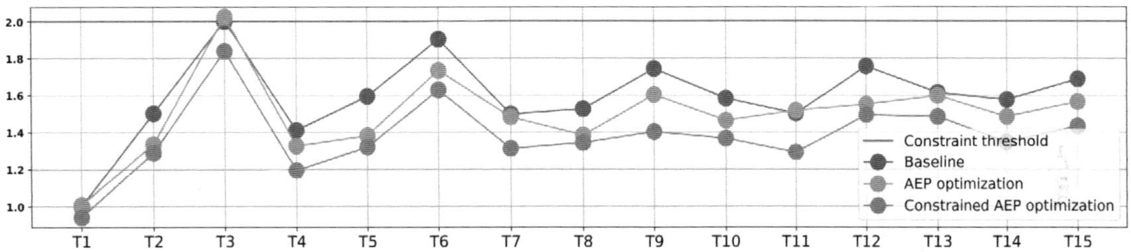

(b) For blade root flap-wise moment. The relative difference w.r.t. the maximum is 10.7%.

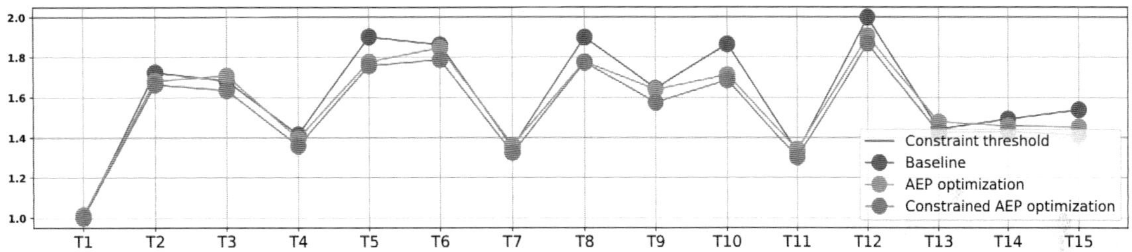

(c) For torsional moment cumulative DEL. The relative difference w.r.t. the maximum is 15.3%.

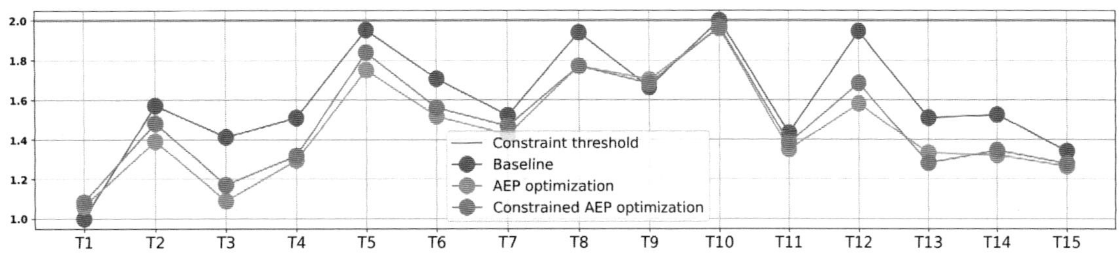

(d) For tower bottom stress cumulative at 350°. The relative difference w.r.t. the maximum is 4.6%.

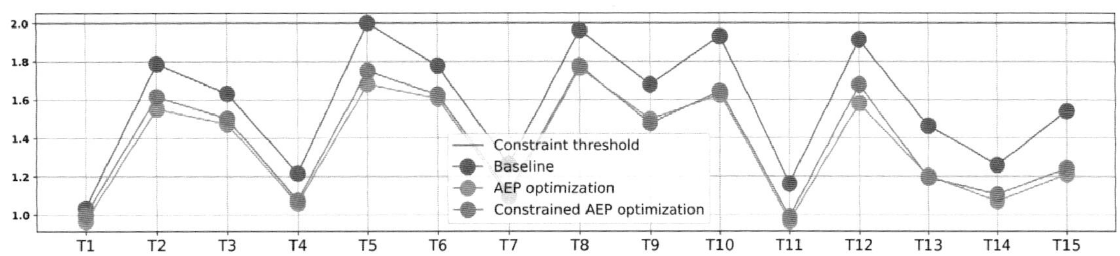

(e) For tower bottom stress at 80°. The relative difference w.r.t. the maximum is 5.7%.

Figure 4: Cumulative DELs for a subset of 15 turbines (for the Jiménez deflection model), normalized w.r.t. the minimum (1.0) and the maximum (2.0) values in the baseline case. The cumulative DELs for the baseline case are represented in blue, those for the AEP optimization methods in orange (without DEL constraints) and in green (with bin-wise constraints). The red lines represent the cumulative DEL threshold that must not be exceeded by those methods. The relative differences w.r.t. the maximum value are indicated to illustrate how close the cumulative DELs are to one another for each DEL channel.

To verify compliance with the global constraints of cumulative DELs (5), a verification step is conducted a posteriori after optimizing all wind configurations for the two considered deflection models. For the Jiménez model, Figures 4a and 4b display the normalized cumulative DELs associated to the two blade root moments (edge-wise and flap-wise), for a representative subset of 15 turbines. Each plot presents the cumulative DELs for three distinct scenarios: the baseline case (in blue), and the AEP optimization methods, both without (in orange) and with (in green) the imposition of bin-wise constraints on edge and flap blade root moment DELs. The red line represents the cumulative DEL threshold that must not be exceeded by those methods. The relative differences with respect to the maximum value are indicated to illustrate how close the cumulative DELs are to one another for each DEL channel. For instance, variations in the blade root edge-wise moment are very small, with the minimum value being only 1.1% lower than the maximum value.

AEP optimization has a positive effect on the fatigue of most turbines, with a slight violation of the flap DEL constraint for Turbine T3. The integration of bin-wise constraints on the blade root DELs significantly reduces the cumulative values for all turbines, including those not represented in the figures. Although the DELs associated with the tower are not directly accounted for, the plots in Figure 4c, 4d and 4e clearly demonstrate that the overarching constraints for those components are satisfied within the AEP optimization framework, with a significant reduction of cumulative DELs for most turbines. When bin-wise constraints on the blade root DELs are incorporated into the optimization problem, the cumulative tower DELs do not change much, which makes sense given that the correspondent local DELs are not accounted for in the optimization process.

Note that Turbines T3, T6, T9, and T12 are the most downstream turbines with respect to the main wind direction, leading to a slightly higher cumulative DEL for the blade root flap-wise moment. Turbines T5, T8, T10, and T12 are located inside the farm, experiencing higher tower-bottom stress.

Similar conclusions are reached regarding the cumulative damage equivalent loads when using the Qian deflection model.

5 Conclusions

Structural load considerations are rarely incorporated into the design of wake steering control strategies, likely due to the challenge of computing accurate solutions in an efficient way. Taking into account cumulative fatigue damage in a rigorous manner inevitably leads to complex optimization problems that must be solved across a large number of wind bin conditions and various load cases. The computational efficiency of surrogate fatigue models is a crucial factor in this context. Indeed, the AEP optimization approach under cumulative DEL constraints we intended to apply to our case study – a small offshore wind farm, for which a fatigue surrogate model accurately predicting five representative DELs had been trained – turned out to be too demanding in terms of computation.

This motivated us to develop an approach for AEP optimization via wake steering under fatigue load constraints, defined and solved bin-by-bin, where the cumulative DELs are verified a posteriori. In spite of the simplified formulation, for the considered case study, this pragmatic approach significantly improved AEP compared to the baseline scenario without wake steering, while maintaining fatigue loads within reasonable limits and distributing them more evenly among the turbines. However, these findings may be site-specific, and the global constraints may not always be verified a posteriori. In such case, a different method would be needed.

A redesign of the computational workflow behind the coupling of the wind farm flow model and the surrogate fatigue model is in progress and has already resulted in a speed improvement of several orders of magnitude. We are confident that this will let us run optimization campaigns of reasonable duration to assess the performance of the AEP optimization under cumulative DEL constraints solved via the Uzawa algorithm.

One key challenge remains the handling of global fatigue constraints in a way that enables fast and practical bin-wise optimization. Addressing this issue would not only streamline offline computations of look-up tables but also pave the way for its implementation online (in real-time). This would offer several advantages even in an open-loop context, such as the ability to account for the operational status of the turbines when computing yaw setpoints.

Finally, for all the considered AEP optimization approaches, further work is envisioned to use more advanced analytical wake modeling configurations, namely by including secondary steering and momentum-conserving superposition.

6 Acknowledgments

This study has been carried out as part of the CAP project, supported under the France 2030 investment plan operated by ADEME, the French Agency for Ecological Transition.

References

Andersson, L. E., Anaya-Lara, O., Tande, J. O., Merz, K. O. & Imsland, L. (2021). Wind farm control - Part I: A review on control system concepts and structures, *IET Renewable Power Generation* **15**(10): 2085–2108.

Ben Gharbia, I., Malisani, P. & Guillemin, F. (2023a). An innovative optimization-based wake steering strategy for production increase and loads mitigation, *WESC 2023*, Glasgow, United Kingdom.

Ben Gharbia, I., Malisani, P. & Guillemin, F. (2023b). Stratégies d'optimisation de la production d'une ferme éolienne sous contrainte de DEL par redirection de sillage, *Technical Report R115/23-0008-IBG*, IFP Energies nouvelles.

Blondel, F. (2023). A momentum-conserving superposition method applied to the super-Gaussian wind turbine wake model, *Wind Energy Science* **8**(2): 141–147.

Blondel, F. & Cathelain, M. (2020). An alternative form of the super-gaussian wind turbine wake model, *Wind Energy Science* **5**(3): 1225–1236.

Blondel, F., Cathelain, M., Joulin, P.-A. & Bozonnet, P. (2020). An adaptation of the super-gaussian wake model for yawed wind turbines, *Journal of Physics: Conference Series* **1618**(6): 062031.

Cadoret, A., Cousin, A., Collet, D., Bonfils, N., Guy, N., Leroy, J.-M., Roguet, E., Caleyron, F., Le Corre, V. & Pfister, J.-L. (2025). Enhancing Structural Health Monitoring with Digital Models, *WindEurope Annual Event 2025*, Copenhagen, Denmark.

Fleming, P. A., Stanley, A. P. J., Bay, C. J., King, J., Simley, E., Doekemeijer, B. M. & Mudafort, R. (2022). Serial-Refine Method for Fast Wake-Steering Yaw Optimization, *Journal of Physics: Conference Series* **2265**(3): 032109.

Gebraad, P., Fleming, P., Ning, S. A. & van Wingerden, J.-W. (2014). Floris.
URL: *https://www.osti.gov/biblio/1262644*

Houck, D. R. (2022). Review of wake management techniques for wind turbines, *Wind Energy* **25**(2): 195–220.

Hulsman, P., Andersen, S. J. & Göçmen, T. (2020). Optimizing wind farm control through wake steering using surrogate models based on high-fidelity simulations, *Wind Energy Science* **5**(1): 309–329.

Ishihara, T. & Qian, G.-W. (2018). A new gaussian-based analytical wake model for wind turbines considering ambient turbulence intensities and thrust coefficient effects, *Journal of Wind Engineering and Industrial Aerodynamics* **177**: 275–292.

Jiménez, Á., Crespo, A. & Migoya, E. (2010). Application of a les technique to characterize the wake deflection of a wind turbine in yaw, *Wind Energy* **13**(6): 559–572.

Liew, J., Riva, R., Friis-Møller, M. & Göçmen, T. (2024). Wind Farm Control Optimisation Under Load Constraints Via Surrogate Modelling, *Journal of Physics: Conference Series* **2767**(9): 092039.

Mc William, M., Bonfils, N., Dimitrov, N. & Dou, S. (2022). Wind farm parameterization and turbulent wind box generation, *Technical report*, DTU and IFP Energies nouvelles.
URL: *https://ifp.hal.science/hal-04033050*

Qian, G.-W. & Ishihara, T. (2018). A new analytical wake model for yawed wind turbines, *Energies* **11**(3): 665.

Ragonneau, T. M. (2023). *Model-Based Derivative-Free Optimization Methods and Software*, PhD thesis, Hong Kong Polytechnic University.

Rasmussen, C. E. & Williams, C. K. I. (2005). *Gaussian Processes for Machine Learning*, The MIT Press.

Sood, I., del Fosse et d'Espierres, C. & Meyers, J. (2025). Quasi–static closed–loop wind–farm control for combined power and fatigue optimization, *Wind Energy* **28**(2).

Zong, H. & Porté-Agel, F. (2020). A momentum-conserving wake superposition method for wind farm power prediction, *Journal of Fluid Mechanics* **889**.

Investigation of the effects of individual pitch control of a wind turbine on loads, power and acoustic emission

Anna Wegner[1*], Ashutosh Sharma[1], Sebastian Mechler[1], Paula Ludwig[1] and Corinna Kraasch[1]

[1] Hydrogen Labs and Field Tests, Fraunhofer Institute for Wind Energy Systems, Bremerhaven, Germany

*E-mail: anna.wegner@iwes.fraunhofer.de

Abstract. Wind turbines have grown larger, causing their rotor blades to sweep across more extensive areas. As wind speed varies with height, the blades experience different loads during each rotation. Most modern turbines are equipped with individual pitch control (IPC) actuators to reduce loads. The acceptance of wind turbines by local communities is increasingly important, as are their acoustic emissions. Previous research found higher noise emissions when a modern turbine operates in IPC mode. This study analyzes the effects of IPC operation on power production, loads, and noise emissions. A modern multi-Megawatt wind turbine is equipped with a measurement system to quantify loads, power and acoustic emission. A direct comparison between IPC and non-IPC mode is not straight forward as the turbine is operated during higher power production more frequently in IPC mode, than during lower power production. However, results show, that the effect of IPC on mechanical loads depend on each individual component of the turbine. Damage equivalent loads (DELs) are either increased or decreased by the use of IPC depending on the component. The acoustic measurements show that the sound pressure level (SPL) is increased in certain ranges of the rotational speed and decreased in others. For the electric power we find higher production during times, when the turbine is operated in IPC mode.

1. Introduction

As wind turbines grow larger, their rotor blades sweep across more extensive areas. Because wind speed varies with height above the ground, blades experience different loads during each rotation. Most modern wind turbines are already equipped with individual pitch control (IPC) actuators, which primarily aim to reduce turbine loads.
With this technique the blade angles are adjusted individually, instead of collectively, to minimize the mechanical loads in above-rated wind speeds. The benefit of IPC is the compensation of asymmetrical loading on the rotor. The individual adjustment of the rotor blade angles is based on measurements taken by sensors attached to the blade roots and the blade bending moments. The IPC system's controllers use the main moments which represent the hub mechanical loading in two orthogonal quantities. The signals are converted to the three-axis system in the IPC controller's output, allowing the pitch angles of each blade to be independently adjusted [1] [2].

Content from this work may be used under the terms of the Creative Commons Attribution 4.0 licence. Any further distribution of this work must maintain attribution to the author(s) and the title of the work, journal citation and DOI.
Published under licence by IOP Publishing Ltd

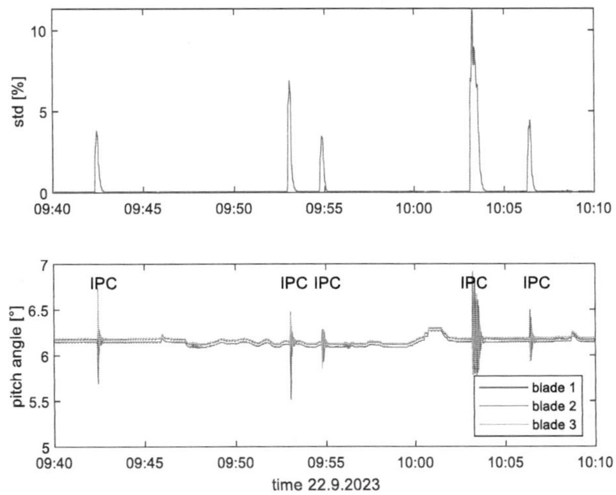

Figure 1. Illustration for the identification of the time intervals, when the turbine is operated in IPC mode. The standard deviation of the pitch angle (upper plot) indicates that the turbine is operating in IPC mode. The pitch angles of all three blades are shown for comparison (lower plot).

With the global expansion of renewable energy, the acceptance of wind turbines by local communities is becoming increasingly important. Consequently, the acoustic emissions of wind turbines are also a critical factor. Previous research [3] has found higher noise emissions when a modern multi-megawatt wind turbine operates in IPC mode compared to non-IPC mode.
In this study, we present an in-depth analysis of the effects of IPC operation on wind turbines, focusing on power production, loads, and noise emissions.

2. Methods

2.1 Test site description

The wind turbine site is located in northern Germany. The site is characterized by flat land with primarily agricultural land. The turbine under test is the south westernmost turbine of a wind farm with approx. 20 other nearby wind turbines. There is no significant elevation in the area, so no site calibration needs to be performed. The most prevalent wind direction is from Southwest, which is also an undisturbed wind sector. To quantify the incoming wind field, a Lidar is located 870 m westerly of the wind turbine. The wind turbine has a rotor diameter of 158 meters, and the hub height of 120 meters. The incoming wind speed was quantified using a ZX300 LiDAR at multiple heights of 38 m, 41 m, 68 m, 94 m, 120 m, 146 m, 172 m, 199 m, 225 m.

2.2 Detection of IPC and non IPC-mode

As no signal was provided by the SCADA system indicating, wether the turbine was operating in IPC mode or not, the pitch angle values were analyzed. For the evaluation of the acoustic data and the power production, the identification of IPC mode was performed as illustrated in Figure 1. For each 40 s time interval the standard deviation of the pitch angles was calculated. For times, when this value increased over a threshold of 0.4%, the turbine was identified to be operated in IPC mode. Based on this identification, power and acoustic datasets were also categorized accordingly into IPC and non-IPC modes.

For the load analysis, datasets of 10-minute duration were used. To distinguish between time intervals, when the turbine was operated in IPC or Non-IPC mode, the moving standard deviation of the pitch angles of rotor blades 1 and 2 was first calculated using a time window of 6 seconds. Subsequently, the number of seconds within the 600-second period in which this standard deviation exceeded the threshold of 0.5 was determined. Datasets were classified as IPC datasets, if the threshold was exceeded for at least 450 seconds. Non-IPC datasets were those in which the threshold was never exceeded. All other datasets were excluded from further analysis. Table 2 shows the number of IPC and Non-IPC datasets as a function of wind speed.

2.3 Acoustic measurement system

The acoustic measurement was carried out in accordance with International Standard IEC 61400-11 [2], using an IMC Cronosflex system with a GRAS low frequency microphone. The measurement microphone was setup 200 m from the base of the turbine (i.e. downwind of the turbine, as per IEC 61400-11) at an elevation of 0 m relative to the base of turbine. The microphone was placed in the centre of a circular, acoustically reflective board. During the measurement period only data points for which the microphone was within 15 degrees of downwind from the turbine were used. The microphone position relative to downwind of the turbine was monitored via the yaw angle output provided from the turbine system. During the time of the measurements, the surrounding land was composed of exposed soil and vegetation. The crop was short and as such the influence on the measurement was considered negligible. There were no nearby reflecting surfaces (houses, barns etc.), as such the influence from reflecting surfaces was negligible. A double windscreen setup was utilised. Data was logged in 10 second interval for post processing (as per the measurement standard) for a period of 4 days (20.09 to 24.09.2023). Figure 2 shows the installation of the system in the field. Output data from the turbine (Power, yaw, RPM, pitch angle, and nacelle wind speed) were obtained as SCADA data signals that were simultaneously acquired with the acoustic and anemometer measurement data. A weighted overall SPL represented by dB (A) were measured as well as 1/3rd Octave band spectrum data was also measured.

2.4 Load measurement system

In accordance with IEC 61400-13 [5], a load measurement system was installed on the wind turbine to measure loads within the blades, nacelle, rotor hub, and at three levels of the tower. The system includes strain gauges, acceleration sensors, fiber-optic strain gauge sensors, rotary encoders, torsion sensors, and inclinometers. Additionally, meteorological sensors, such as temperature, pressure, and humidity sensors, were installed to account for temperature compensation on the wind turbine. The load measurement system comprises data loggers and Q.Station devices from Gantner, installed in the rotor hub, tower base, and nacelle. The recorded signals are stored on an external hard drive located in the cabinet within the nacelle. Load data was synchronously measured alongside other types of data. The dataset underwent filtering based on various checks, including the availability of channels such as yaw, wind speed at nacelle height, rotor position, and speed. Calibration of the strain sensors were done using gravity calibration with slow pitching and rotor rotation maneuvres.

Journal of Physics: Conference Series **3025** (2025) 012018 doi:10.1088/1742-6596/3025/1/012018

Figure 2 placement of meteorological system as well as wooden board downstream of the wind turbine for the acoustic measurements.

2.5 Power measurement system

In accordance with IEC 61400-12-1 [6], a power measurement system for measuring current and voltage is installed on the wind turbine. The turbine system is a double-fed asynchronous

Table 1. Overview of identification of data sets of IPC and non-IPC with the two different methods.

Wind speed bin	Data sets IPC 10 sec	Data sets non-IPC 10 sec	Data sets IPC 10 min	Data sets non-IPC 10 min
5-6	1141	80730		1010
6-7	6852	70329	11	745
7-8	20346	53314	29	241
8-9	32444	16352	166	83
9-10	27041	16352	238	27
10-11	20310	8123	166	3
11-12	14270	4153	47	
12-13	7837	1687	18	
13-14			3	

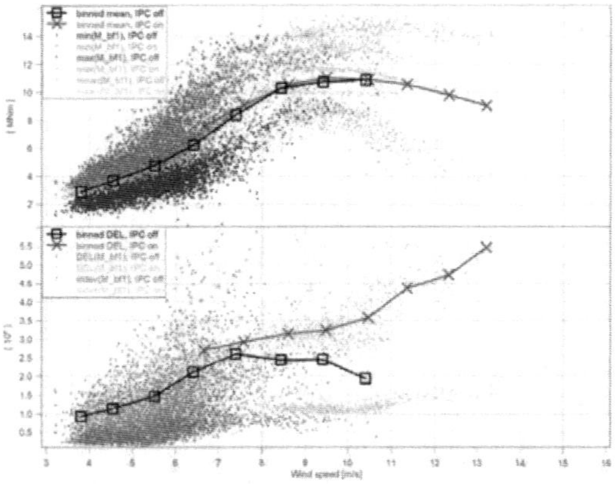

Figure 3 Bending moment and damage equivalent loads (DELs) for flapwise movement at the blade root.

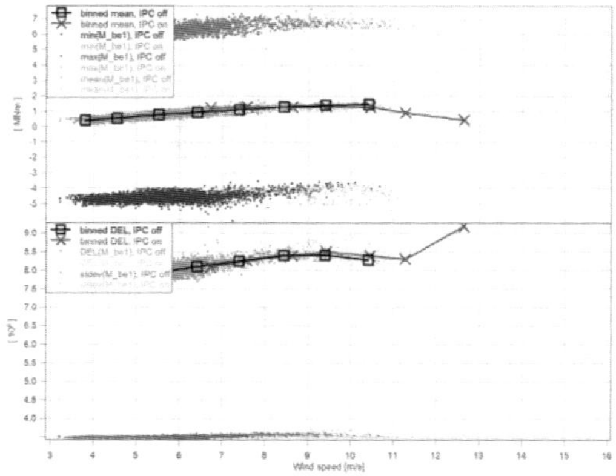

Figure 4 Bending moment and damage equivalent loads (DELs) for edgewise movement at the blade root.

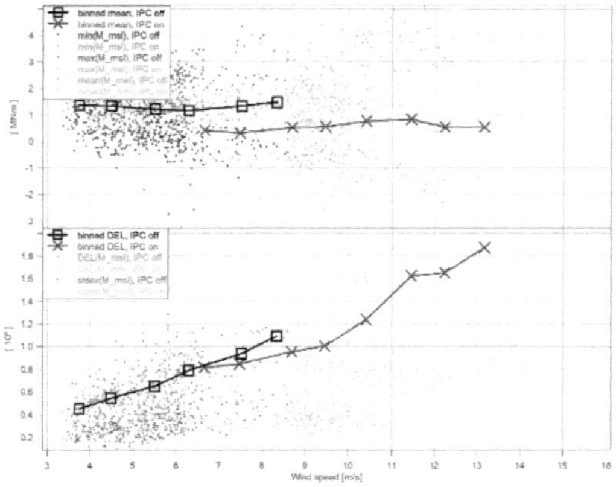

Figure 5 Bending moment and damage equivalent loads (DELs) for lateral direction of the main shaft.

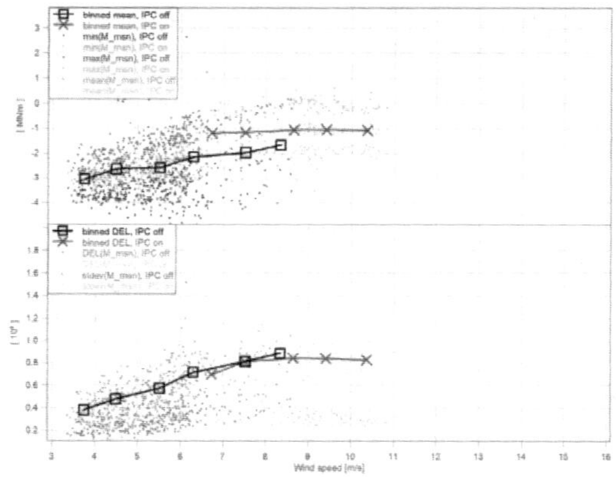

Figure 6 Bending moment and damage equivalent loads (DELs) for normal direction of the main shaft.

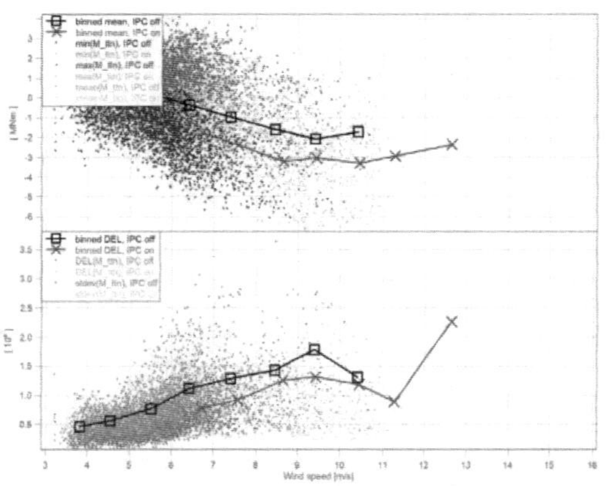

Figure 7 Bending moment and damage equivalent loads (DELs) for normal direction at the tower top.

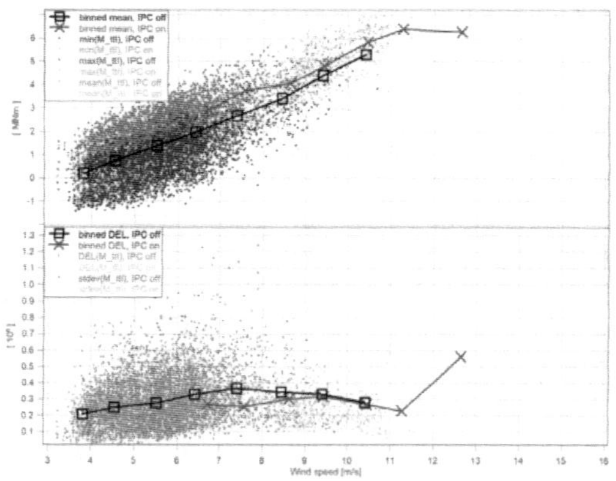

Figure 8 Bending moment and damage equivalent loads (DELs) for lateral direction at the tower top.

Table 2. Summary of electric power production and SPL during IPC mode and non IPC mode for certain RPM and wind speeds, respectively.

Electric Power [kW]		
Wind speed	IPC	Non-IPC
5.5	1228.47	713.57
6	1476.67	947.14
6.5	1797.95	1238.27
7	2152.70	1589.67
7.5	2504.75	1984.61
8	2899.89	2439.76
8.5	3287.97	2906.33
9	3652.76	3312.18
9.5	4014.60	3687.38
10	4359.18	4120.80
10.5	4664.79	4516.86
11	4899.91	4804.60
11.5	5061.41	5008.95
12	5168.57	5134.56
12.5	5252.26	5226.37
13	5285.35	5230.95

Sound pressure level [dB (A)]		
Rpm	Non-IPC	IPC
5.5	46.97	48.00
6	49.08	48.14
6.5	48.82	49.99
7	49.84	49.88
7.5	51.24	51.78
8	52.25	53.63
8.5	52.99	55.43
9	54.22	54.83
9.5	56.10	55.88
10	56.50	55.09

machine, as the measurement takes place before the step-up transformer and therefore also before the grid connection point, the stator and rotor drive train must be measured. The voltage is measured at six locations, including both the stator phase (6 kV) and the rotor phase (690 V), each equipped with measurement technology. The 690 V is measured directly using the Gantner A128 SEB measuring module, while the 6 kV is reduced via a voltage divider (HST6) with a transformation ratio of 1:2000 and then measured using the Gantner A111 measuring module. Current measurements are conducted similarly at two different points. The current is measured using Rogowski coils, specifically the PEM UK CWT15 model.

All times, when the turbine was not operated in normal power production or the wind direction outside the valid wind sector were excluded from the analysis.

3. Results

In general, with higher rotational speed the turbine is operated in IPC mode more frequently. This is consistent using both methods to identify the IPC mode.

3.1 Acoustic emissions

Acoustic measurements were conducted with the help of a single microphone measurements according to the standard IEC61400-11 [2] over a span of 4 days (20.09 till 24.09.2024) with a total of 4289 data points, each data point averaged over a 10 second interval. The microphone was approximately 200 m behind the wind turbine. Table 2 presents the SPL from single microphone data during times when the turbine was operating in IPC mode and non-IPC mode across various RPM bins. IPC on was considered only for the times when it was detected on for more than at least 10 seconds.

We find higher acoustic emissions during times when the turbine was operating in IPC mode and operating in region 2 (i.e. below rated RPM of the wind turbine). For region 3 i.e. at rated RPM the SPL levels were lower for the IPC On cases as compared to IPC off. The single microphone measurements also indicate a higher number of occurrences of wind turbine under IPC mode at higher RPM and getting less predominant at lower RPM (see Table 2).

3.2 Mechanical Loads

The effect of IPC on the mechanical loads is not uniform throughout the turbine, i.e. both loads and damage are reduced or increased during IPC mode depending on the respective measurement channel.

When looking at the blade root (Figure 3 and Figure 4), we find that bending moments are similar, when the turbine in operated in IPC mode and non-IPC mode for both edgewise and flapwise bending. However, flapwise DELs are higher when the turbine is operated in IPC mode compared to non-IPC mode. Bending moments of the main shaft are higher when the turbine is in IPC mode compared to non-IPC mode for the lateral direction, whereas for the normal direction the observation is vice versa with higher values for the bending moment when IPC mode is on (Figure 5 and Figure 6). The DELs in lateral direction are lower during IPC mode in lateral direction, whereas for normal direction no significant difference can be observed. For tower top normal direction both bending moments and DELs are lower when the turbine is operating in IPC mode (Figure 7 and Figure 8). For the lateral direction DELs are lower and bending moments are higher during IPC mode operation. Previous studies show a decrease in DELs, when IPC is applied (e.g.[7]). These discrepancies have to be evaluated in detail. Reasons could be, that during the presented study, the turbulence intensity and wind shear is not investigated, which could have an influence on the results. Additionally, the fact, that the turbine is controlled using IPC mode more often for higher wind speeds than for lower wind speeds might bias the results.

3.3 Electrical Power

Table 2 shows the results from the evaluation of electric power, when the turbine is in IPC mode or non-IPC mode. The results show, that the power production is higher, when the turbine is in IPC mode. This effect is visible especially in region 2 (below rated RPM).

4. Conclusion

This study presents an investigation of the effect of IPC on acoustic emission, mechanical loads and power. The results show, that the use of IPC has both advantages and disadvantages. Here, we see a large potential in the optimization of IPC in a more comprehensive controller development not only for the individual wind turbine, but also for an entire wind park. The influence of IPC on the noise of the wind turbine holds the potential to optimize the turbine controller during night times on noise emissions. On the other hand during day times, when noise regulations are less strict other parameters could be optimized. Concerning the mechanical loads, the different effects of IPC on different components offer potential with regard to optimizing the service life of wind turbines. Here IPC offers significant potential for stressing or relieving components that are already more or less severely damaged by controlling the turbine.

References

[1] E. A. Bossanyi, „Wind Turbine Control for Load Reduction," *Wind Energy,* Bd. 6, p. 229–244, 2003.

[2] R. B. Morim, F. de Morais Carnielutti, L. D. da Rosa, G. Ricardo Hubner, C. M. Franchi, C. Eduardo de Souza und H. Pinheiro, „Analysis of Wind Turbine Power Generation with Individual Pitch Control," in *IEEE PES Innovative Smart Grid Technologies Conference - Latin America (ISGT Latin America)*, Gramado, Brazil, 2019.

[3] A. Sharma, C. Kraasch und A. Wegner, „Acoustic evaluation: microphone array measurements on a multi-megawatt individual pitch control wind turbine," *INTER-NOISE and NOISE-CON Congress and Conference Proceedings,* Bd. 3376, Nr. 12, pp. 2993 - 3992, 2024.

[4] International Electrotechnical Commission, Wind turbines - Part 11 Acoustic noise measurement techniques, 2019.

[5] International Electrotechnical Commission, *Windenergieanlagen: Teil 13 Messung von mechanischen Lasten DIN EN 61400-13 VDE 0127-13:2023-02,* 2023.

[6] International Electrotechnical Commission, *IEC 61400-12-1:2022 Wind energy generation systems - Part 12-1: Power performance measurements of electricity producing wind turbines,* 2022.

[7] D. Ossmann, P. Seiler, C. Milliren und A. Danker, „Field testing of multi-variable individual pitch control on a utility scale wind turbine," *Renewable Energy,* Bd. 170, pp. 1245-1256, 2021.

Impact of Innovative Maintenance Technologies on Offshore Wind Farm Performance

Christian Schläger, Torge Lorenz and David Baumgärtner

Fraunhofer Institute for Wind Energy Systems IWES, Bremerhaven, Germany

E-mail: schlaeger@solar.uni-hannover.de

Abstract. Cost reduction within the operations & maintenance (O&M) phase plays a key role in reducing the levelized cost of energy generated by offshore wind farms. While innovative maintenance technologies offer promising solutions, they also present significant financial risks for wind farm operators, as it is challenging to assess the profitability of technology integration. The simulation-based assessment of innovative concepts can facilitate the identification of their potential, support the integration of new technologies, and help prioritize research and development efforts. In this study, we comparatively evaluate the potential of various innovative O&M technologies using the O&M simulation software OffshoreTIMES. Innovations in four different application areas are investigated: (1) novel crew transfer vessel (CTV) designs, (2) crane-less technologies, (3) inspection drones, and (4) rotor blade repair robots. To address diverse technological development stages, each technology is evaluated in a conservative near-present and an optimistic future scenario. Additionally, combined technology scenarios are simulated in order to gain insights into the interactions between technologies. Our findings demonstrate that innovative maintenance concepts can significantly enhance availability and reduce O&M costs. Among the assessed concepts, crane-less technologies emerge as the most promising. The potential of next-generation CTVs varies greatly depending on site conditions, with notable benefits in milder weather environments. While automation of rotor blade maintenance using inspection drones and repair robots improves performance, their impact is smaller, and the transition to autonomous operation offers only marginal additional benefits. Integrated scenarios show that the technologies are not in competition with each other; rather, they are complementary, resulting in a cumulative performance benefit.

1 Introduction

The offshore location and associated harsh environmental conditions pose a number of significant challenges to offshore wind energy projects that are especially noticeable during their operations & maintenance (O&M) phase, such as turbine access restrictions during bad weather [1, 2], increased maintenance requirements [2–4], health and safety risks [4–6] and very costly equipment [7, 8]. Direct O&M costs, i.e. all costs incurred during the O&M phase of an offshore wind farm, typically account for around 25 to 30 % of the total project cost [9, 10]. In addition, downtime due to turbine failure results in lost revenue, commonly referred to as indirect O&M costs. Reducing total O&M costs, the combination of both, plays a key role in reducing the levelized cost of electricity (LCOE) of offshore wind energy and is therefore attracting considerable research interest [2].

One promising way to tackle offshore challenges and reduce costs is the integration of innovative maintenance technologies. Some of these are being increasingly used in industry already, e.g. drones for visual rotor blade inspections or robots for automated leading edge repairs. The integration of such

Content from this work may be used under the terms of the Creative Commons Attribution 4.0 licence. Any further distribution of this work must maintain attribution to the author(s) and the title of the work, journal citation and DOI.
Published under licence by IOP Publishing Ltd

innovations is associated with high financial risks for wind farm operators due to high operating or investment costs. A key challenge lies in assessing the impact a particular technology will have on wind farm performance and, consequently, whether its integration or development will be profitable.

Previous research primarily focused on developing and showcasing innovations in O&M, whereas some recent studies have begun to address the challenge of quantifying technology potential using an activity-based cost modeling approach. Kabbabe Poleo et al. [11] utilized this methodology to explore the effect of drone inspections on LCOE across differing levels of autonomy. Koltsidopoulus et al. [12] calculated the benefits of introducing various robotic systems in the future based on a projected technology roadmap. Their emphasis, though, was on the integration of such technologies as a whole, rather than comparing them with each other. In another study by the Offshore Wind Innovation Hub [13], cost modeling was used to assess and compare the potential of various O&M innovations. However, the study did not specify the cost assumptions underlying its results, limiting the transparency of the methodology.

Cost modeling is an appropriate approach for producing initial estimates; however, it fails to fully capture the complexity of the system. Limited logistical resources and weather-related constraints necessitate more accurate modeling, achievable through the detailed integration of all O&M activities into a comprehensive time-series simulation. Simulation tools for O&M are an active area of research, supporting the entire spectrum of decision-making, with applications ranging from fleet optimisation [14–16], site comparison [17], future scenario analysis [18], to environmental assessment [19]. A detailed literature review on this topic is presented by Seyr et al. [3].

In consideration of the capabilities of O&M simulation software, its application to innovative maintenance technologies represents a logical extension; however, this approach has received limited attention in the literature. One example is the study by Netland et al. [20], in which the authors evaluate the concept of a remote inspection robot located inside the nacelle. However, the practical relevance of this concept to the industry appears to be limited. Another example is the work of Rinaldi et al. [21] on tow-to-port maintenance for floating wind farms.

A review of the literature reveals several gaps in assessing the potential of innovative maintenance technologies. Studies often either rely on simplified cost modeling or, when using comprehensive O&M simulation software, focus on a single technology. Additionally, inherent challenges in O&M modeling contribute to further limitations. One key challenge is the uncertainty in input variables due to limited data access, particularly regarding turbine reliability and costs [3]. This affects the models informative value, making it difficult to quantify the potential of innovations, especially when considered in isolation.

In this study, we aim to address those challenges through a comparative analysis of innovative O&M technologies using the OffshoreTIMES simulation software. To this end, we investigate innovations in four distinct application areas: (1) novel crew transfer vessel (CTV) designs that mitigate weather-related transfer restrictions, (2) crane-less technologies for replacing large turbine components, (3) inspection drones for automating rotor blade surveys, and (4) robotic blade crawlers for minor rotor blade repairs.

2 Innovative Maintenance Technologies

2.1 Next-generation CTVs

Personnel transfer from a wave-exposed vessel to the foundation structure is one of the most hazardous activities in the offshore wind industry due to the difficulty and high frequency of this task [22]. The risk increases with wave height, making its viability both site- and time-dependent, which significantly limits potential maintenance windows for CTV crews. In the industry, the significant wave height h_s is the key parameter for the access/no-access decision, typically stipulated in contractual agreements with maintenance service providers [23]. For conventional CTVs, this threshold ranges from 1.0 m to 2.0 m, depending on vessel type and size [1]. The widespread use of motion-compensating gangways in service operation vessels (SOVs) allows for a significantly higher wave tolerance, making SOVs the default choice in harsher environments.

Enabling easier and safer turbine access in rough seas for smaller CTVs is the objective of various innovative ship and access system designs. One approach downscales the gangway principle from SOVs to CTVs, allowing motion-compensated transfers without direct contact between the vessel and turbine foundation [24]. This solution is advantageous as it can be retrofitted onto conventional CTVs or used for temporary fleet modifications, though it is limited to larger CTVs with sufficient free deck area. Other concepts enhance the conventional ‘Bump and Jump’ method, where the CTV is propelled against the foundation structure, establishing friction-based contact [1]. Some concepts dampen wave-induced motions through innovative hull technologies, such as hydraulically decoupled catamaran hulls [25] or actively controlled hydrofoils [26]. Another approach employs a bow-mounted gripper mechanism for a form-fit connection to the foundation, eliminating non-vertical relative movement [27]. However, this requires modifications to both the vessel and the foundation structure, limiting its practicality.

These concepts offer solutions for improving turbine access for CTVs in harsh conditions, with concept owners generally claiming to enable personnel transfer in significant wave heights of 2.0 to 2.5 m. Beyond enhancing comfort and safety during transfers, these innovations could reduce maintenance costs by enabling cost-effective CTVs to handle more tasks, allowing for more flexible and responsive operations. However, this comes with increased system complexity. Conventional catamaran CTVs achieve near-total availability due to their simplicity, while advanced designs may introduce reliability concerns.

2.2 Up-tower cranes

The limited predictability of wind and wave conditions makes offshore heavy lift operations more challenging than their onshore counterparts [2]. Wind turbines are typically equipped with small service cranes capable of lifting tools and smaller replacement parts; however, larger components such as gearboxes, generators, or rotor blades necessitate the use of specialized crane equipment. For these operations, heavy lift vessels (HLVs) are conventionally employed, but their utilization presents significant logistical and economic challenges due to the exceptionally high charter costs, which can be up to ten times greater than those for comparable onshore lifting equipment [8]. Consequently, these vessels are the cause of a significant proportion of direct O&M costs despite their limited deployment time.

A variety of crane-less concepts have been developed as cost-effective alternatives to traditional heavy-lift equipment, collectively referred to as up-tower cranes in this study. The objective of these systems is to enable medium to major component replacements without the necessity for HLVs. A key feature of these systems is that the crane base is positioned at the top of the turbine, simplifying the lifting process and enabling smaller designs. The crane can be installed directly on top of the nacelle [28, 29], or alternatively, it can climb onto the turbine tower [30]. Crane installation on top of the nacelle is typically performed in a modular manner, with components being installed successively. In principle, up-tower cranes are suitable for use with existing turbines. However, their applicability is subject to several constraints, including space requirements, weight limitations, and compatibility, which restricts their use for certain turbine types or components. Some manufacturers provide compatibility guidelines for various common turbine models [28]. A general principle that can be deduced is that the larger and more powerful the turbine, the fewer components are replaceable with existing crane-less systems.

The integration of up-tower cranes into an offshore O&M concept necessitates the transportation of equipment and personnel to the turbine. Smaller crane components can be transported by CTVs, while larger elements and spare parts require a larger vessel. This can be a simple barge with tugboats, as demonstrated by a manufacturer [28]. These vessels have much lower charter costs, are more readily available than HLVs, and place fewer demands on the service port, simplifying O&M planning. However, barges with tugboats are more susceptible to wave-induced motion and thus have lower weather-related operating limits, which may tighten time constraints on replacement operations.

2.3 Drone inspections

The rotor constitutes the most maintenance-intensive component group of a wind turbine [31], making regular rotor blade inspections an essential part of offshore maintenance strategies. Inspections aim to detect damage that could potentially propagate further or impair the aerodynamic performance of the rotor, thereby reducing energy production. A range of non-destructive testing (NDT) methods are available for this purpose, including optical, thermographic, X-ray, and ultrasonic techniques. Conventionally, rotor blade inspections and repairs are performed by rope-access technicians, which is resource-intensive, hazardous and highly susceptible to weather constraints [5, 32].

An emerging approach to rotor blade maintenance involves the use of unmanned aerial vehicles to automate optical inspections, usually employing off-the-shelf hardware. Equipped with high-resolution cameras, drones circumnavigate the rotor blade capturing image data while maintaining a safe clearance of a few meters [4]. Typically, inspections are conducted under visual line-of-sight (VLOS) by a pilot on the deck of a CTV or on the transition platform. While conventional rotor blade inspections require significant turbine downtime, drone-based inspections can substantially reduce the standstill period [32]. In practice, the flight time for an optical drone inspection is approximately 30 minutes per turbine, with further reductions anticipated through increased process automation [11]. In contrast, conventional rope-access inspections typically require several hours.

All relevant drone and camera systems are notably compact, facilitating transport by CTVs or SOVs and enabling seamless integration into existing O&M strategies. However, the primary focus of drones remains on optical inspection, and they are typically not equipped to perform other NDT procedures. Consequently, they cannot fully substitute conventional rotor blade inspections.

2.4 Blade-robots

A particularly frequent damage type on rotor blades is leading edge erosion, in which continuous environmental exposure degrades the blade's leading edge, impairing turbine performance and incurring significant economic consequences [2, 33]. The corresponding repair process, leading edge repair, stands out as a particularly well-suited application for robotic maintenance due to its repetitive nature.

Several robotic concepts for rotor blade maintenance are being developed, with the aim of facilitating both NDT and repair operations, thereby reducing the necessity for rope-access activities. In this work we refer to those concepts collectively as blade-robots. Most commercial designs feature a modular architecture, allowing for the integration of different tools and adaptability to various maintenance tasks. The mechanisms for task operation and movement on the rotor blade vary significantly across concepts. Examples include rope-stabilized contact with the leading edge, using anchors at the nacelle and ground [34], gripping onto the leading edge of a horizontally positioned rotor blade [35], and the use of robotic legs with suction cups for free movement along the blade surface [36].

Currently, blade-robot operations are typically conducted within VLOS by trained pilots. However, similar to drones, blade-robots have the potential for increased autonomy as robotic systems continue to improve in reliability and performance. A conceivable future concept involves deploying blade-robots via cargo drones. For instance, a blade-robot and its transport drone could be stationed at an offshore substation or an SOV, featuring autonomous takeoff, landing, and charging capabilities. The deployment and retrieval process with a cargo drone has already been demonstrated under laboratory conditions for one of the blade-robots [37].

3 Methodology – Simulation of O&M Activities

3.1 OffshoreTIMES O&M Simulation Software

The aim of this study is to examine the innovative maintenance concepts that are presented in the previous section in appropriate scenarios, which will allow their impact on wind farm performance to be quantified on a comparative basis. For this purpose, we conduct several simulations using OffshoreTIMES, a comprehensive O&M simulation software developed by Fraunhofer IWES. In the following, a brief overview of the most relevant aspects for this work is provided.

OffshoreTIMES depicts the complete 25-35 years long O&M phase of an offshore wind farm in a high-resolution time-series simulation with a discrete time step of 5 minutes. The simulation provides a high level of detail, which includes for example the position and status of individual turbine components, spare parts, technicians and vessels. Hourly metocean data is used by the model to consider weather restrictions of marine operations and to compute wind turbine energy production, including inner-park turbine wake modeling with FOXES [38].

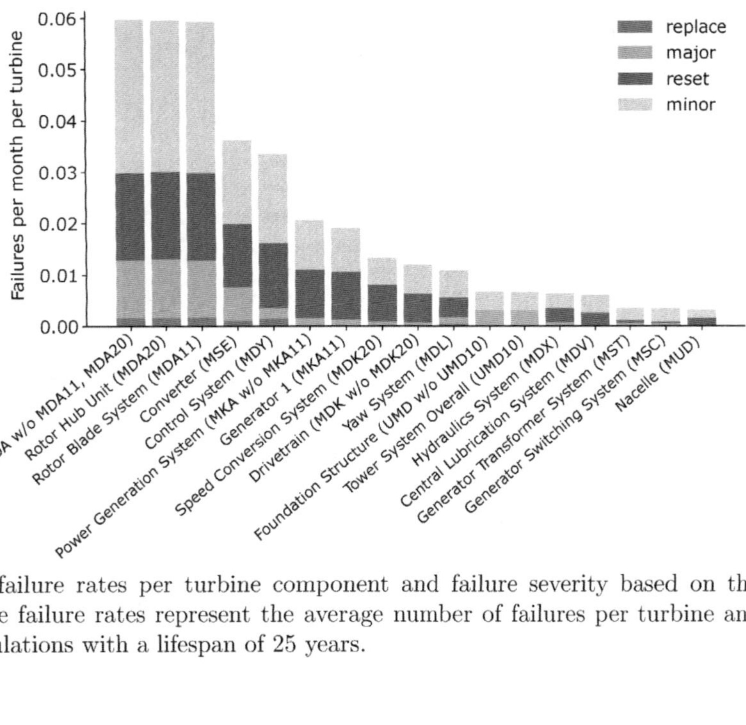

Figure 1: Observed failure rates per turbine component and failure severity based on the employed reliability model. The failure rates represent the average number of failures per turbine and month as recorded over 50 simulations with a lifespan of 25 years.

The OffshoreTIMES code allows for a flexible configuration of the simulated crew-logistics strategy (including ports, storages, technicians and different types of vessels) and the wind turbine reliability model. The reliability model plays a central role in the simulation, as it defines the demand for maintenance and inspection activities and describes which logistical resources are necessary to carry out the maintenance tasks. At any simulation time step, turbine component failures may occur stochastically according to the associated failure rates. In this work, we employ a reliability model developed in previous work [39], that describes each wind turbine consisting of 17 individual failure components. According to this model, component failures are specified by different severities (reset, minor, major, replacement). For every failure component, each failure severity comes with its own likelihood of occurrence and repair formula, specifying the personnel, time, material and vessel types that are required in order to remedy the failure. Figure 1 illustrates the failure rates of the 17 individual turbine components, categorized by severity.

The built-in post-processing computes several key performance indicators (KPIs) to evaluate the performance of the O&M strategy, such as direct and indirect O&M costs, turbine availability, and the utilization of logistical resources. In order to achieve statistically robust results from the stochastically driven simulations, a series of Monte Carlo iterations is carried out. Figure 2 illustrates the convergence behavior for one KPI, the energy-based availability. The plot shows the relative deviation of the KPI mean value as a function of the number of iterations, normalized to the mean value after 450 iterations. By definition, the deviation at n = 450 is zero, as all values are calculated relative to this reference. The analysis indicates that 50 iterations per simulation scenario are sufficient to achieve convergence, with a relative error for energy-based availability remaining below 0.05 %. A similar behaviour is observed for other KPIs, such as total O&M costs, which exhibit a relative error of approximately 0.3 %. Therefore in this work, we perform 50 Monte Carlo iterations for each simulation scenario.

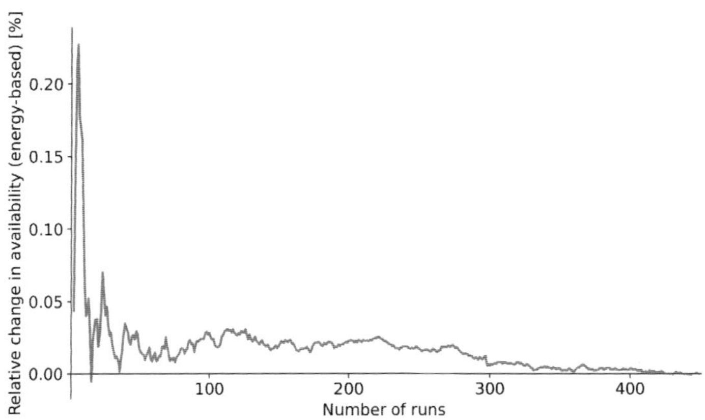

Figure 2: Convergence behavior of the Monte Carlo simulation for the energy-based availability

3.2 Reference Scenario

To examine the innovations analyzed in this study under realistic and representative conditions, the Borkum Riffgrund 2 wind farm in the German Bight is selected as a reference scenario. This wind farm, commissioned in 2019, consists of 56 MHI Vestas V164-8.0MW turbines, with a total installed capacity of 448 MW. It is located approximately 58 km from the nearest service port and therefore represents a medium-sized offshore wind farm at a moderate distance from shore. Turbine coordinates are obtained from the Global Offshore Wind Farm Database of 4C Offshore [40]; the power curve data is sourced from the manufacturer's specifications [41], and the thrust coefficient curve is taken from a comparable reference turbine of equal rated power [42]. Metocean data is sourced from the ERA5 data set [43] with an hourly resolution.

In the simulation, three ships are available as logistical resources: a permanently chartered SOV, a temporarily chartered HLV as needed, and a temporarily chartered CTV to assist during the annual inspection campaign in summer. In the reference scenario, the SOV is utilised for all daily maintenance operations. The SOV employs a staggered shift system, with up to five worker groups beginning their 12-hour shifts at an interval of one hour. It is assumed that the SOV returns to its service port in Emden, Germany, at least every 14 days for personnel changes and storage loading with spare parts. The motion-compensated gangway installed on the SOV allows for transfer operations up to $h_s = 3.5\,m$ [1]. Furthermore, the SOV has a daughter craft with eight seats, that can be used in favourable weather

IOP Publishing
doi:10.1088/1742-6596/3025/1/012019

conditions as an additional resource. In practice, the operator clusters some maintenance operations across several adjacent wind farms; this aspect is not considered in our analysis but indirectly accounted for by a reduced SOV personnel capacity and charter rate. The exchange of large turbine components requires an HLV, which is located in Esbjerg, Denmark and can be chartered as needed. Due to the high financial risk associated with bad weather, chartering of the HLV is restricted to the months from February to August. Additionally, a CTV can be chartered to meet seasonal inspection demands, which is operated from the nearby service port in Norddeich, Germany. The modeling parameters for the ships operating in the reference scenario are listed in Table 2 in the appendix.

Similar to ships, OffshoreTIMES allows both permanent employment and temporary hiring of technicians. In this study, temporarily hired technicians are exclusively used for the inspection campaign and crewing the temporarily chartered HLV. This assumption is reasonable as in practice, many inspection activities are carried out by service providers, especially when using innovative technologies such as inspection drones or blade-robots. Permanent technicians prioritize corrective maintenance and assist with inspections when capacity allows. Inspection tickets for each failure component are generated in May based on the annual or biennial frequency dictated by the reliability model. The inspection campaign typically concludes within one to two months. The number of temporarily hired inspection technicians is constrained by the crew capacity of the deployed ships. The count of permanent technicians is fixed at ten skilled workers available for deployment, a configuration that has proven suitable in test simulations. In practice, SOV personnel follow a two-week shift rotation, meaning that each technician in our simulation corresponds to two real-world technicians taking alternating shifts. Personnel costs for technicians are set at 800 €/d for permanent employees and 1000 €/d for temporary hires. For corrective repairs, it is assumed that spare parts are readily available in the onshore inventory at any time. Furthermore, the SOV storage is systematically loaded with small repair parts that are required on a regular basis.

3.3 Modeling of Innovations

This paper explores innovations across four application areas. To address for diverse technological development stages, each innovation is evaluated in a conservative near-present and an optimistic future scenario with advanced technology. Alongside the individual innovation scenarios, simulations of combined scenarios are also performed to understand the interaction between technologies at their respective maturity levels. In the following paragraphs, each of the four innovations is presented, detailing how the technology is integrated into OffshoreTIMES and highlighting the differences between the innovation scenario and the reference case. An overview of all simulation scenarios is given in Table 1.

Table 1: Description of the simulation scenarios

Scenario Name	Description
Reference scenario	The reference scenario – detailed in Section 3.2.
Next-gen CTV	Daily maintenance needs are met by two CTVs instead of the SOV, with maximum wave tolerances of $h_s = 2.0$ m.
Next-gen CTV upscaled	Daily maintenance needs are met by two CTVs instead of the SOV, with maximum wave tolerances of $h_s = 2.5$ m.
Up-tower crane	5 of 9 large components are replaced using the crane barge instead of an HLV.
Up-tower crane upscaled	All 9 large components are replaced using the crane barge instead of an HLV.
Drone pilot	Additional pilot-operated drone inspections are conducted, improving early fault detection in the rotor blade system.
Drone autonomous	Additional autonomously conducted drone inspections are performed, improving early fault detection in the rotor blade system.
Blade-robot pilot	Minor failures in the rotor blade system can be repaired faster and with less personnel using a blade-robot.
Blade-robot autonomous	Minor failures in the rotor blade system can be repaired faster and without personnel using a blade-robot and a cargo drone.
Present technologies	Up-tower crane, drone and blade-robot are deployed together in their conservative configuration.
Future technologies	Up-tower crane, drone and blade-robot are deployed together in their optimistic configuration.

Innovative CTV designs aim to increase wave tolerance compared to conventional CTVs, thereby enabling maintenance operations to be planned and executed more reliably. This principle translates into a modified parameterisation of the deployed vessels. Since an SOV is used for routine operations in the reference scenario, it is replaced by two CTVs in the innovation scenarios. Conventional CTVs in catamaran configuration typically operate up to wave heights of $h_s = 1.5$ m [17]. In the present and future innovation scenarios, we assume next-generation CTVs to enhance the wave tolerances to $h_s = 2.0$ m and 2.5 m, respectively. These values are frequently claimed by technology developers. The other selected ship parameters are similar to those of the inspection CTV, as listed in Table 3 in the appendix. Increased charter rates and potentially lower ship reliability are not included in the scenarios. The inspection CTV, which is used only during the summer months, retains its low wave tolerance from the reference case.

Up-tower cranes aim to enable large component replacements without relying on an HLV, thereby offering more cost-effective and flexible operations. The reliability model describes each turbine with 17 distinct failure components, with the nine largest conventionally requiring an HLV for replacement operations. In the innovation scenarios, the repair formulas for these large replacements are modified so that a crane barge is used instead. However, as discussed in Section 2.2, space and weight limitations mean that not all components can necessarily be replaced using up-tower cranes. In the conservative innovation scenario, only five of these nine components are considered replaceable using an up-tower crane, whereas in a future scenario, all nine components are assumed to be replaceable in this way. Personnel requirements during replacement operations remain unchanged. For simplicity, the simulation aggregates the barge, tugboats, the up-tower crane system, and any additional auxiliary equipment into a single vessel. The weather and wave restrictions for the up-tower crane and barge operation are assumed to be $h_s = 2$ m, $h_{max} = 2.5$ m, and an average wind speed at 10 m height of $u_{10\,m} = 12$ m/s. The ship parameters for the crane barge are listed in Table 3 in the appendix. In contrast, the HLV can operate under much harsher conditions, with wave heights up to $h_s = 3.5$ m and $h_{max} = 6$ m. The resulting difference from the reference case primarily lies in the substantially lower costs and more flexible charter options of the crane barge, which, however, also entails stricter weather restrictions.

Drone inspections aim to automate optical rotor blade surveys. In the reliability model, each turbine component is assigned a condition score that degrades over time and is refreshed following repair or replacement. Before a component ultimately fails, it typically exhibits early indications of damage that inspections can detect once its condition falls below a certain threshold. Consequently, maintenance can be scheduled to repair emerging faults before complete failure occurs, thereby preventing turbine downtime and revenue loss. As optical inspections of the rotor blades cannot fully replace conventional inspections of this component, which also include other NDT methods, we model drone inspections as an additional measure without altering the conventional inspection schedule. In the innovation scenarios, drone inspections are performed every 145 days, increasing the probability of early fault detection in the rotor blade system. During drone inspections, emerging faults are detected with a probability of detection (POD) of 40 %. In the conservative scenario, drones are operated on-site in VLOS by a team of two pilots transferred via CTV or SOV. In the future scenario, autonomous drones are instead operated from an offshore base, thus eliminating the need for ship transfers. In both cases, the duration of drone inspections is 60 minutes.

Blade-robots aim to automate repetitive tasks on rotor blades that are conventionally performed by rope-access technicians, enabling both inspections and repairs. The predominant application of such systems at present is the automation of leading edge repair, which in our simulation is most closely represented by repairs of rotor blade failures with minor severity. In the reference case, a minor rotor blade repair takes 9 hours and requires two technicians. In contrast, the innovation scenarios assume that blade-robot automation reduces this repair time to 3 hours. Similar to the drone-based approaches, two operational modes are considered: a conservative, pilot-operated scenario where blade-robot operations are conducted in VLOS by two pilots transferred by CTV or SOV, and a future scenario in which a cargo drone deploys and retrieves the blade-robot autonomously, thereby eliminating the need for ship transfers and personnel.

4 Results

Figure 3 provides an overview of two main KPIs across all scenarios, expressed as deviations from the median of the reference scenario. The energy-based availability is an important indicator of the technical performance of the O&M strategy, defined as the ratio of electricity produced to the theoretical maximum in the absence of turbine failures. In the reference scenario, availability is 94.03 %. The conservative and optimistic next-generation CTV scenarios yield absolute availability reductions of 7.05 % and 3.44 %, respectively, indicating that an SOV-based logistics concept remains far superior in Borkum Riffgrund 2. All other technology innovations lead to improvements, with optimistic variants mostly outperforming their conservative counterparts. The most notable improvement is observed in the optimistic up-tower crane scenario, which achieves an absolute increase of 0.77 %, whereas the conservative up-tower crane configuration does not result in a significant enhancement, compared to the reference scenario. Drone inspections lead to an absolute availability increase of approximately 0.35 % in both scenarios, with little additional benefit observed from transitioning to autonomous operations. The blade-robot shows a smaller impact, achieving only a 0.13 % increase in availability under autonomous operation.

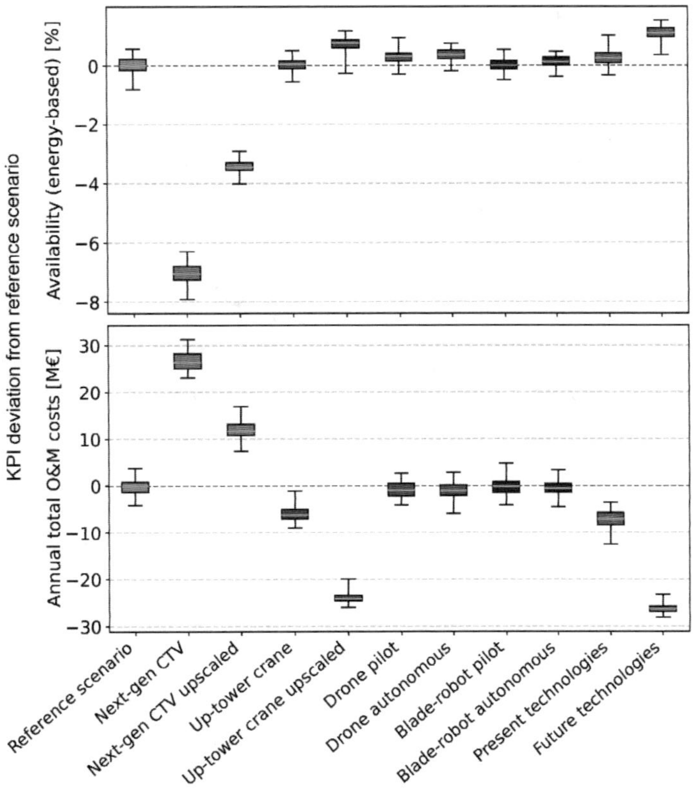

Figure 3: Results box plot of the KPIs energy-based availability and total O&M costs in all simulation scenarios. For each, the absolute KPI deviation from the median of the reference scenario is shown. The orange line marks the median, the box represents the first and third quartiles, and the whiskers indicate the full range of results. The box colours indicate the technology used in the scenario.

Furthermore, the annual total O&M costs are compared in Fig. 3, which is the most significant parameter for an economic analysis of the O&M strategy. Revenue losses are calculated using a fixed electricity price of 19.4 €/MWh. Overall, the trends in total O&M costs are consistent with those observed for availability. The deployment of up-tower cranes in the optimistic configuration leads to the most substantial cost savings, with an estimated reduction of 24.0 M€/a. Even in the conservative up-tower crane scenario, cost reductions of 5.8 M€/a are achieved, which surpass the cost savings from both drone inspections and blade-robot repairs. While the integration of drone inspections and robotic rotor blade repairs does reduce O&M costs, their impact is less pronounced. In the autonomous scenario, drones and blade-robots result in annual cost reductions of approximately 0.7 M€/a and 0.4 M€/a, respectively.

In addition to the individual innovation scenarios, we investigate combined technology scenarios, in which innovations are used concurrently within the wind farm. For those scenarios, next-generation CTVs are excluded due to their inferior performance. The results demonstrate further improvements for both KPIs, with overall gains close to the sum of the individual innovation-related improvements. This leads to an estimated 1.11 % increase in availability and a cost reduction of 26.1 M€/a in the future technologies scenario. These findings highlight the complementary nature of the individual technological innovations, indicating that they do not adversely interfere with one another.

Having outlined the overall results of the main study, we now turn to a more detailed examination of two of the four innovations, namely next-generation CTVs and drone inspections. The performance of next-generation CTVs in the wind farm under investigation is unsatisfactory. In order to assess whether this outcome is influenced by local conditions, additional scenarios are simulated for the case of Baltic 2, an alternative wind farm in the Baltic Sea located closer to its CTV port (32 km instead of 54 km) and comprising 80 turbines rated 3.6 MW. The Baltic Sea generally experiences milder weather than the German Bight, making CTVs a common choice. For this reason, employment of conventional CTVs with $h_s = 1.5$ m is also part of the comparison. Vessel efficiency can be evaluated by analysing daily vessel status reports. The distribution of status messages from the primary vessel utilised for daily maintenance operations over the entire simulation period is presented in Fig. 4 for the reference scenario and the CTV-related scenarios in both wind farms.

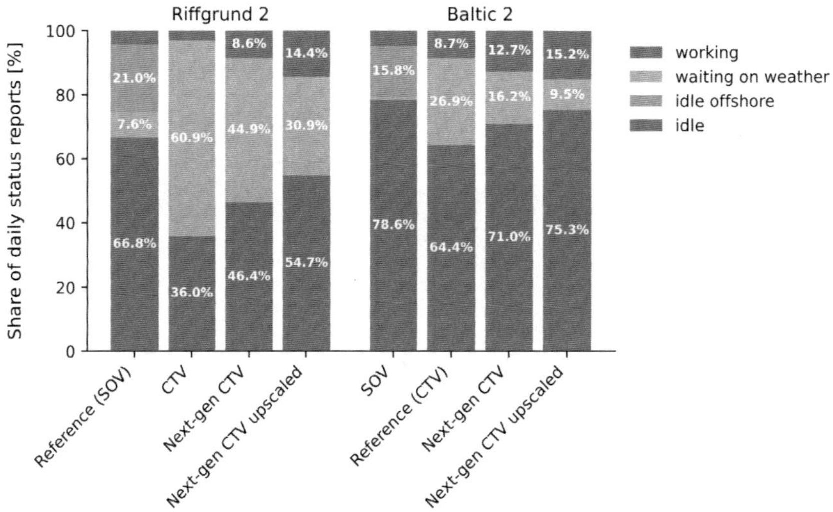

Figure 4: Distribution of daily status reports for the primary vessel (SOV or primary CTV) in different simulation scenarios

In the reference scenario, the SOV operates with only 7.6 % weather-induced downtime. However, both next-generation CTV scenarios experience multiple times higher weather-related downtime. In contrast, conventional CTVs perform relatively well in the Baltic Sea, achieving 92.1 % wind farm availability with only 26.9 % waiting on weather reports. The introduction of next-generation CTVs significantly reduces weather-related waiting time, leading to substantial absolute availability improvements of 1.5 % and 2.3 %, compared to conventional CTV use. This marks one of the highest availability gains observed in our study, surpassed only by the optimistic up-tower crane scenario. Additionally, these improvements translate into considerable O&M cost reductions of 4.3 M€/a and 6.7 M€/a, respectively. These findings highlight the strong site-specific potential of next-generation CTVs, particularly in the milder weather conditions of the Baltic Sea.

A key advantage of detailed O&M modeling over simpler activity-based cost models is its ability to yield insights beyond financial metrics, such as determining the most efficient method for technology integration and identifying the required capabilities for economic viability. Optical drone inspections serve as a representative example. In modeling this innovation, two key parameters are considered: the probability of detection (POD) and the inspection frequency. The POD is determined by the selected drone technology, which encompasses factors such as drone control, camera quality, and effectiveness of image data analysis, while the inspection frequency can be directly influenced through scheduling. A sensitivity analysis is conducted to evaluate how variations in these parameters influence wind farm performance.

Figure 5a illustrates the percentage of rotor blade damages that are repaired early, i.e. after damage has been detected during inspections and before component failure occurs. This metric highlights the practical benefit that drone inspections contribute to the O&M concept, and enables a quantitative evaluation of the trade-off between the required operational effort and the practical benefits achieved. For instance, the data indicate that a configuration with a POD of 30 % and an inspection interval of 50 days results in a similar share of early repairs as one with a POD of 50 % and an inspection interval of 90 days. Consequently, investing in drone equipment or service providers capable of achieving a POD of 50 % could reduce the inspection frequency by half while maintaining comparable benefits.

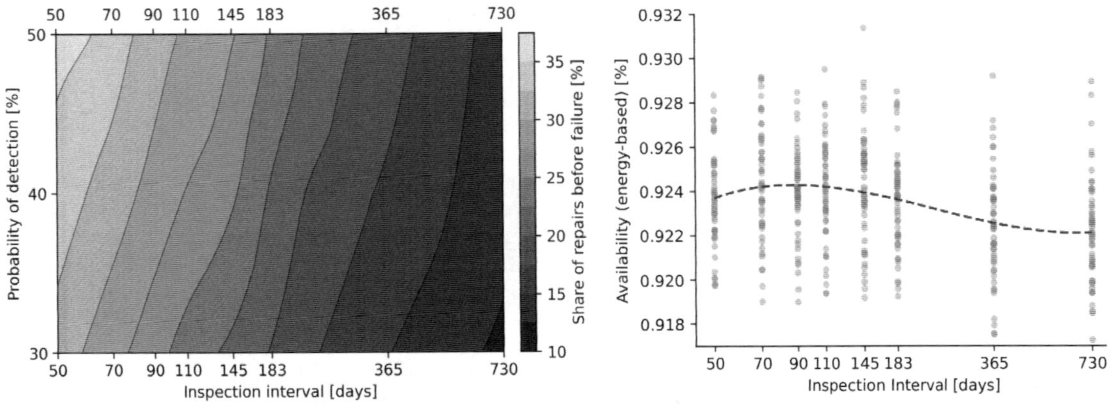

(a) Share of rotor blade damages repaired before component failure as a function of inspection interval and POD. Cubic interpolation is performed between scenarios.

(b) Availability at different inspection intervals and POD of 40 %, showing individual Monte Carlo iterations and a cubic polynomial fit.

Figure 5: Sensitivity study of pilot-operated drone inspections. The inspection intervals (log-scaled) vary as indicated in the plots, while the POD is set to 30, 40, and 50 %. For each of the 24 scenarios, 50 Monte Carlo iterations are performed.

Increasing the inspection frequency too much can overwhelm the logistical framework and raise costs, while too few inspections may result in insufficient early detection of faults. Thus, the goal for an effective O&M strategy is to identify an optimal inspection frequency that strikes a balance between maximizing early fault detection and manageable logistical demands. This relationship is illustrated in Fig. 5b, where the availability results from all Monte Carlo iterations for a POD of 40 % and different inspection intervals are shown, along with a polynomial fit. Based on our model parameters, such as POD, reliability model, and logistics concept, we find that this expected maximum is at approximately 90 d. However, the influence of drone inspections on availability is relatively minor, compared to the variability introduced by stochastic effects. This analysis demonstrates, how O&M modeling supports decision-making by quantifying the operational impact of new technologies and identifying optimal integration strategies.

5 Discussion

The outcomes of O&M modeling are highly sensitive to input parameters, as demonstrated by the impact of next-generation CTVs on wind farm availability. As the primary ship type is changed, the innovation's performance is strongly influenced by the underlying ship modeling assumptions for both SOV and next-generation CTVs. Our study indicates that next-generation CTVs achieve suboptimal performance in the German Bight when compared to SOVs; yet they show substantial potential in the Baltic Sea. This is intuitive, as SOVs can maintain operations under harsher conditions due to their higher wave tolerance. However, the SOV performance may be overestimated in our model due to an overly high assumption for its operational limits in terms of h_s. Specifically, we use a wave height limit of $h_s = 3.5$ m, as provided by the gangway manufacturer for the access system installed on the SOV deployed in Riffgrund 2 [1]. In contrast, our analysis of automatic identification system (AIS) data has revealed that most SOVs become inoperative at wave heights already as low as 2.5 to 3 m [44]. This underlines the importance to complement manufacturer specifications with empirical operational data to accurately capture real-world performance constraints.

The reliability model plays a central role in determining the demand for maintenance operations, making the simulation results highly sensitive to the assumed failure rates. The model employed in this study is derived from older and smaller wind turbines, which limits its direct applicability to the 8 MW turbines of Borkum Riffgrund 2. Being aware of this simplification is particularly important when interpreting results such as the optimal 90-day drone inspection interval, as these depend strongly on the underlying reliability model. Consequently, conducting more detailed analyses would necessitate the employment of customised reliability models that reflect the specific failure behaviour of a given wind farm.

When assessing the impact of innovations on total O&M costs, the results are highly dependent on the underlying cost modeling parameters. Key influencing factors in this regard include vessel charter rates, the costs associated with implementing new technologies, and the revenue model. However, cost estimates in O&M modeling often suffer from significant uncertainties, due to limited data access [3].

This uncertainty is most pronounced for innovations that fundamentally alter vessel logistics, such as next-generation CTVs and up-tower cranes. Since these technologies introduce the largest structural changes to O&M operations, their economic impact is also the most sensitive to cost assumptions. With more precise cost or technology data, the approach can provide an accurate estimate of the impact of innovative O&M technologies for a given wind farm scenario.

Additionally, our analysis does not explicitly account for the investment or operational costs of the new technologies themselves. For instance, we assume charter costs for next-generation CTVs to be comparable to conventional CTVs, and we apply estimated costs for barge and tugboat operations, as well as for personnel conducting drone inspections. However, we do not include specific cost estimates for the up-tower crane, the inspection drone, or the blade-robot. This decision is motivated by the lack of precise cost data and the desire to assess the technologies from a broader perspective, as integration costs are highly case-specific. Consequently, the results should be interpreted with the understanding that the projected O&M benefits must outweigh the yet-to-be-determined technology costs for these innovations to be economically viable.

6 Conclusion

In this study, we conducted simulations using the O&M simulation software OffshoreTIMES to evaluate innovative maintenance technologies across four application areas: next-generation CTVs, up-tower cranes, drone inspections, and blade-robots. In order to ensure accurate modeling, a concise literature and industry overview was provided, outlining the current status of these innovations. Each technology was assessed under both conservative near-present and optimistic future scenarios, as well as in combined technology configurations, using a real wind farm as the reference scenario.

Our results show that innovative maintenance concepts have the potential to increase wind farm availability and reduce O&M costs. However, their effectiveness varies significantly across different technologies. Next-generation CTVs demonstrated a strong dependence on site-specific conditions, performing suboptimally in our reference wind farm in the German Bight but showing notable improvements in an alternative Baltic Sea location. The integration of up-tower cranes resulted in the greatest increase in availability and cost savings. This is particularly evident in the optimistic technology scenario, where full component compatibility eliminates the requirement for HLVs entirely. This demonstrates that the technology potential is highly dependent on lifting compatibility. The automation of rotor blade maintenance using inspection drones and blade-robots also contributed to performance improvements, though their impact was relatively limited. Notably, the transition from pilot-operated to fully autonomous operations yielded only minor additional benefits in both applications. Among all assessed technologies, blade repair robots exhibited the lowest potential, as they only address a small subset of maintenance operations. The additional integrated scenarios demonstrate that the combination of multiple technologies results in benefits that are approximately equal to the sum of the individual technology benefits, with no adverse impact on one another. Instead, the technologies complement each other, enhancing overall performance.

This study demonstrates the potential of detailed O&M modeling as a powerful tool for evaluating the impact of innovative maintenance technologies on wind farm performance. By leveraging a comparative approach, we illustrate how this can provide meaningful insights even with limited empirical data. Furthermore, O&M modeling offers a structured approach to evaluating the role of new technologies in existing maintenance strategies. For instance, our sensitivity study on inspection drone parameters identifies an optimal inspection interval of 90 days, highlighting how O&M modeling can guide technology integration.

7 Acknowledgements

This work has been carried out as part of the research project SOBeKo, which was funded by the German Federal Ministry for Economic Affairs and Climate Action (BMWK) under grant number 03EE3001A [45]. The authors would like to thank the project partners for their support and cooperation.

A Modelling parameters

Table 2: Modelling parameters of the vessels deployed in the reference scenario. The crew capacity and charter rate of the SOV are reduced to indirectly account for the clustering of maintenance activities with adjacent wind farms.

Vessel type	SOV	SOV daughter	HLV	CTV (inspection)
Crew capacity	20	8	105	24
Shift mode	day	day	day & night	day
Maximum offshore duration	14 d	1 d	-	1 d
Transit speed	13 kn	23 kn	9.6 kn	23 kn
Transit speed inside wind farm	13 kn	23 kn	4.8 kn	11.5 kn
Fuel consumption (transit)	1.000 L/h	150 L/h	1.000 L/h	320 L/h
Fuel consumption (idle)	350 L/h	-	120 L/h	120 L/h
h_s (operation)	3.5 m	1.2 m	3.5 m	1.5 m
h_s (transit)	10.0 m	1.2 m	3.5 m	1.5 m
h_{max} (operation)	4.0 m	1.5 m	6.0 m	1.8 m
h_{max} (transit)	10.0 m	1.5 m	6.0 m	1.8 m
$u_{10\,m}$	-	-	10 m/s	-
Charter mode	permanent	permanent	temporary	temporary
Minimum charter time	-	-	14 d	-
Mobilisation time	-	-	30 d	-
Mobilisation costs	-	-	2, 500, 000 €	-
Demobilisation costs	-	-	2, 500, 000 €	-
Charter costs per day	9, 100 €	-	250, 000 €	7, 500 €
Personnel transfer duration	15 min	5 min	15 min	15 min
Service port (shore distance)	Emden (84 km)	SOV	Esbjerg (209 km)	Norddeich (58 km)

Table 3: Modelling parameters of the vessels deployed in innovation scenarios

Vessel type	CTV (primary)	CTV (secondary)	Crane Barge
Innovation Scenarios	Next-gen CTV	Next-gen CTV	Up-tower crane
Crew capacity	12	26	50
Shift mode	day	day	day & night
Maximum offshore duration	1 d	1 d	-
Transit speed	23 kn	23 kn	9.6 kn
Transit speed inside wind farm	11.5 kn	11.5 kn	4.8 kn
Fuel consumption (transit)	250 L/h	320 L/h	1.000 L/h
Fuel consumption (idle)	120 L/h	120 L/h	120 L/h
h_s (transit & operation)	2.0 m | 2.5 m	2.0 m | 2.5 m	2.0 m
h_{max} (transit & operation)	2.3 m | 2.8 m	2.3 m | 2.8 m	2.5 m
$u_{10\,m}$	-	-	12 m/s
Charter mode	permanent	permanent	temporary
Minimum charter time	-	-	1 d
Mobilisation time	-	-	5 d
Charter costs per day	3, 500 €	5, 000 €	25, 000 €
Personnel transfer duration	15 min	15 min	15 min
Service port (shore distance)	Norddeich (58 km)	Norddeich (58 km)	Emden (84 km)

References

[1] Hu B and Yung C 2020 Offshore Wind Access Report 2020 Tech. Rep. TNO 2020 R11992 TNO URL `https://repository.tno.nl/islandora/object/uuid%3Ad6cefbf6-2916-42fc-bb3d-b2d713938e09` Accessed: March 2025

[2] Ren Z, Verma A S, Li Y, Teuwen J J and Jiang Z 2021 Offshore Wind Turbine Operations and Maintenance: A State-of-the-Art Review*Renewable & Sustainable Energy Reviews* **144** 110886

[3] Seyr H and Muskulus M 2019 Decision Support Models for Operations and Maintenance for Offshore Wind Farms: A Review*Applied Sciences* **9** 278

[4] Liu Y, Hajj M R and Bao Y 2022 Review of Robot-Based Damage Assessment for Offshore Wind Turbines*Renewable & Sustainable Energy Reviews* **158** 112187–112187

[5] Khalid O, Hao G, Desmond C J, Macdonald H, McAuliffe F D, Dooly G and Hu W 2022 Applications of Robotics in Floating Offshore Wind Farm Operations and Maintenance: Literature Review and Trends *Wind Energy* **25**

[6] Tveiten C K, Albrechtsen E, Heggset J, Hofmann M, Jersin E, Leira B and Norddal P K 2011 HSE challenges related to offshore renewable energy Tech. Rep. A18107 SINTEF Technology and Society URL `https://www.sintef.no/globalassets/upload/teknologi_og_samfunn/sikkerhet-og-palitelighet/final-report-hse-challenges-offshore-renewable-energy.pdf/` Accessed: March 2025

[7] Dalgic Y, Lazakis I and Turan O 2013 Vessel charter rate estimation for offshore wind O&M activities *Developments in Maritime Transportation and Exploitation of Sea Resources IMAM 2013* ed Soares C and Peña F (CRC Press) pp 899–907 ISBN 978-1-138-00124-4

[8] van Bussel G and Zaaijer M B 2001 Reliability, Availability and Maintenance aspects of large-scale offshore wind farms, a concepts study *Proceedings of MAREC 2001 Marine Renewable Energies Conference* (Newcastle, United Kingdom) pp 119–126 ISBN 1-902536-43-6

[9] BVG Associates 2019 Guide to an offshore wind farm: Updated and extended Tech. rep. BVG Associates URL `https://guidetoanoffshorewindfarm.com/` Accessed: March 2025

[10] Stehly T, Beiter P and Duffy P 2020 2019 Cost of Wind Energy Review Technical Report NREL/TP-5000-78471 National Renewable Energy Laboratory URL `https://research-hub.nrel.gov/en/publications/2019-cost-of-wind-energy-review` Accessed: March 2025

[11] Kabbabe Poleo K, Crowther W and Barnes M 2021 Estimating the Impact of Drone-based Inspection on the Levelised Cost of Electricity for Offshore Wind Farms*Results in Engineering* **9** 100201

[12] Koltsidopoulos A, Wavell D and Gray A 2021 Quantifying the impact of Robotics in Offshore Wind Tech. rep. Offshore Wind Innovation Hub URL `https://ore.catapult.org.uk/resource-hub/analysis-reports/quantifying-impact-robotics-offshore-wind` Accessed: March 2025

[13] Offshore Wind Innovation Hub 2020 Operations & Maintenance: Cost Drivers Tech. rep. ORE Catapult & Knowledge Transfer Network Glasgow URL `https://offshorewindinnovationhub.com/industry_insight/operations-maintenance-cost-drivers/` Accessed: March 2025

[14] Sperstad I B, Stålhane M, Dinwoodie I, Endrerud O E V, Martin R and Warner E 2017 Testing the Robustness of Optimal Access Vessel Fleet Selection for Operation and Maintenance of Offshore Wind Farms*Ocean Engineering* **145** 334–343

[15] Dalgic Y, Dinwoodie I, Lazakis I, Mcmillan D and Revie M 2015 The Influence of Multiple Working Shifts for Offshore Wind Farm O&M Activities – Strathow-OM Tool *Design and Operation of Offshore Wind Farm Support Vessels 2014* (London, UK: Royal Institution of Naval Architects)

[16] Endrerud O E V, Liyanage J P and Keseric N 2014 Marine logistics decision support for operation and maintenance of offshore wind parks with a multi method simulation model *Proceedings of the Winter Simulation Conference 2014* (Savanah, GA, USA: IEEE) pp 1712–1722

[17] Hu B, Swamy S K and Hermans K W 2018 Including Realistic Vessel and Human Behaviour in Simulations of Offshore Wind Farm Operations TNO 2018 R10937 ECN Part of TNO URL `https://repository.tno.nl/islandora/object/uuid:7156668e-d901-4f84-aef0-8e3b9dd59795` Accessed: March 2025

[18] Hofmann M and Sperstad I B 2014 Will 10 MW Wind Turbines Bring Down the Operation and Maintenance Cost of Offshore Wind Farms*Energy Procedia* **53** 231–238

[19] Garcia-Teruel A, Rinaldi G, Thies P R, Johanning L and Jeffrey H 2022 Life Cycle Assessment of Floating Offshore Wind Farms: An Evaluation of Operation and Maintenance*Applied Energy* **307** 118067

[20] Netland Ø, Sperstad I B, Hofmann M and Skavhaug A 2014 Cost-Benefit Evaluation of Remote Inspection of Offshore Wind Farms by Simulating the Operation and Maintenance Phase*Energy Procedia* **53** 239–247

[21] Rinaldi G, Thies P and Johanning L 2020 Improvements in the O&M modelling of floating offshore wind farms *Proceedings of the 4th International Conference on Renewable Energies Offshore (RENEW 2020)* (Lissabon, Portugal: CRC Press (Taylor & Francis)) pp 481–487

[22] McAuliffe F D, Murphy J, Lynch K, Desmond C J, Norbeck J A, Nonås L M, Attari Y, Doherty P, Sorensen J D, Giebhardt J, Lopez C, Hildebrandt S, Potestio S and Godavarthi V S 2017 Driving Cost Reductions in Offshore Wind: The Leanwind Project Final Publication Tech. rep. LEANWIND consortium URL `https://www.leanwind.eu/wp-content/uploads/LEANWIND-final-publication.pdf` Accessed: March 2025

[23] McMorland J, Collu M, McMillan D and Carroll J 2022 Operation and Maintenance for Floating Wind Turbines: A Review *Renewable & Sustainable Energy Reviews* **163** 112499–112499

[24] Ampelmann L-type – Motion Compensated Gangway URL `https://www.ampelmann.nl/systems/l-type` Accessed: March 2025

[25] Wallaby Boats GmbH Wallaby boats: Your partner for high performance daughter crafts, CTV, pilot boats and small workboats. URL `https://www.wallaby-boats.de/home` Accessed: March 2025

[26] BAR Technologies Ltd Foil Optimised Stability System (FOSS) URL `https://www.bartechnologies.uk/work-boats/foil-optimised-stability-system-foss/` Accessed: March 2025

[27] Mobimar Ltd WIND FARM CREW TRANSFER VESSEL URL `https://www.mobimar.com/commercial-vessels/trimaran-concept/wind-farm-crew-transfer-vessel` Accessed: March 2025

[28] Liftra ApS Denmark LT1200 – Liftra Self-Hoisting Crane: Craneless Replacement of Major Components URL `https://liftra.com/en/products/selfHoistingCrane` Accessed: March 2025

[29] Kenz Figee Group BV Up-Tower Cranes: Leading Supplier URL `https://kenzfigee.com/up-tower-cranes/` Accessed: March 2025

[30] KoalaLifter Wind Craneless Systems URL `https://koalalifter.com/` Accessed: March 2025

[31] Carroll J, McDonald A and McMillan D 2016 Failure Rate, Repair Time and Unscheduled O&M Cost Analysis of Offshore Wind Turbines: Reliability and Maintenance of Offshore Wind Turbines *Wind Energ.* **19** 1107–1119

[32] Shafiee M, Zhou Z, Mei L, Dinmohammadi F, Karama J and Flynn D 2021 Unmanned Aerial Drones for Inspection of Offshore Wind Turbines: A Mission-Critical Failure Analysis *Robotics* **10** 26

[33] Verma A S, Castro S G P, Jiang Z and Teuwen J J E 2020 Numerical Investigation of Rain Droplet Impact on Offshore Wind Turbine Blades under Different Rainfall Conditions: A Parametric Study *Composite Structures* **241** 112096

[34] SIA Aerones Engineering Robotic Wind Turbine Care Systems URL `https://aerones.com/` Accessed: March 2025

[35] BladeRobots Next Generation Automated Leading Edge Maintenance URL `https://bladerobots.com/` Accessed: March 2025

[36] BladeBUG Ltd Advanced robotics for turbine maintenance URL `https://bladebug.co.uk/` Accessed: March 2025

[37] Jiang Z, Jovan F, Moradi P, Richardson T, Bernardini S, Watson S, Weightman A and Hine D 2023 A Multirobot System for Autonomous Deployment and Recovery of a Blade Crawler for Operations and Maintenance of Offshore Wind Turbine Blades *Journal of Field Robotics* **40** 73–93

[38] Schmidt J, Vollmer L, Dörenkämper M and Stoevesandt B 2023 FOXES: Farm Optimization and eXtended Yield Evaluation Software *Journal of Open Source Software* **8** 5464 ISSN 2475-9066

[39] Faulstich S, Berkhout V, V Berkhout, Mayer J and Siebenlist D 2016 Modelling the Failure Behaviour of Wind Turbines *Journal of Physics: Conference Series (JPCS)* **749** 012019

[40] 4C Offshore Ltd Global Offshore Wind Farms Database URL `https://www.4coffshore.com/windfarms/` Accessed: March 2025

[41] Vestas Wind Systems A/S Vestas V164-8.0 MW URL `https://pdf.archiexpo.com/pdf/vestas/vestas-v164-80-mw/88087-134417.html` Accessed: March 2025

[42] Desmond C, Murphy J, Blonk L and Haans W 2016 Description of an 8 MW Reference Wind Turbine *J. Phys.: Conf. Ser.* **753** 092013

[43] Hersbach H, Bell B, Barrisford P, Horányi A, Muñoz Sabater J, Nicolas J, Peubey C, Radu R, Rozum I, Schepers D, Simmons A, Soci C, Dee D and Thépaut J N 2023 ERA5 hourly data on single levels from 1940 to present. Copernicus Climate Change Service (C3S) Climate Data Store (CDS)

[44] Baumgärtner D and Kaczenski J 2024 Innovative Benchmarking of Offshore Wind O&M Strategies via AIS Data Analysis *J. Phys.: Conf. Ser.* **2875** 012033 ISSN 1742-6596

[45] Fraunhofer Institute for Wind Energy Systems (IWES) Joint project: SOBeKo - Electricity market-oriented optimisation of operating concepts for wind farms; subproject: Future operating concepts and strategies for wind farms. Funded by BMWK. Reference number: 03EE3051A. URL `https://www.enargus.de/detail/?id=1200953` Accessed: March 2025

Journal of Physics: Conference Series **3025** (2025) 012020 doi:10.1088/1742-6596/3025/1/012020

Examining the Potential for Floating Offshore Wind Farms in Irish Waters to Achieve the 37 GW Offshore Wind Energy Goal by 2050

Krzysztof Szczepaniec[1], Fergal O'Rourke[2] and Peter Ryan[3*]

[1,2,3] Centre for Renewables and Energy at Dundalk Institute of Technology, School of Engineering, Dundalk, Ireland

*E-mail: peter.ryan@dkit.ie

Abstract. Significant legislative initiatives and ambitious targets characterise Ireland's transition to offshore renewable energy. Despite ambitious goals, the full potential of Irish waters has not been thoroughly explored. This study evaluates the capacity of Irish waters to host floating offshore wind farms, taking into account recent policy developments and strategic objectives. A comprehensive methodology was developed, incorporating technical, legislative, and industry best practices, which led to identifying an area of interest spanning 58,486 km^2. Bathymetric analysis indicates that approximately 80% of the total power capacity of 197 GW can be deployed in water depths ranging from 60 m to 200 m. This area of interest can accommodate nearly five times the most ambitious target (Target III) and almost twenty times the lowest target (Target I). The area's assessment of wind conditions and energy yield revealed exceptional potential. The gross energy yield per wind turbine ranges from 80.8 GWh/a to 88.6 GWh/a, averaging 84.8 GWh/a.

1. Introduction

Efforts to combat climate change are crucial, but energy diversification and access to affordable energy sources have also become key factors in the transition to renewable energy.

For these initiatives to succeed, legislation and policy must evolve. Like many other EU coastal states, Ireland has started legislative changes supporting the transition to offshore renewable energy (ORE).

In the end of 2022, seven offshore wind projects with a total capacity of approximately 4.5 GW have received maritime area consent [1]. The leading projects utilise fixed-bottom technology while floating offshore wind (FOW) projects are still in the early stages of development.

The latest Irish policy on ORE aims to install 20 GW by 2040 and 37 GW, or 50 GW, in the "stretch wind target" by 2050. The 27 GW out of 37 GW target and 40 GW out of 50 GW target of ORE are allocated to FOW [2,3].

When selecting locations for offshore wind farms, it is vital to consider technical factors, comply with applicable laws and policies, and mitigate potential conflicts with stakeholders and the environment. Adopting a holistic approach can reduce project risks and enhance public perception, leading to expedited project completion and lower costs.

Content from this work may be used under the terms of the Creative Commons Attribution 4.0 licence. Any further distribution of this work must maintain attribution to the author(s) and the title of the work, journal citation and DOI.
Published under licence by IOP Publishing Ltd

Nonetheless, there is a lack of diligent inventory of Irish waters that would consider a thorough review of legislation and policy and wind resource characteristics, projected energy output and seabed geophysical properties in viable deployment zones. Finally, the actual capacity of Irish waters to host floating offshore wind farms (FOWF) in the context of the ambitious policy remains unknown.

2. State-of-the-Art

The maximum depth viable for installing fixed-bottom offshore wind turbine generators (WTGs) is between 60 m and 80 m [4,5]; however, about 80% of the sites with the best wind resources are located in deeper waters [6]. Locating the wind farms in these areas will increase energy output and efficiency; however, it poses technical and commercial challenges.

Much of the existing literature on site selection for FOWF tends to focus on a specific method, a single aspect of the process, or a limited number of criteria rather than developing a holistic approach that could be adopted in real-world scenarios.

One of the notable examples of site selection for FOWF in Irish waters is a study conducted by Martinez and Iglesias [7]. The main criterion of site feasibility in this study is LCOE. The researchers also considered several factors restricting the search area, such as bathymetry, distance from shore, and some environmental aspects.

Another example of the site selection for FOWF off the Irish coast was conducted by Loughney et al. [8]. This study aims to apply the Multiple Attribute Decision Analysis (MADA) approach to identify the best site for a FOWF. In order to narrow the search area, researchers considered several factors and finally chose the area of Shannon Foynes Bay off the Galway for a detailed study.

The application of the Multi-Criteria Decision Making method (MCDM) for site selection in Irish waters was also investigated by Deveci et al. [9]. In contrast, this study focused on the rank of five of the most advanced fixed-bottom offshore wind projects; therefore, the project locations are known and fixed. The rank criteria include economic indices, chosen technical aspects and some environmental and social factors.

Another notable example is the study conducted by Diaz and Guedes Soares [10]. Researchers first developed the GIS research tool for FOWF site selection based on exclusion criteria and, in further studies [11–13], supplemented this tool with evaluation criteria and application of MCDMs for final site rank.

While there are commendable examples of research related to site selection for FOWFs, there remains an opportunity to enhance the methodology for practical implementation. Many existing studies face challenges, including limited connections to relevant legislation and policies and the need for more clearly defined objectives that align with the utilised criteria. Furthermore, incorporating tools and data sets recognised and widely used in the wind industry would enhance the methodology. Addressing these aspects could significantly improve the applicability of these methods in real-world scenarios and allow for the estimation of the real potential of the waters to host FOWF.

3. Methodology

The methodology is divided into two sections. Firstly, it presents a delineation of an area of interest (AOI) that is feasible for FOWF deployment. Secondly, assumptions are presented to estimate the AOI's power capacity and energy yield.

3.1 Area of interest

The methodology must incorporate the legislative, policy, and technical requirements and employ tools and data recognised in the wind industry. In order to define a set of criteria, a hierarchical decomposition of the main objective into sub-objectives linked to relevant legislation and policy has been proposed. This creates a robust framework and justifies consideration of each criterion's function (exclusion or evaluation).

The main objective is to evaluate the potential of Irish waters within the context of the goals of FOWF development. Eight sub-objectives representing a broad spectrum of development aspects have been proposed to achieve this primary objective. Finally, each sub-objective has been linked to policy and legislation, informing the selection of criteria. As a result, 41 criteria were identified and considered in the study. Data processing and spatial analysis were conducted using GIS software. Consequently, the final exclusion map presenting the AOI has been derived. The sub-objectives and criteria together with the reference to relevant policy and legislation, are presented in Table 1.

Table 1. Exclusion criteria, connection to legislation and policy, buffer zones and sources of data.

Sub-objectives	No.	Criteria	Policy and legislation	Data source
1. Legislative	1.1	EEZ	[14–16]	[17]
	1.2	Maritime Spatial Planning (MSP)	[18–20]	[15,21]
	1.3	OREDP and OREDP II	[15,16]	[15,16,21]
2. Offshore infrastructure and licenses	2.1	Existing and consented ORE	[19]	[22,23]
	2.2	ORE Test sites	[15,20]	[22]
	2.3	Gas and oil platforms and licenses	[15,19]	[22]
	2.4	Underwater pipelines	[15,19,20]	[21,22]
	2.5	Underwater Power Cables and Telecommunication	[15,20]	[22,24–27]
	2.6	Exploration wells	[15,19]	[28]
	2.7	Meteorological equipment, navigation buoy	[15]	[28]
3. Military zones	3.1	Military training zones	[20]	[22]
4. Navigation and fishing, ship routes	4.1	Cargo ships	[19,20]	[22]
	4.2	Passenger ferries		[22]
	4.3	Tankers		[22]
	4.4	Fishing		[22]
	4.5	Other: sailing, dredging, service, tug and tow, pleasure, high-speed craft, military		[22]
5. Sea usage and exploitation	5.1	Waste, sewage, munition, dredge	[15,20]	[22,29]
	5.2	Anchorage area	[20]	[15,30]
	5.3	Traffic Separation Scheme (TSS)	[20]	[15,30]
	5.4	Licensed aquaculture AQ.	[19]	[28]
6. Environment	6.1	Special Protection Areas (SPA)	[15,20,31]	[32]
	6.2	Special Areas of Conservation (SAC)		[32]
	6.3	Proposed and existing Natural Heritage Areas (NHA), (pNHA)		[32]
	6.4	OSPAR Declining Threatened Habitats		[22]
	6.5	Important Bird Areas (IBA)		[33]

Sub-objectives	No.	Criteria	Policy and legislation	Data source
	6.6	ObSERVE High-Density Areas		[15,34]
	6.7	Herring Spawning Grounds (HSG)	[15,20]	[28]
	6.8	Marine Protected Areas (MPAs)	[15,20]	[22]
	6.9	RAMSAR Convention	[15,20]	[22]
7. Heritage and social	7.1	UNESCO World Heritage Sites and Global Geoparks	[15,20]	[35]
	7.2	UNESCO Dublin Bay Biosphere		[36]
	7.3	Shipwrecks	[20]	[22,37]
	7.4	Blue Flag Beaches and Green Coast Beaches	LTA [20]	[22]
	7.5	Seascape and landscape	LTA	[21]
8. Technology	8.1	Bathymetry	[15]	[22]
	8.2	Sea substrate	NR	[22,37]
	8.3	Distance to shore	NR	[28]
	8.4	Wind speed	NR	[6]
	8.5	Seabed slope	NR	[22]
	8.6	Landslide susceptibility	NR	[22]
	8.7	Submarine gas and fluid emissions	NR	[22]

NR- no reference to legislation and policy, LTA- to be considered in Lower Tier Assessment

3.2 Power density, capacity of AOI and forecasted energy yield

The theoretical power density (PD) calculation is defined as capacity per square kilometre, and the estimation of energy yield in the AOI is based on the exemplary WTG IEA-15 MW offshore reference turbine [38].

The average PD of European offshore fixed-bottom wind farms is 6.1 MW/km^2 [4,39]. The PD of existing FOWF, Hywind Scotland, Hywind Tampen and Kincardine is 9.8 MW/km^2, 8.6 MW/km^2, and 12.5 MW/km^2, respectively. Due to the small-scale and pilot nature of existing FOWFs, the PD of these wind farms may not correlate with the PD of future large-scale FOWFs; therefore, it should be considered only as indicative.

In the latest Leasing Round 5 for Celtic Sea FOW conducted by Crown Estate the minimum allowed PD of FOWFs in the proposal is 4 MW/km^2 [40]. Assuming that the lowest commercially acceptable PD is 4 MW/km^2, the proposed wind farm layout has been adjusted to meet this requirement. A 10-by-10 grid of WTGs, with a total capacity of 1.5 GW, has been used as a reference site layout. In order to achieve the required minimum power density, the spacing between WTGs was set to 8.8 RD. The layout assumptions are presented in Figure 1.

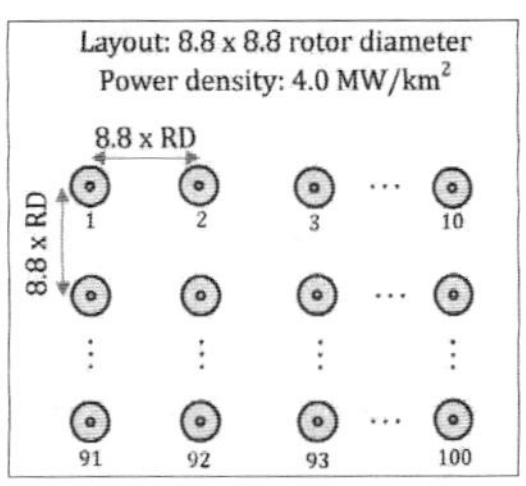

Figure 1. Layout assumption adopted in the study.

The energy yield estimation is based on the power curve of the IEA-15 MW wind turbine and the Weibull distribution of the ten-year (2008 to 2017) average annual wind speed at hub height (HH). The long-term wind data with a spatial resolution of 250 m x 250 m has been obtained from the Global Wind Atlas [6], and the vertical extrapolation to HH has been performed using the WAsP model [41].

4. Results and discussion

This section is organised into several subsections that present results obtained from executing the various steps of the methodology.

4.1 Area of interest

The resultant AOI covers 58,486 km^2, 13.7% of the initially investigated area of the Irish EEZ. The final AOI is a product of applying pertinent exclusion criteria, as detailed in Table 1. Figure 2 presents the AOI highlighting suitable locations for deploying FOWF within the case study area in chosen bathymetric ranges and visible seas' boundaries.

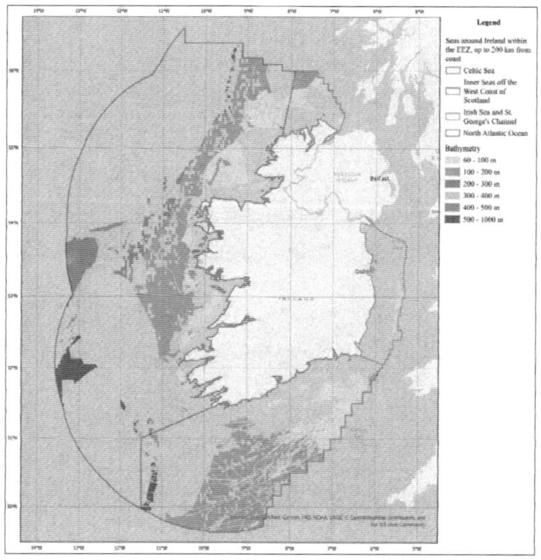

Figure 2. The final extent of the AOI, considering chosen bathymetry ranges and the boundaries of seas around Ireland, defined within the EEZ, a distance of 200 km from shore and constrained by relevant exclusion criteria.

The spatial coverage of the AOI within a particular sea is summarised in Table 2.

Table 2. The spatial coverage of the AOI within a specific sea, along with the percentage of that coverage relative to the total AOI and the total sea surface area.

Area	Area of interest [km^2]	Percentage of the AOI in a specific sea compared to the total AOI	Percentage of the AOI in a specific sea compared to the total sea surface area
AOI- All seas	58,486	100.0%	13.7%

Area	Area of interest [km²]	Percentage of the AOI in a specific sea compared to the total AOI	Percentage of the AOI in a specific sea compared to the total sea surface area
AOI- Atlantic Ocean	35,453	60.6%	10.8%
AOI- Inner Seas off the West Coast of Scotland	1,515	2.6%	26.8%
AOI- Irish Sea	395	0.7%	3.8%
AOI- Celtic Sea	21,124	36.1%	25.7%

The Atlantic Ocean comprises 60.6% of the AOI, while the Celtic Sea encompasses 36.1%. The Inner Seas off the West Coast of Scotland account for 2.6%, and only 0.7% is in the Irish Sea. The most notable reduction in available space occurs in the Irish Sea, where a mere 3.8% of the sea area within the Irish EEZ is suitable for FOWF deployment.

The bathymetric analysis indicates that 25.7% of the AOI is located at 60 m to 100 m depths, while 53.6% resides within 100 m to 200 m depths. Therefore, nearly 80% of the potential power capacity can be installed in waters up to 200 m deep.

4.2 Wind conditions and energy yield in the AOI

The long-term annual mean wind speed at a height of 100 m within the AOI ranges from 9.7 m/s to 11.0 m/s, with an average of 10.2 m/s and a median of 10.1 m/s. The highest mean wind speed, recorded at 10.4 m/s, occurs in the Atlantic Ocean, while the lowest, at 9.9 m/s, is observed in the Irish Sea. Wind speed distribution in the AOI with marked seas' boundaries in the study area is presented in Figure 3.

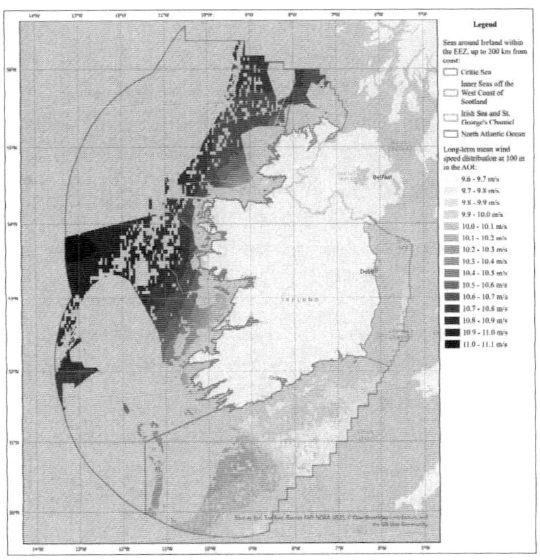

Figure 3. Long-term annual wind speed distribution in the AOI at 100 m and the boundaries of seas around Ireland, within the EEZ and 200 km distance from the coast. Based on wind data obtained from Global Wind Atlas covering the period 2008 to 2017 [6].

The region with the most abundant wind resources is located in the western part of the Atlantic Ocean, specifically between latitudes 53°N and 54°N, extending westward from longitude 13°W. In this area, the long-term mean annual wind speed at a height of 100 m reaches 11 m/s. Additionally, excellent wind conditions can be found in the Inner Seas off the West Coast of Scotland, where the mean wind speed is 10.3 m/s, with a median of 10.1 m/s, a maximum of 10.7 m/s, and a minimum of 9.9 m/s. The Celtic Sea exhibits slightly milder wind conditions compared to the Atlantic Ocean and the Inner Seas off the West Coast of Scotland.

The annual energy yield forecast has been based on long-term annual wind distribution across AOI. Results illustrate the forecasted annual gross energy yield per WTG. A summary of the energy yield characteristics in the AOI and in specific sea regions is presented in Table 3.

Table 3. Forecasted energy yield per WTG in the AOI and in specific sea regions.

Parameter	AOI All	AOI - Atlantic Ocean AOI	AOI - Inner Seas off the West Coast of Scotland	AOI - Irish Sea	AOI - Celtic Sea
AEP Mean [GWh/a]	84.8	86.2	85.7	81.5	82.8
AEP Min. [GWh/a]	80.8	82.3	83.2	80.9	80.8
AEP Max. [GWh/a]	88.6	88.6	86.6	83.7	84.8
AEP Median [GWh/a]	84.9	86.3	85.8	81.4	82.8
CF Mean	64.5%	65.6%	65.2%	62.1%	63.0%
CF Min.	61.5%	62.6%	63.3%	61.6%	61.5%
CF Max.	67.4%	67.4%	65.9%	63.7%	64.6%
CF Median	64.6%	65.7%	65.3%	62.0%	63.1%

The energy yield within the AOI ranges from 80.8 GWh/a to 88.6 GWh/a, with an average of 84.8 GWh/a. The highest mean energy yield is anticipated in the AOI of the Atlantic Ocean, reaching 86.2 GWh/a, while the lowest mean output is found in the Irish Sea at 81.5 GWh/a. Significant energy yields are also projected for the Inner Seas off the West Coast of Scotland and the Celtic Sea, averaging 85.7 GWh/a and 82.8 GWh/a, respectively. Although the Irish Sea has the lowest forecasted energy yield, it still holds considerable potential, with a mean capacity factor (CF) of 62.1%.

4.3 Potential of AOI, total installed power capacity

The AOI can host over 13,000 IEA-15 MW reference WTGs, totalling an impressive power capacity of 197 GW. The potential total installed capacity in a particular sea is presented in Table 4.

Table 4. The potential of the installed power capacity by the sea and the share of specific sea regions in the total capacity.

AOI within the sea	Installed power capacity [GW]	Percentage of installed capacity
AOI- All seas	196.6	100.0%
AOI - Atlantic Ocean	119.2	60.6%
AOI - Inner Seas off the West Coast of Scotland	5.1	2.6%
AOI - Irish Sea	1.3	0.7%
AOI - Celtic Sea	71.0	36.1%

The largest FOWF can be situated in the Atlantic Ocean and the Celtic Sea. The Irish Sea and the Inner Seas off the West Coast of Scotland can only support 3.3% of the overall installed capacity. The power capacity in chosen bathymetric ranges has been presented in Figure 4.

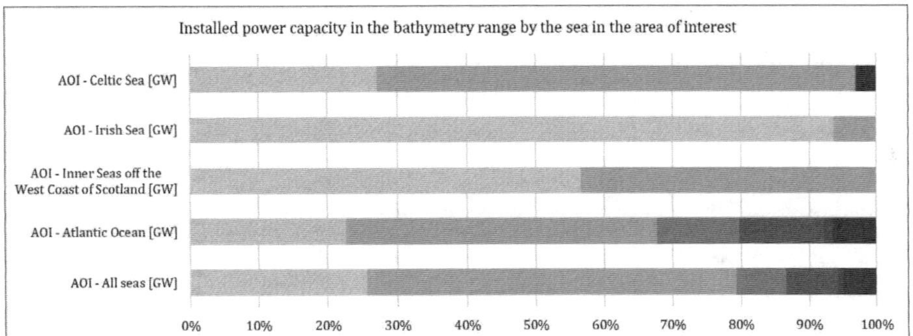

	AOI - All seas [GW]	AOI - Atlantic Ocean [GW]	AOI - Inner Seas off the West Coast of Scotland [GW]	AOI - Irish Sea [GW]	AOI - Celtic Sea [GW]
60 - 100 m	50.5	27.1	2.9	1.2	19.3
100 - 200 m	105.3	53.6	2.2	0.1	49.5
200 - 300 m	14.5	14.3	0.0	0.0	0.1
300 - 400 m	14.7	14.7	0.0	0.0	0.0
400 - 500 m	1.6	1.6	0.0	0.0	0.0
500 - 1000 m	10.0	7.9	0.0	0.0	2.1

Figure 4. Installed power capacity in bathymetry range by the sea in the AOI.

The bathymetric assessment shows that 50.5 GW of installed power capacity can be deployed in water depths ranging from 60 m to 100 m, while waters with a bathymetry range between 100 m and 200 m can accommodate 105.3 GW. However, regions with deeper waters, specifically those between 200 m and 1000 m, can only support 40.8 GW of installed power capacity.

The ambitious plans project the deployment of 20 GW of ORE by 2040, primarily to meet domestic demand, with more ambitious forecasts estimating 37 GW or even 50 GW by 2050. Approximately 10 GW out of the 20 GW target, 27 GW out of the 37 GW target, and 40 GW out of the 50 GW target will be achieved through the deployment of FOWF [2,3]. The highest targets anticipate significant energy exports to the UK and the EU.

The potential installed power capacity across the entire AOI and within specific marine regions and their bathymetry are detailed in Table 5.

Table 5. The potential of the installed power capacity by the sea and the specific bathymetry for three considered targets.

Targets	Bathymetry in AOI	AOI - All seas	AOI - Atlantic Ocean	AOI - Inner Seas off the West Coast of Scotland	AOI - Irish Sea	AOI - Celtic Sea
I - 10 GW	60 - 1000 m	1966%	1192%	51%	13%	710%
	60 - 100 m	505%	271%	29%	12%	193%
	100 - 200 m	1053%	536%	22%	1%	495%
	200 - 300 m	145%	143%	0%	0%	1%
	300 - 400 m	147%	147%	0%	0%	0%
	400 - 500 m	16%	16%	0%	0%	0%
	500 - 1000 m	100%	79%	0%	0%	21%
II - 27 GW	60 - 1000 m	728%	441%	19%	5%	263%
	60 - 100 m	187%	100%	11%	5%	71%
	100 - 200 m	390%	198%	8%	0%	183%
	200 - 300 m	54%	53%	0%	0%	0%
	300 - 400 m	54%	54%	0%	0%	0%
	400 - 500 m	6%	6%	0%	0%	0%
	500 - 1000 m	37%	29%	0%	0%	8%
III - 40 GW	60 - 1000 m	492%	298%	13%	3%	178%
	60 - 100 m	126%	68%	7%	3%	48%
	100 - 200 m	263%	134%	6%	0%	124%
	200 - 300 m	36%	36%	0%	0%	0%
	300 - 400 m	37%	37%	0%	0%	0%
	400 - 500 m	4%	4%	0%	0%	0%
	500 - 1000 m	25%	20%	0%	0%	5%

The AOI has the capacity to accommodate nearly five times the most ambitious target (Target III) and almost twenty times the lowest target (Target I). Targets I and II can be realised by deploying FOWF within the Atlantic Ocean at depths ranging from 60 m to 100 m. By extending operations to depths of up to 200 m, all three targets can be successfully attained within these waters. In the Celtic Sea, Target I is achievable within the 60 m to 100 m bathymetric range; however, by expanding the operational depth to 200 m, all three targets can also be met. Furthermore, there is considerable potential in the Inner Seas off the West Coast of Scotland, where approximately half of Target I can be realised through deployment in bathymetry ranging from 60 m to 200 m. In contrast, the Irish Sea offers the least potential for installed power capacity, with only 3.3% to 13.3% of the respective targets being attainable; nonetheless, nearly all deployments can occur at depths between 60 m and 100 m.

5. Conclusions

This study addresses the research question regarding the realistic potential of Irish waters to accommodate FOWFs, considering recent policy developments and the country's strategic ambitions.

The findings reveal that the AOI constitutes 13.7% of the Irish EEZ. Notably, approximately 60% of the AOI is situated in the Atlantic Ocean, while the AOI in the Irish Sea accounts for only 0.7% of the total area. Approximately 80% of the AOI lies in relatively shallow waters with a bathymetry range of 60 m to 200 m.

The primary factors contributing to the reduction in the available area include bathymetry, distance from the coast, environmental constraints, shipping routes, and fishing activities.

The annual average wind speed in AOI exceeds 7 m/s at a height of 100 m, which is considered the minimum required annual average wind speed to ensure the project's financial viability [4,15].

The gross annual energy yield in this area ranges from 80.9 GWh/a to 88.6 GWh/a, with a mean value of 84.8 GWh/a. This translates to a mean CF of 64.5%, with minimum and maximum values of 61.5% and 67.4%, respectively. For context, the world's top-performing FOWF, Hywind Scotland, achieved a CF of 54% over five years of operation [42]. The average CF for offshore wind farms in the UK is approximately 40% [43].

The AOI can sustain a significant number of WTGs, amounting to a total installed power capacity of 197 GW, more than 16 times the total installed power capacity of all conventional and renewable energy sources operating in Ireland in 2022 [44]. The Atlantic Ocean can support up to 119.2 GW of installed power capacity, while the Celtic Sea has the potential for 71 GW. In contrast, the Irish Sea and the Inner Seas off the West Coast of Scotland can each accommodate 1.3 GW and 5.1 GW, respectively.

The most ambitious target of deploying 40 GW of FOWFs can be met by utilising relatively shallow waters with a bathymetry of 60 m to 100 m.

In the context of surplus energy export, the most favourable AOI appears to be located in the Inner Seas off the West Coast of Scotland. This region benefits from exceptional wind conditions, advantageous bathymetry, and proximity to Northern Ireland and Scotland. The Celtic Sea also holds significant potential for energy export, particularly with the possibility of interconnections to the UK and France. Additionally, the AOI in the Irish Sea is valuable from a commercial perspective due to its closeness to the high-energy demand area of Dublin, along with its favourable bathymetry. The AOI of the Atlantic Ocean offers the most significant potential because of its vast area and highest energy yield. Nonetheless, risks related to challenging sea conditions may not compensate for energy gain. Given the considerable distance to neighbouring countries, FOWF in this area is anticipated to primarily cater to domestic demand, with any surplus energy being exported through interconnections and green hydrogen initiatives.

References

[1] Department of the Environment Climate and Communications, Minister Ryan grants Ireland's first Maritime Area Consents for Offshore Wind Energy projects, (2023). https://www.gov.ie/en/press-release/976eb-minister-ryan-grants-irelands-first-maritime-area-consents-for-offshore-wind-energy-projects/ (accessed August 27, 2024).

[2] Sustainable Energy Authority of Ireland, Offshore Renewable Energy Technology Roadmap Executive Summary, 2024.

[3] Department of the Environment Climate and Communications, The Future Framework Policy Statement, 2024.

[4] P. Beiter, W. Musial, A. Smith, L. Kilcher, R. Damiani, M. Maness, S. Sirnivas, T. Stehly, V. Gevorgian, M. Mooney, G. Scott, A Spatial-Economic Cost-Reduction Pathway Analysis for U.S. Offshore Wind Energy Development from 2015-2030, 2016. www.nrel.gov/publications.

[5] B. Cerfontaine, D. White, K. Kwa, S. Gourvenec, J. Knappett, M. Brown, Anchor geotechnics for floating offshore wind: Current technologies and future innovations, Ocean Engineering 279 (2023). https://doi.org/10.1016/j.oceaneng.2023.114327.

[6] Technical University of Denmark and the World Bank Group, Global Wind Atlas, (2023). https://globalwindatlas.info/en/about/introduction (accessed September 19, 2023).

[7] A. Martinez, G. Iglesias, Site selection of floating offshore wind through the levelised cost of energy: A case study in Ireland, Energy Convers Manag 266 (2022). https://doi.org/10.1016/J.ENCONMAN.2022.115802.

[8] S. Loughney, J. Wang, M. Bashir, M. Armin, Y. Yang, Application of a multiple-attribute decision-analysis methodology for site selection of floating offshore wind farms off the West coast of Ireland, in: Developments in Renewable Energies Offshore - Proceedings the 4th International Conference on Renewable Energies Offshore, RENEW 2020, CRC Press/Balkema, 2021: pp. 389–398. https://doi.org/10.1201/9781003134572-45.

[9] M. Deveci, U. Cali, S. Kucuksari, N. Erdogan, Interval type-2 fuzzy sets based multi-criteria decision-making model for offshore wind farm development in Ireland, Energy 198 (2020). https://doi.org/10.1016/j.energy.2020.117317.

[10] H. Díaz, C. Guedes Soares, An integrated GIS approach for site selection of floating offshore wind farms in the Atlantic continental European coastline, Renewable and Sustainable Energy Reviews 134 (2020). https://doi.org/10.1016/j.rser.2020.110328.

[11] H. Díaz, C. Guedes Soares, A novel multi-criteria decision-making model to evaluate floating wind farm locations, Renew Energy 185 (2022) 431–454. https://doi.org/10.1016/j.renene.2021.12.014.

[12] H. Díaz, A.P. Teixeira, C. Guedes Soares, Application of Monte Carlo and Fuzzy Analytic Hierarchy Processes for ranking floating wind farm locations, Ocean Engineering 245 (2022). https://doi.org/10.1016/j.oceaneng.2021.110453.

[13] H. Díaz, S. Loughney, J. Wang, C. Guedes Soares, Comparison of multicriteria analysis techniques for decision making on floating offshore wind farms site selection, Ocean Engineering 248 (2022). https://doi.org/10.1016/j.oceaneng.2022.110751.

[14] United Nations Convention, United Nations Convention on the Law of the Sea, (1982).

[15] Department of the Environment Climate and Communications, DRAFT Offshore Renewable Energy Development Plan II, 2023.

[16] Government of Ireland, Offshore Renewable Energy Development Plan, 2014.

[17] Flanders Marine Institute, Marine Regions, (2024). https://www.marineregions.org/ (accessed February 26, 2024).

[18] European Parliament and of the Council, Directive 2014/89/EU establishing a framework for maritime spatial planning, 2014.

[19] Government of Ireland, Maritime Area Planning Act 2021, (2021).

[20] Government of Ireland, National Marine Planning Framework, 2021.

[21] Marine Institute, Ireland's Marine Atlas, (2016). https://atlas.marine.ie/ (accessed September 19, 2023).

[22] European Marine Observation and Data Network, European Marine Observation and Data Network (EMODnet), (2024). https://emodnet.ec.europa.eu/en (accessed February 26, 2024).

[23] Offshore Market Intelligence & Marine Cable Consulting, Offshore Market Intelligence & Marine Cable Consulting | 4C Offshore, (2024). https://www.4coffshore.com/ (accessed March 6, 2024).

[24] Greenlink Interconnector, Greenlink Interconnector | energy infrastructure | Ireland and Wales, (2024). https://www.greenlink.ie/ (accessed February 26, 2024).

[25] The Kingfisher Information Service – Offshore Renewable & Cable Awareness project, Homepage | KIS-ORCA, (2019). https://kis-orca.org/ (accessed February 26, 2024).

[26] EirGrid, Interconnection | Industry | EirGrid, (2024). https://www.eirgrid.ie/industry/interconnection (accessed February 26, 2024).

[27] EirGrid, Celtic Interconnector, la première liaison France-Irlande, (2020). https://www.celticinterconnector.eu/ (accessed February 26, 2024).

[28] Marine Institute, Marine Institute, (2022). https://www.marine.ie/ (accessed February 26, 2024).

[29] Environmental Protection Agency, Environmental Protection Agency, Ireland (EPA) Geoportal, (2024). https://gis.epa.ie/Home (accessed February 26, 2024).

[30] UK Hydrographic, UK Hydrographic Office - GOV.UK, (2024). https://www.gov.uk/government/organisations/uk-hydrographic-office (accessed February 26, 2024).

[31] Department of Housing Local Government and Heritage, National Marine Planning Framework Strategic Environmental Assessment (SEA) Statement, 2021.

[32] National Parks & Wildlife Service, National Parks & Wildlife Service, (2024). https://www.npws.ie/ (accessed February 26, 2024).

[33] BirdLife International, BirdLife International - BirdLife is the world leader in Bird Conservation, (2024). https://www.birdlife.org/ (accessed February 26, 2024).

[34] Department of the Environment Climate and Communications, gov - ObSERVE Programme, (2024). https://www.gov.ie/en/publication/12374-observe-programme/ (accessed July 25, 2024).

[35] United Nations Educational Scientific and Cultural Organization (UNESCO), Ireland - UNESCO World Heritage Convention, (2024). https://whc.unesco.org/en/statesparties/ie (accessed February 26, 2024).

[36] Dublin Bay Biosphere, Home | Dublin Bay Biosphere, (2024). https://www.dublinbaybiosphere.ie/ (accessed February 26, 2024).

[37] Infomar, Home | Infomar, (2024). https://www.infomar.ie/ (accessed February 26, 2024).

[38] E. Gaertner, J. Rinker, L. Sethuraman, F. Zahle, B. Anderson, G. Barter, N. Abbas, F. Meng, P. Bortolotti, W. Skrzypinski, G. Scott, R. Feil, H. Bredmose, K. Dykes, M. Shields, C. Allen, A. Viselli, Definition of the IEA Wind 15-Megawatt Offshore Reference Wind Turbine Technical Report, 2020. www.nrel.gov/publications.

[39] P. Enevoldsen, M.Z. Jacobson, Data investigation of installed and output power densities of onshore and offshore wind turbines worldwide, Energy for Sustainable Development 60 (2021) 40–51. https://doi.org/10.1016/j.esd.2020.11.004.

[40] V. Joncheray, Information Memorandum Celtic Sea Floating Offshore Wind, 2023.

[41] DTU Wind and Energy Systems, WAsP - WAsP, (2024). https://www.wasp.dk/software/wasp (accessed February 28, 2024).

[42] Equinor, Equinor leads the global floating offshore wind industry, (2025). https://www.equinor.co.kr/en/news/equinor-leads-the-global-floating-offshore-wind-industry (accessed February 4, 2025).

[43] Equinor, Hywind Scotland remains the UK's best performing offshore wind farm - equinor.com, (2024). https://www.equinor.com/news/archive/20210323-hywind-scotland-uk-best-performing-offshore-wind-farm (accessed September 15, 2024).

[44] EirGrid, Ireland Capacity Outlook 2022-2031, 2022.

doi:10.1088/1742-6596/3025/1/012021

Field investigation into mooring Load Reduction Device performance in hurricane conditions at an offshore test site

Tom Doyle[1], **Danny Golden**[1], **Syed Hasan**[1,2]

[1]Dublin Offshore Technologies Ltd, Dublin, Ireland
[2]University College Dublin, Dublin, Ireland

E-mail: tom.doyle@dublinoffshore.ie

Abstract. Floating Offshore Wind (FOW) is the next frontier in clean and renewable wind energy exploitation of particular significance to the European economy. To keep the FOW competitive against other resources, the Levelised Cost of Energy (LCoE) should keep up with projections in the next decade for the commercial floating wind farms to be economically viable. One of the contributors to LCoE is the cost of the mooring system for the Floating Offshore Wind Turbine (FOWT). This paper presents a field investigation of a quarter scale Mechanical Compliance Device (MCD) added to an inline heavy chain mooring attached to a test platform. The MCD (i.e. Dublin Offshore's Load Reduction Device aka LRD) is aimed at peak load mitigation in the mooring line of an FOWT, while allowing for geometric compliance during operational conditions. This field campaign happened in Q4 2020 at Smart Bay Test Site (SBTS), Ireland. The test platform survived Hurricane Epsilon and Storm Aiden during the test period. The tests validated the Load Extension Curve of the selected MCD, demonstrating its effectiveness in mitigating peak loads. These tests along with prior scaled tank testing show that the LRD is a prime candidate for reducing the demand for a high Minimum Breaking Load for the mooring line. This leads to substantial reduction in the line specification, thus, keeping capital as well as operating costs down. Moreover, the construction of LRD is simple, robust and comprised of easy to source materials.

1 Introduction

Floating Offshore Wind (FOW) has been forecasted to achieve tremendous growth in installed capacity from 121MW (2022) to reaching 264GW by 2050. The driving factor for this growth is due to more consistent and stronger winds available offshore with about 80% of offshore wind resource in water depths deeper that 60m [8]. This is evidenced by the fact that high load factor ($> 65\%$) have been achieved for pre-commercial wind farms [16]. Commercial farms ($> 200MW$ capacity) are forecasted to be online by 2030.

The Levelised Cost of Energy (LCoE) for FOW is expected to close the gap with onshore wind, plummeting from less than €75/MWh in 2030 to below €50/MWh by 2050 [6]. The fall in LCoE is due to numerous factors [8, 16] like

- Innovations in FOWT platform design, materials and manufacturing.
- Availability of appropriate port and vessel lift infrastructure for installation and maintenance.

Content from this work may be used under the terms of the Creative Commons Attribution 4.0 licence. Any further distribution of this work must maintain attribution to the author(s) and the title of the work, journal citation and DOI.
Published under licence by IOP Publishing Ltd

- High Voltage dynamic cables for export
- Increasing efficiency of installation, operation and maintenance techniques
- Better economies of scale with larger wind farms

Amongst the many technical challenges that need to be surmounted for achieving these LCoE projections, is the reduction in the cost of station keeping of FOWTs. The mooring system of an offshore floating platform is the subsystem required for station keeping. Concepts and standards from Floating Oil & Gas (O&G) platforms have been adopted to FOWTs. However, since the energy resource for O&G exploitation is different, several differences in functional aims emerge.

Unlike Floating O&G structures, an offshore FOWT is designed to exploit the wind forces head on rather than avoiding it. Even in operating conditions, large windage area of the turbines impose tremendous load at large heights. E.g., the reference IEA 15MW turbine imposes close to 250T of maximum rotor thrust at the Rotor and Nacelle Assembly (RNA) situated about 150m from the still waterline [10]. Additionally, a FOWT platform should avoid resonance due to 1P and 3P frequencies of the wind turbine. Rotor control strategies have also been adopted by wind turbine manufacturer as negative damping is experienced when the turbine runs above rated load. The station keeping for dynamic export cables is more sensitive to excursion, especially in shallow waters for a FOWT than say a flexible riser for an O&G platform. In general, compared with O&G, a FOWT has a smaller payload with a higher center of gravity leading to stability issues [8]. The FOWT platform and associated mooring system are designed to compensate all of these demands.

The sizing of the mooring lines for an FOWT is largely a function of peak loads on the mooring system during storm survival conditions. Typically, this means high stiffness properties to counter higher loads at lower excursion. However, a FOWT will spend most of its design life under revenue generating operating conditions. High stiffness would imply higher loads during operating conditions as well, adversely affecting fatigue life of the mooring line. It is, therefore, desirable to have sufficient compliance in the mooring system for operational conditions while allowing some sort of peak load mitigation during survival conditions within the limits of station keeping excursions [17]. Compliance, herein, is defined as ability of the platform to follow the wave orbital motions at sea surface.

To avoid the development of entirely new types of mooring lines, readily available solutions from O&G can be adopted with modifications. The requirements for FOWT platform moorings have guided development of innovative solutions acting as an inline ancillary to a mooring line. Passive and active solutions are being developed [2, 5, 12, 13, 14]. In this paper, we discuss field testing of one of these ancillaries i.e. Dublin Offshore's Load Reduction Device (LRD). The main advantage of LRD is its simple, cheap, robust and proven construction. Also, it does not require specialised installation aids or maintenance equipment.

1.1 Operating Principle
LRD has been described as a "ballasted pendulum" [9] consisting of a buoyancy chamber counteracted by a solid ballast and integrated into mooring line (Figure 1a). The LRD is connected inline with the mooring via two arms hinged to LRD body transferring mooring tension to it. The working principle of the LRD is based on providing a restoring moment (due to the offset between the buoyancy and the ballast) in opposition to the overturning moment created by the mooring tension as LRD rotates. This arrangement results in a bespoke 3-phase Load Extension Curve (LEC) with a soft stiffness response in Phase 2 that shifts the peak loads in the mooring to larger extensions compared with the baseline mooring (Figure 1b). As compliance is exhausted for out of design survival conditions, further excursion is met with highly stiff Phase 3 properties keeping the FOWT within station keeping limits. The offset between the hinge points of LRD provides the geometric compliance as seen in a typical LEC. Extensive numerical modeling and tank testing (scale 1:60) has already been published to design and validate the behaviour of LRD [7, 11].

1.2 Research Objectives
To the best knowledge of authors, the described field test presents the first instance of a successful deployment of a Mechanical Compliance Device (MCD) for mooring in the sea. No wind turbines were deployed on the test platform (in this case a Data Buoy) due to limitations of the test platform. However, the field tests remain of value to the mooring system and LRD dynamics. LRD design was slightly undersized relative to field requirements to allow validation of Phase 3 LEC in storm survival conditions. The objectives of the field test for LRD were as follows:

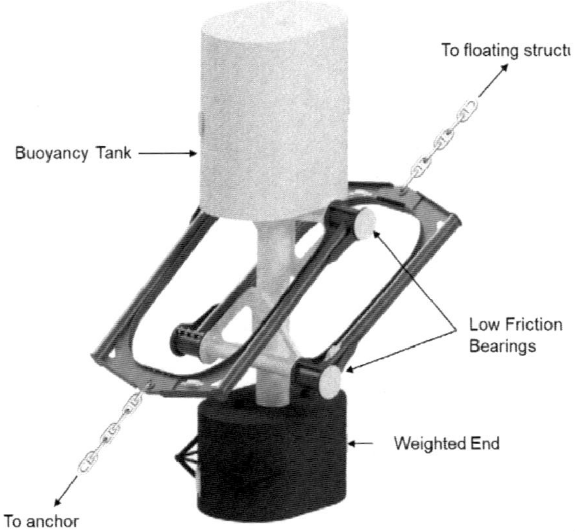

(a) A prototype LRD Version 2

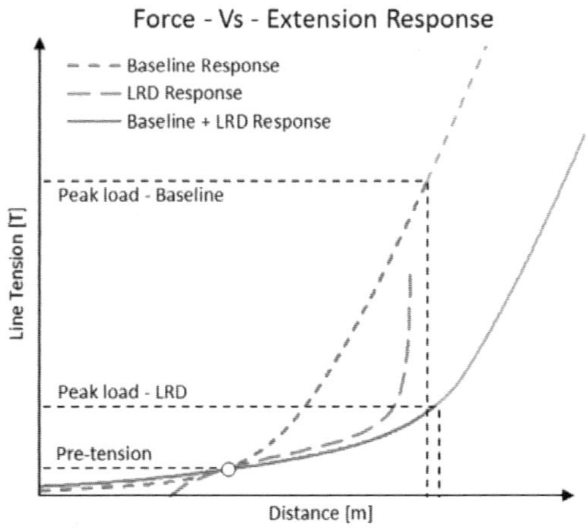

(b) Figurative load extension curve demonstrating LRD effectiveness

Figure 1

1. Verify the scalability of the LRD
2. Validate LRD operational performance
3. Verify numerical models used in design

2 Field setup

The location for the field test was chosen to be the Smart Bay Test Site (SBTS) in the Galway Bay during Q4 2020. SBTS is an equipped observatory located 1.5km offshore with a minimum depth of 20m and is suitable for quarter scale tests (Figure 2a). The metocean conditions at SBTS are mainly oriented along North East-South West axis, approximately. The western side is exposed to the Atlantic Ocean contributing to swells while localised wave climate is a function of fetch-limited wind, coastal geography and bathymetry. Figure 2b shows the 10-year mean annual scatter table. Extensive site characterisation studies of SBTS can also be found in [1, 3, 4]

The test setup consisted of the a two-point moored Data Buoy (DB), an adjacent moored wave buoy (DWR4) and a subsea observatory along with an optical/electrical backhaul to a shore station. The DB acted as a test platform and was instrumented for providing its location, wind conditions and pressure. The anchor positions of the DB mooring were marked by float buoys for identification and recovery. DWR4 relayed the sea state condition in real time to the shore station. Details of the test equipment is provided in Table 2.

2.1 Mooring Layout & Instrumentation

Figure 4 shows the mooring arrangement at the site for the test platform (i.e. the Data Buoy). The mooring arrangement consisted of two catenary legs oriented along the NE-SW axis. The western leg of the mooring line (henceforth, called Primary Line) was oriented weatherward while eastern leg (henceforth, called Back Line) was leeward. The anchor positions on each leg were recorded post-deployment and prior to recovery. The Primary Line was anchored via a drag embedment anchor while the Back Line was supported by sinkers. The Primary Line was integrated with an instrumented quarter scale LRD. Accelerometers were attached to the LRD at four locations. Two of these were attached to upper and lower parts of main LRD body between the buoyancy and ballast with aim of providing LRD orientations. The other two accelerometers were attached to the arms of LRD providing the declination of the western mooring leg.

(a) Illustrative layout of test site and its bathymetry (Courtesy: Marine Institute)

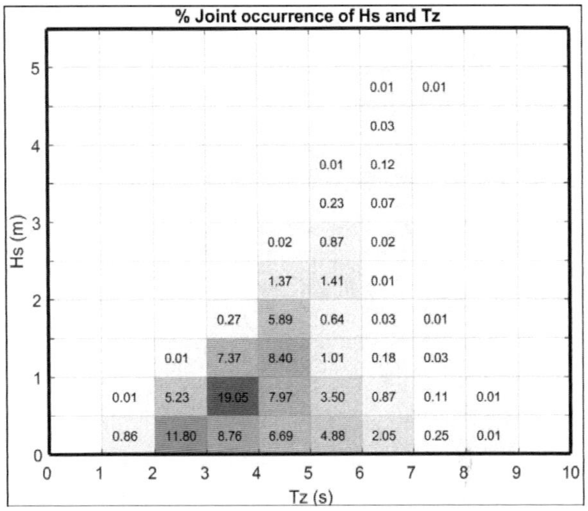

(b) 10-year mean scatter (2006-2015) at Smart Bay Test Site (SBTS) [15]

Figure 2

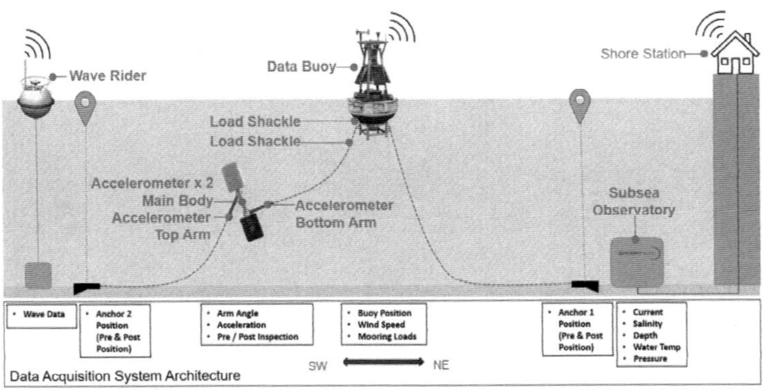

Figure 3: Field set up and data acquisition at SBTS

Data Set	Data Source	Sampling
Wind Speed	Data Buoy	1Hz measurement, output averaged over 10mins
Wind Direction		
Relative Pressure		
Wave Height	DWR4	Logged every 30min
Wave Period		
Wave Direction		
Current Speed	Subsea Observatory	10min time series
Current Direction		
Water Depth		
Water Temperature		Average hourly temperature time series

Table 1: Environmental data acquisition details

Load shackles on top part of the Primary Line provided an interface to record the line tension at the upper part of the mooring leg. Table 3 provides the details of the mooring arrangement. The submerged load shackles were of SHK-B Bow type Cable Load Shackle rated to 12Te.

2.2 Data Acquisition

Table 1 shows the details of environmental data acquisition telemetry during the test campaign. The DWR4 buoy was programmed to transmit the statistical wave parameters every 30 minutes. Additionally, the raw time series for horizontal and vertical displacement at 128Hz was also harvested from the DWR4 taking advantage of the proximity of DWR4 to the backhaul of the subsea observatory.

The data acquisition related to GPS positioning of DB(i.e. the test platform) and load cells for mooring line tension were also done remotely. The DB position was estimated to be accurate within 2.5m. The line tensions were sampled at 10 Hz. The GPS position of the anchors and sinkers were noted after installation and recovery. The three-axis accelerometer data for each of the four accelerometers attached to LRD was sampled at 12Hz and stored locally for post recovery processing (See Figure 4). Careful calibration and time alignment was undertaken for the load shackles and accelerometers before deployment.

Equipment	Model	Specification	Value
Data Buoy (Test Platform)	Mobilis DB8000	Dry Weight	3Te
		Buoyancy	8Te
		Max Diameter	3m
		Height	7.38m
		Draft	2.27m
Wave Buoy	Datawell Directional Waverider 4	Dry Weight	~ 0.35Te
		Diameter	~ 0.9m
Mechanical Compliance Device	LRD Version 2	Scale	$\frac{1}{4}$
		Dry Weight	~ 3Te
		Height	~ 4m
		Offset between hinges	~ 1.27m
		Safe Working Load (SWL)	5Te
		Minimum Breaking Load (MBL)	≥ 20Te
		Rated Water Depth (operation)	20m

Table 2: Test equipment details

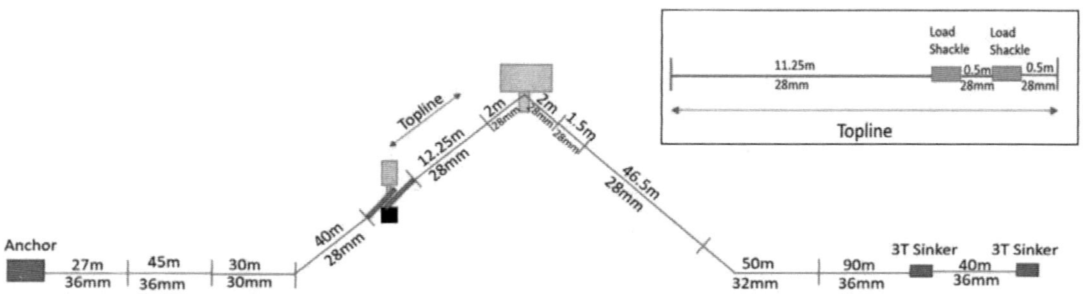

Figure 4: Mooring general arrangement at SBTS

Mooring Leg	Line Type	Length
Primary Line (Western)	Data Buoy	
	28mm Bridle	2 m
	28mm Chain	1.5 m
	LRD	
	32mm Studless Chain (incl. load shackles)	10.75 m
	28mm Studless Chain	40 m
	30mm Studless Chain	30 m
	36mm Studless Chain	70 m
	Drag Embedment Anchor	
Back Line (Eastern)	Data Buoy	
	28mm Bridle	2 m
	28mm Chain	1.5 m
	28mm Studless Chain	46.5 m
	32mm Studless Chain	50 m
	36mm Studless Chain	90 m
	3Te Sinker	
	36mm Studless Chain	40 m
	3Te Sinker	

Table 3: Layout and specification of the mooring system

(a) Installation of LRD at SBTS

(b) Underwater snapshot of deployed LRD

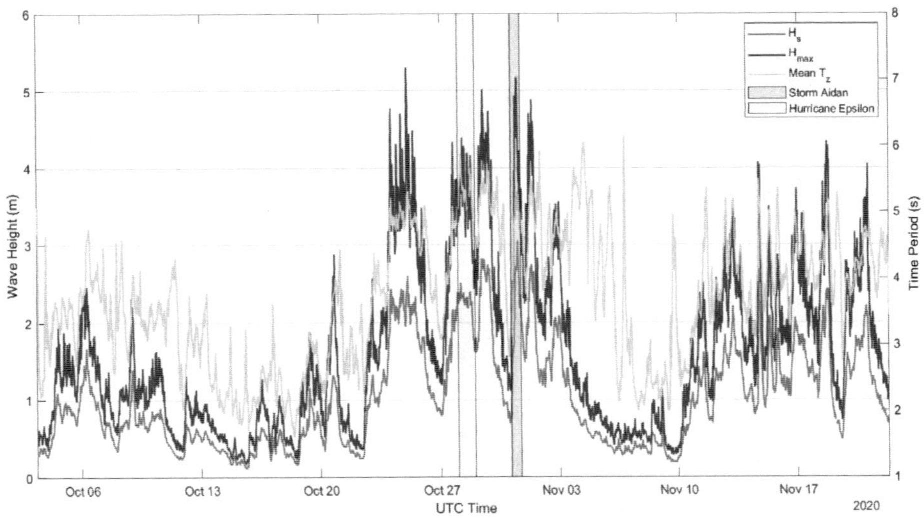

(c) In-situ wave climate during the test period

Figure 5

2.3 LRD Installation & Recovery

The LRD-integrated mooring system was successfully installed in October 2020 (Figure 5a and 5b). The installation period was intended to last 1000 hours however due to adverse weather conditions the system remained on site for 1200 hours, until the next available weather window. The test period coincided with Hurricane Epsilon and Storm Aiden in addition to periods of relative calm resulting in sea states across a broad spectrum being observed. A maximum significant wave height (H_s) of 3.05m was recorded, corresponding to a maximum wave height (H_{max}) of approximately 5.5m (Figure 5c). This equates to a 22m wave at full scale and therefore represented a true test of the mechanical system. The mechanical integrity of the system was successfully demonstrated with no damage or degradation evident in test data or during the post recovery mechanical inspection. Field test data was successfully recovered with real time data available from the load shackles and wave rider. Accelerometer data was recovered from standalone sensor pods at the end of the test period.

Visual inspection of the LRD was carried out following its recovery (Figure 6). The results of the visual inspection indicated that the LRD has behaved as intended. No damage or degradation of performance was evident over the test duration. The inspection is summarised in Table 4.

Figure 6: LRD post-recovery

Item	Inspection Results
Mooring Integrity	No failures observed in the chain, connectors, shackles or LRD
LRD structural Integrity	No signs of yielding, cracking or overload. No signs of water ingress in buoyancy tank.
LRD mechanical Damage	No evident damage to fixings, bearings or fenders Negligible bearing wear. Mounting of accelerometers was still effective.
Corrosion & Anodes	No degradation to coating with a minor crack, later determined to be related to coating only (unrelated to LRD body)
Marine Growth	Soft growth observed on most surfaces of LRD

Table 4: Results of post deployment mechanical inspection

3 Results & Discussion

The acceleration of LRD arms and body was used to determine the orientation of LRD arms and body relative to the true vertical as described in [11]. The principle assumption is that LRD behaves like a pendulum in the plane of the mooring line. Figure 7a shows a 1-minute sample of the correlation between LRD arms and its body.

3.1 Load Extension Validation

A time series (1-minute sample) comparison of the recorded line tension and the LRD extension is presented for October, shown in Figure 7b. Looking at the time series it is evident that there is correlation between high levels of line tension and high levels of LRD extension.

One of the objectives of the field campaign was to provide validation to the LEC for the LRD. Hence, the force and extension data were split in to 1-minute bins for the duration of the deployment. For each bin the maximum and minimum force and extension are recorded. The maxima and minima are then plotted as shown in Figure 7c demonstrating clear alignment with the predicted LEC.

3.2 Numerical Validation

The as-built design numerical model in Orcaflex was fed with 128Hz raw displacement time series from DWR4 wave buoy. The predicted peak load in the mooring line was then compared with as-measured load shackle data at the test site (Figure 8). Strong correlation between the two data sets is evident from the comparison and therefore, also demonstrates high fidelity in the numerical model in predicting peak load events.

4 Conclusion

This paper presents the results of 1:4 scale offshore testing of Dublin Offshore's LRD deployed for 1200 hours in October-November 2020. The collected load and acceleration data was used to characterise the behaviour of the LRD and verify that the device can be effectively scaled up from the 1:60 scale models previously assessed during tank testing campaigns. The deployment duration coincided with Hurricane Epsilon and Storm Aiden resulting in a highly valuable dataset across normal operating and survival conditions. The acceleration data was used to determine the LRD orientation and the mooring line declination angle thereby allowing the LRD extension to be calculated. The results of testing show good agreement between the LRD quasi-static stiffness response curve and the experimental measurements and similarly show good agreement between the mooring behaviour predicted in the numerical models in comparison with the measured performance across a range of sea states. It is concluded that, in addition to demonstrating the mechanical survivability of the design, the 1:4 scale deployment has demonstrated that the LRD can be scaled up and transferred into real world open ocean conditions with no loss

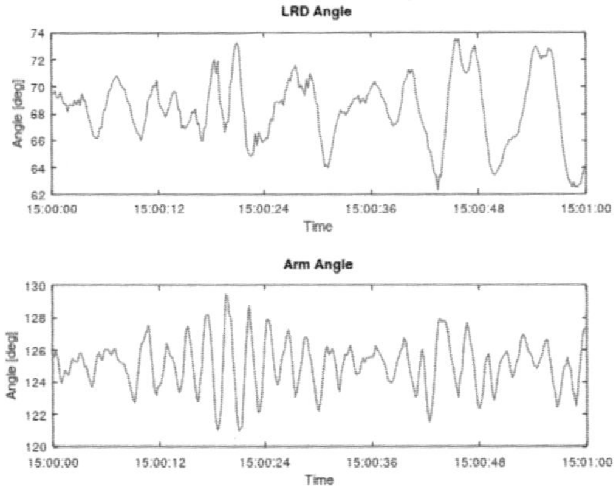

(a) LRD body orientation compared to the arm angle for a 1-minute sample in October, 2020

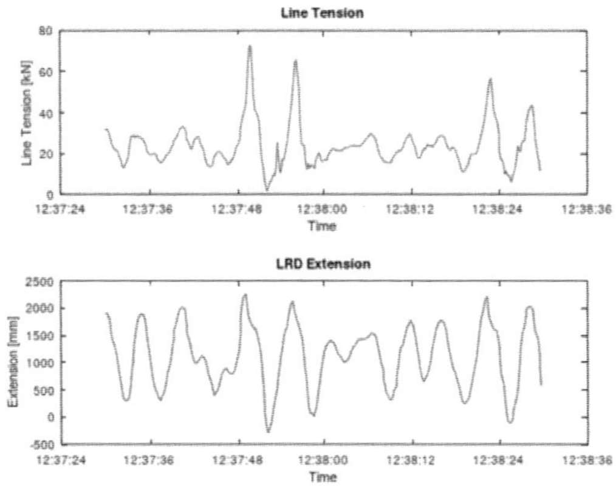

(b) Upper line tension compared to the calculated LRD extension for a 1-minute sample on October 29, 2020

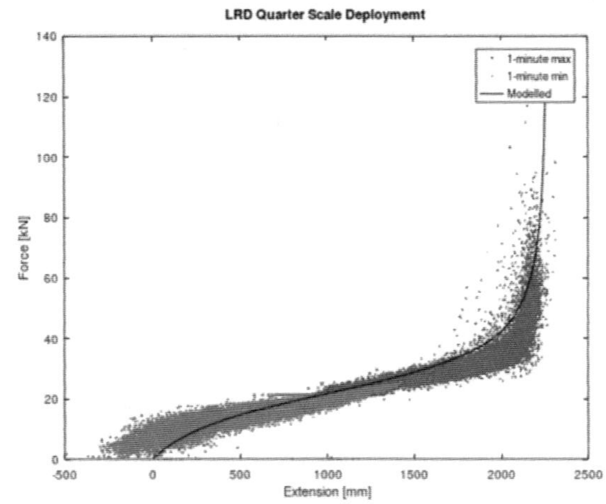

(c) Correlation of test data with the design LRD Load Extension Curve

Figure 7

(a) Numerical modelling of the test site in Orcaflex

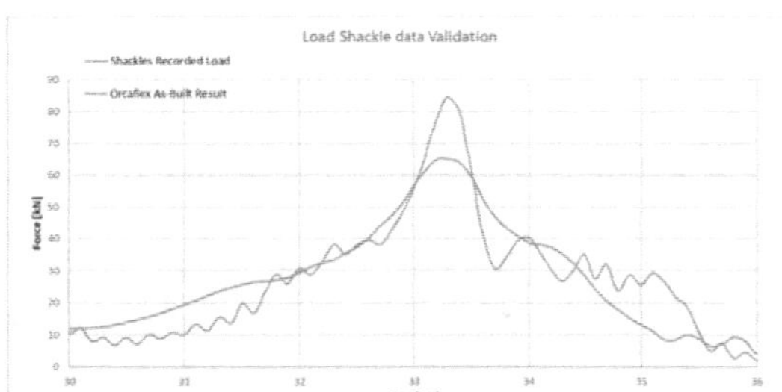

(b) Validation of shackle load with the Orcaflex model

Figure 8

of performance and is therefore a viable alternative mooring architecture for station keeping of FOW platforms. Further campaigns in deeper waters are planned in near future.

5 Acknowledgments

The design and manufacture of LRD was supported by the Floating Wind Technology Accelerator Competition (FLWTAC) which was funded by the Scottish Government and managed by the Carbon Trust's Floating Wind Joint Industry Project (FLW JIP). The field campaign described here received support from Ocean DEMO, an Interreg NWE Project. This publication has emanated from research supported in part by a grant from Research Ireland under grant number 23/IRDIFB/12108. For the purpose of Open Access, a CC BY public copyright licence is applied to any Author Accepted Manuscript version arising from this submission.

References

[1] Iulia Anton and Stephen Nash. *Wave Resource Characterisation at the Galway Bay Marine and Renewable Energy Test Site.* Tech. rep. National University of Ireland, Galway, 2020, pp. 1–107.

[2] Iwan Aryawan et al. "Potential Mooring System Optimization Using Polymer Spring Component – Floating Offshore Wind Turbine Application". In: *Proceedings of the Annual Offshore Technology Conference* 2023-May (2023). ISSN: 01603663. DOI: 10.4043/32410-MS.

[3] Reduan Atan et al. "Assessment of wave characteristics and resource variability at a 1/4-scale wave energy test site in Galway Bay using waverider and high frequency radar (CODAR) data". In: *Ocean Engineering* 117 (2016), pp. 272–291. ISSN: 00298018. DOI: 10.1016/j.oceaneng.2016.03.051.

[4] B. Cahill. *Resource Characterisation of the Galway Bay 1/4 Scale Wave Energy Test Site.* Tech. rep. University College Cork, 2012, pp. 1–74.

[5] P. Casaubieilh et al. "Performance improvements of mooring systems for wave energy converters". In: *Renewable Energies Offshore - 1st International Conference on Renewable Energies Offshore, RENEW 2014* (2015). Seaflex, pp. 897–903. DOI: 10.1201/b18973-126.

[6] Det Norske Veritas. *Floating Wind: The Power to Commercialize.* Dec. 2020. URL: https://www.dnv.com/focus-areas/floating-offshore-wind/commercialize-floating-wind-report/.

[7] T. Doyle, S. A. Hasan, and D. Golden. "Tank testing of Load Reduction Device for Floating Offshore Wind Moorings". In: *Proceedings of Offshore Technology Conference 2025.* OnePetro, Submitted.

[8] Emma C. Edwards et al. "Evolution of floating offshore wind platforms: A review of at-sea devices". In: *Renewable and Sustainable Energy Reviews* 183 (May 2023), p. 113416. ISSN: 18790690. DOI: 10.1016/j.rser.2023.113416.

[9] Oscar Festa, Susan Gourvenec, and Adam Sobey. "Comparative analysis of load reduction device stiffness curves for floating offshore wind moorings". In: *Ocean Engineering* 298 (Apr. 2024). ISSN: 00298018. DOI: 10.1016/j.oceaneng.2024.117266.

[10] Evan Gaertner et al. *Definition of the IEA 15 MW Offshore Reference Wind Turbine.* 2020, pp. 1–44. URL: https://www.nrel.gov/docs/fy20osti/75698.pdf.

[11] Danny Golden et al. "An investigation into the use of load reduction devices integrated with high-modulus synthetic mooring lines for semi-submersible station keeping". In: *Journal of Ocean Engineering and Marine Energy* (Apr. 2024). ISSN: 2198-6444. DOI: 10.1007/s40722-024-00323-3.

[12] Tessa Gordelier et al. "A novel mooring tether for highly-dynamic offshore applications; mitigating peak and fatigue loads via selectable axial stiffness". In: *Journal of Marine Science and Engineering* 3 (4 2015). Exeter Tether, pp. 1287–1310. ISSN: 20771312. DOI: 10.3390/jmse3041287.

[13] Darren Hayes, Danny Golden, and Tom Doyle. *FLWTAC Load Reduction Device.* June 2021. URL: https://ctprodstorageaccountp.blob.core.windows.net/prod-drupal-files/documents/resource/public/DublinOffshore_LRD.pdf.

[14] Jamie F. Luxmoore et al. "Analytical performance assessment of a novel active mooring system for load reduction in marine energy converters". In: *Ocean Engineering* 124 (2016), pp. 215–225. ISSN: 00298018.

[15] Marine Institute. *Galway Bay Marine and Renewable Energy Test Site. Foreshore Lease Application. Environmental Report.* Tech. rep. Feb. 2016, pp. 1–280. URL: https://tethys.pnnl.gov/sites/default/files/publications/marine_institute-spiddal-environmental_report.pdf.

[16] Dan Kyle Spearman et al. *Floating Wind Joint Industry Project Phase II Summary Report.* July 2020. URL: https://www.carbontrust.com/our-work-and-impact/guides-reports-and-tools/floating-wind-joint-industry-programme-phase-ii-summary-report.

[17] Philipp R. Thies, Lars Johanning, and Paul McEvoy. "A novel mooring tether for peak load mitigation: Initial performance and service simulation testing". In: *International Journal of Marine Energy* 7 (2014), pp. 43–56. ISSN: 22141669. DOI: 10.1016/j.ijome.2014.06.001.

Optimal day-ahead trading and power control for a hybrid wind-hydrogen plant with multi-agent reinforcement learning

Stijn Ally[1,2,3], **Timothy Verstraeten**[1,2], **Ann Nowé**[1] **and Jan Helsen**[2,3]

[1]Faculty of Engineering, Vrije Universiteit Brussel, Brussels, Belgium
[2]Artificial Intelligence Lab Brussels, Vrije Universiteit Brussel, Brussels, Belgium
[3]Flanders Make@VUB, Vrije Universiteit Brussel, Brussels, Belgium

E-mail: stijn.ally@vub.be

Abstract. Hybrid wind-hydrogen plants have multiple revenue sources, subject to uncertainties and trade-offs. As a consequence, maximizing their overall profitability is a challenging optimization problem. Since electricity is typically traded in advance of its actual power generation, weather forecasts play a crucial role in the power trading strategy. Additionally, the trading and control strategies of other market players influence the imbalance of the public grid and, as a result, have an impact on the revenue that can be generated by grid balancing.

To address this challenge, we propose a novel approach based on multi-agent reinforcement learning (RL), with two RL agents: one for day-ahead power trading, and a second for real-time power control of the wind farm and the electrolyzer. The RL system is trained using real-world data from a large offshore wind farm in the Belgian North Sea.

Results demonstrate the effectiveness of this approach, with the RL agents collaboratively maximizing the total operational profit of the hybrid plant, achieving a significantly higher profitability compared to conventional methods.

Keywords: hybrid wind-hydrogen plant, multi-agent reinforcement learning, power trading, power control, offshore wind farm

1 Introduction

The power output of a wind farm is influenced by numerous factors and fluctuates due to variable weather conditions. Electricity is typically traded ahead of its actual generation. Consequently, inaccuracies in weather forecasts can result in the wind farm being unable to meet its contracted energy, or, conversely, being able to produce more energy. Such deviations from the contracted energy may incur imbalance costs or represent missed revenue opportunities. An additional source of revenue is grid balancing, where a wind farm intentionally provides a surplus or shortage of energy, in order to help balancing the public grid. However, the grid imbalance also exhibits stochastic behavior, as it depends on the actions of numerous other electricity producers and consumers.

In addition to wind energy, green hydrogen is another puzzle piece for the transition towards a carbon-free energy production. Although electrolyzers are currently not yet financially competitive with traditional hydrogen production methods (such as steam methane reformation of natural gas or coal gasification), in the future they may become increasingly advantageous when combined with offshore wind farms [1, 2], especially when the share of wind power in the total energy mix continues to grow and

Content from this work may be used under the terms of the Creative Commons Attribution 4.0 licence. Any further distribution of this work must maintain attribution to the author(s) and the title of the work, journal citation and DOI.
Published under licence by IOP Publishing Ltd

financial mechanisms favor green hydrogen production. Surplus wind power can be used for hydrogen production, avoiding curtailment of the wind farm power production. Moreover, the extra flexibility of the hybrid plant may be advantageous to provide additional grid balancing services.

For a hybrid wind-hydrogen plant owner, selling more electricity in the day-ahead power market, maximizing revenues through grid balancing and producing more hydrogen, are three conflicting objectives. Given the uncertainties associated with weather forecasts, wind farm power capability, grid imbalance and the operational status of the electrolyzer stacks, finding an optimal power trading and control strategy is a challenging optimization problem.

Optimal trading and control strategies for renewables are active fields of research. Power systems can have a complex structure, which may vary across countries, resulting in different optimization settings. In research often only a part of the overall optimization problem is addressed [3, 4, 5, 6, 7, 8]. Our work goes beyond aforementioned studies on several aspects. We integrate power control for both the wind farm and the electrolyzer into the day-ahead power trading strategy. Moreover, we account for the impact of the hybrid plant imbalance on the imbalance tariffs imposed by the transmission system operator (TSO). Furthermore, we leverage long-term real-world data, taking the relationships between weather forecasts, actual weather conditions, wind farm power production and power market dynamics into account.

We propose a novel application of multi-agent reinforcement learning for the optimal trading and control of a hybrid wind-hydrogen plant. Two RL agents collaborate to maximize the total operational profit of the plant. One agent performs trades on the day-ahead power market, while the other dynamically controls the wind farm and electrolyzer in near real-time in order to maximize revenues from grid balancing and hydrogen production.

In order to create an adequate RL environment for the training of these RL agents, our work involves also the development of a data-driven model predicting the imbalance tariff based on some key parameters of the Belgian power system. Furthermore, a deep-learning wind farm power model, developed in prior work [9], simulates the wind farm power production with high accuracy. In addition, an electrolyzer model is developed, which accounts for essential operating characteristics of industrial-scale electrolyzers.

The proposed approach is validated using an extensive real-world dataset from a large offshore wind farm located in the Belgian North Sea, combined with pricing data from the day-ahead EPEX spot power market and balancing data from the Belgian TSO (Elia).

Results demonstrate that the RL agents effectively learn to jointly maximize the operational profit of the hybrid plant, achieving significantly higher profitability compared to some baseline algorithms.

2 Methodology

2.1 Reinforcement learning

Reinforcement learning (RL) is a branch of machine learning that focuses on enabling an intelligent agent to make optimal decisions in a dynamic environment to maximize cumulative rewards [10]. RL can handle large amounts of data, including stochastic data, and optimize control actions based on multiple criteria. A model-free RL agent learns the dynamics of the environment and a mapping between states and actions from direct interaction with that environment or from collected historical data, without the need for prior knowledge about the environment.

Many RL algorithms exist, each with different characteristics and application areas. For this work, the Soft Actor Critic (SAC) [11] algorithm is selected. This algorithm can handle continuous action spaces (which is required both for trading and power control), and discrete and continuous observation spaces. Actor-critic methods combine the benefits of value based and policy-based methods. Policy-based RL is effective in learning stochastic policies. At the same time, value based RL excels in sample efficiency and stability. A central feature of SAC is entropy regularization. The policy is trained to maximize a trade-off between the expected return and entropy, a measure of randomness in the policy. This objective has a number of advantages. First, the policy is incentivized to explore more widely, while giving up on clearly unpromising avenues. Second, the policy can capture multiple modes of near-optimal behavior. Lastly, it considerably improves learning speed.

2.2 RL environment

The proposed methodology is based on an RL system with two RL agents. Figure 1 gives an overview of the RL environment. The first RL agent, denoted as the DA-agent, trades electricity on the day-ahead (DA) power auction, based only on information available at the moment of trading (i.e. before 12 a.m. at the day preceding the day of electricity generation), such as the weather forecast. The second RL agent, denoted as the RT-agent, sets the power setpoints of the wind farm and the electrolyzer, based on real-time (RT) available information, controlling so the hydrogen production rate and the grid balancing.

The imbalance tariff applied by the TSO depends on the grid imbalance, which, on its term, is also influenced by the imbalance of the hybrid plant itself. The two RL agents have as objective to optimize jointly the total profit generated by the hybrid wind-hydrogen plant.

The DA-agent sells electricity for every 15-minute block of the next day. It is assumed that the quantity of traded power by the DA-agent does not influence significantly the day-ahead power prices, as the maximum production capacity of a wind farm is much smaller than the total power traded in the day-ahead auction [4]. Consequently, on the day-ahead market a wind farm is thus a price-taker. On the other hand, the RT-agent sets every minute the power setpoints of the wind farm and the electrolyzer. As the imbalance of an offshore wind farm can be substantial compared to the total imbalance of the Belgian grid, the influence of the farm imbalance on the imbalance tariff cannot be ignored.

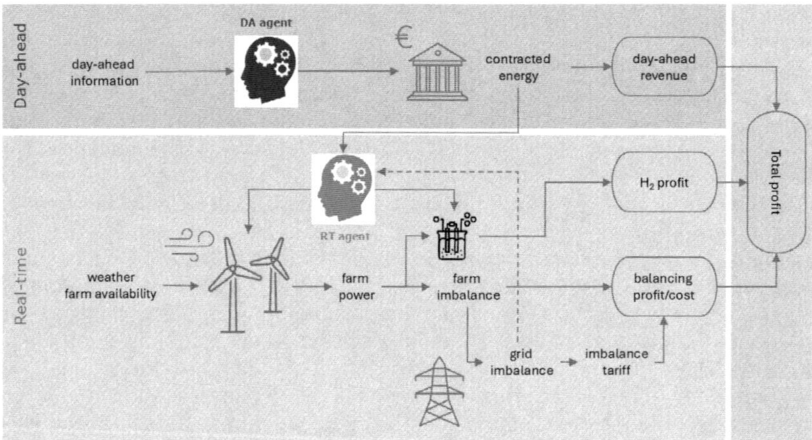

Figure 1: Overview of the RL environment with the two RL agents. The DA-agent trades electricity on the day-ahead auction based solely on information available at the moment of trading, generating a day-ahead revenue. The RT-agent updates dynamically the power setpoints of the wind farm and the electrolyzer, controlling the hydrogen production and grid balancing. The two RL agents have as objective to maximize collaboratively the total profit.

As can be seen from Figure 1, the day-ahead action of the DA-agent (i.e. the amount of electricity contracted on the day-ahead auction) has an influence on the actions to be taken by the RT-agent, as well as on the maximum profit that the RT-agent can achieve the next day. In contrast, the RT-agent has no influence on the day-ahead revenue generated by the DA-agent. However, for the DA-agent it is important to take into account the control behavior of the RT-agent and its possible future profit, in order to be able to trade the optimal amount of energy. Therefore, the reward function for the DA-agent $R_k^{DA\text{-}agent}$ at quarter-hour time step k comprises, in addition to the reward related to the day-ahead revenue R_k^{DA}, also rewards related to the hydrogen production $R_k^{H_2}$ and grid balancing R_k^B:

$$R_k^{DA\text{-}agent} = R_k^{DA} + R_k^{H_2} + R_k^B, \tag{1}$$

whereas the reward function of the RT-agent $R_k^{RT\text{-}agent}$ comprises only rewards related to the hydrogen production and grid balancing:

$$R_k^{RT\text{-}agent} = R_k^{H_2} + R_k^B. \tag{2}$$

The rewards related to the day-ahead revenue R_k^{DA}, hydrogen production $R_k^{H_2}$ and the balancing of the public grid R_k^B are calculated respectively as the contracted power multiplied with the day-ahead electricity price, the mass of the produced hydrogen multiplied with the hydrogen price and the imbalance of the hybrid plant multiplied with the imbalance tariff.

Note that the reward related to grid balancing can be rewarded only at the end of each 15-minute time block, during which the RT-agent must perform 15 pairs of continuous actions. This reward is thus very sparse. In order to increase the learning speed and favor convergence towards a smooth power control policy for the wind farm, a penalty (negative reward) R_t^O is added based on the change of the power setpoint for the wind farm ΔSP_t^{WF} at each 1-minute time step t:

$$R_t^O = -\left(\frac{\Delta SP_t^{WF}}{P_R^{WF}}\right)^2 \times R_{max}^O, \forall t \text{ with } t \in [1, 12], \tag{3}$$

with P_R^{WF} the rated wind farm power and R_{max}^O the maximum oscillation penalty.

2.3 Data

The training of the RL agents is performed with historical data, because in live conditions, untrained agents could cause major financial losses during the initial training phase and provoke unhealthy behavior of the physical systems. The historical data is composed of real-world data: wind speed and direction forecast data for the location of the offshore wind farm at a height of 100m (i.e. hub-height) from a commercial weather forecast provider, wind measurement data from the offshore wind farm, weather measurement data from a weather station in the vicinity of the offshore wind farm, EPEX spot electricity prices and balancing data from the Belgian TSO (Elia).

2.4 System models

2.4.1 Wind farm power model As the RT-agent controls, and thus modifies, the power production of the wind farm, the historical power measurement data of the wind farm cannot be used as such. Instead, an accurate wind farm power model is used to predict the farm power production based on wind inflow measurements and the power setpoint set by the RT-agent. The wind farm power model has been generated by deep learning based on long-term SCADA data of the wind farm [9].

2.4.2 Electrolyzer model Due to the lack of operational data from relevant large-scale electrolyzer plants (as these are still under development), in contrast to the data-driven wind farm power model, a discrete rule-based electrolyzer model was developed. The model takes typical technical characteristics into account, such as a limited power range and ramping constraints [12, 13].

2.4.3 Imbalance tariff model In Belgium, every balance responsible party (BRP) is responsible for balancing its power injection and offtake on a quarter-hourly basis. However, when BRP's are unable to do so, the Belgian TSO takes action to restore the balance within the Belgian control area by activating balancing services to reduce the residual system imbalance (SI) between the total electricity generation and consumption. Each BRP who is imbalance is consequently imposed an imbalance tariff proportional to the resources activated by Elia to manage the global grid imbalance, calculated as shown in Tabel 1,

Table 1: Tariffs for maintaining and restoring the residual system balance, imposed to individual balance responsible parties [14]

Imbalance of the BRP	System Imbalance (SI)	
	Positive	Negative or zero
Positive Negative	*MDP*-α	*MIP*+α

where *MDP* is the marginal price of downward activation by Elia, *MIP* is the marginal price of upward activation by Elia, and α is a parameter that offers an additional incentive in the event of a structural imbalance in the Belgian control area:

$$\alpha = 0\ €/MWh, \text{ if } |SI| \leq 150MW \tag{4}$$

$$\alpha = \left(a + \frac{b}{1 + \exp\frac{c-x}{d}}\right) \times cp, \text{ if } |SI| > 150MW \tag{5}$$

doi:10.1088/1742-6596/3025/1/012022

with a, b, c and d constant parameters, $x = average\,(|SI_t|, |SI_{t-1}|)$ and cp a function of MIP, MDP and SI. So, in Belgium, a single pricing system is applied. Passive balancing, also known as net imbalance volume (NIV) chasing, is allowed and even incentivized in case of a large residual system imbalance. This balancing strategy is referred to as passive balancing, as there is no actual trading and no counterparty.

Due to the large power capability of an offshore wind farm, the imbalance of a single wind farm can have a significant impact on the imbalance in the Belgian grid control area and, by consequence, also on the net control volume (NCV) activated by Elia. Consequently, MDP, MIP and α can be impacted, and so the imbalance tariff as well. In order to calculate the impact on MDP and MIP, one could consider to recalculate the settling of all reserve market auctions based on all individual auction bids. However, to be able to do so, the data of all bids should be available, and this calculation is computationally too expensive to get a tractable training duration for the RL agents. Therefore, data-driven models for MDP and MIP are developed based on historical data, as functions of the NRV and the day-ahead spot price. The imbalance tariff can then be calculated by the formulas given in Table 1 and equations (4) and (5).

2.5 RL agents

Both the DA-agent and RT-agent are implemented as SAC agents, using the open-source library Stable-Baselines3 (SB3) [15].

2.5.1 DA-agent The observation space of the day-ahead trading agent is limited to data available day-ahead at the moment of trading. As, at the one hand, the time resolution of the spot EPEX power auction for Belgium is one hour, and, at the other hand, the power balancing for the Belgian control area is done on a quarter-hour basis, the discrete time step of the DA-agent is selected to be 15 minutes, with four sequential time blocks having the same electricity price.

The observation space of the DA-agent comprises the normalized values of the following variables: day-ahead wind speed forecast, cosine and sine of day-ahead wind direction forecast, day-ahead electricity price and hydrogen price.

The action of the DA-agent is the quantity of sold power, limited to positive values capped by the rated power of the wind farm.

It may be surprising that, although the day-ahead electricity price is not known to the DA-agent at the moment of trading, it is still provided as input to the DA-agent. As argued higher, the trading volume of one single wind farm does not significantly influence the price settlement of the day-ahead market. The action of the trading agent represents consequently the trading volume by the agent for a given electricity price. In case the agent outputs significantly varying trading volumes for different electricity prices, multiple price-quantity bids can be submitted on the EPEX spot market across a range of relevant electricity prices. Consequently, once the auction is settled, the correct amount of power will be contracted at the final market price.

2.5.2 RT-agent In contrast to the DA-agent, the real-time power control agent has access to real-time data. The RT-agent updates every minute the power control setpoints of the wind farm and the electrolyzer. The 1-minute time step is significantly shorter than the 15-minute time-block which is used for the application of the imbalance tariff, and so the quasi real-time controller can optimize the plant imbalance during this period.

The observation space of the RT-agent comprises the normalized values of the following variables: contracted energy volume (contracted by the DA-agent), day-ahead electricity price, hydrogen price, maximum wind farm power capability, sequence number of the actual 1-minute step within the 15-minute balancing period, power setpoints of the wind farm and hydrogen unit at previous minute, power of the electrolyzer at previous minute and the (non-validated) system imbalances at the three preceding minutes. Note that the maximum wind farm power capability, observed by the RT-agent, is inferred by the wind farm power model (see §2.4) from the actual wind speed, wind direction and air density measurements.

The RT-agent has two actions: the delta (increase or decrease) of the power setpoints of the wind farm and electrolyzer unit compared to the respective setpoints at the preceding minute.

2.6 RL models training methodology

As the objective is to find the optimal combined actions of the DA-agent and RT-agent, and as the actions of the agents are influencing each other, the two agents have to be trained jointly. The training sequence of the two agents is described in Algorithm 1.

Algorithm 1: Training sequence of RL agents

initialise $P_k^{CE}, \forall k \in K$, based on a heuristic rule;
for $l = 1, 2, ..., L$ **do**
    train RT-agent for N^{RT} training steps, with P_k^{CE} in the observation space;
    train DA-agent for N^{DA} training steps, with RT-agent as power controller;
    predict $P_k^{CE}, \forall k \in K$ with DA-agent;
end

K is the quantity of 15-minute time steps in the training data set, L the number of training loops in the joint training phase, N^{RT} and N^{DA} the number of 1-minute and 15-minute training steps per training loop for respectively the RT-agent and DA-agent, and P_k^{CE} the power contracted by the DA-agent for 15-minute time step k.

2.7 Baseline algorithms

As comparison basis for the RL system proposed in this paper, three alternative rule-based algorithms are implemented.

For each of these algorithms, on the day-ahead power market, the day-ahead forecast of the farm power [9] is contracted whenever selling electricity is more profitable than producing hydrogen. Otherwise, this amount is reduced by the maximum continuous rating of the electrolyzer.

2.7.1 Plant power balancing In this approach, the power setpoints of the wind farm and electrolyzer are adjusted in real-time to ensure the hybrid plant matches the day-ahead contracted electricity as closely as possible, thereby minimizing the imbalance of the hybrid plant. Any excess wind power is utilized to maximize hydrogen production.

2.7.2 Maximum production In this approach, the power setpoint of the wind farm is fixed at its rated capacity, maximizing electricity production regardless of the day-ahead contracted power. When hydrogen production is more profitable than selling electricity, the electrolyzer is ramped-up to its maximum capacity at its maximum ramp-up speed. Otherwise, the electrolyzer is ramped-down to 0 MW as quickly as possible.

2.7.3 Perfect weather forecast In this hypothetical scenario, at the moment of day-ahead trading, the weather conditions and farm power capability are perfectly known for every minute of the following day, and electricity is contracted accordingly. In real-time, the power control of the wind farm and electrolyzer are performed identically as for approach §2.7.2. It is important to note that for this scenario the perfect weather forecast shall exclusively be known by the plant owner and not by the other market players. Otherwise, this information would affect the day-ahead bidding of all players, thereby influencing day-ahead prices and grid imbalance.

3 Results and discussion

3.1 Data

The data set spans over a long-term period from December 2021 up to June 2023. In order to create independent training and validation data sets, both including data with various weather conditions and electricity price regimes, the data set is split up in one-week-blocks which are attributed randomly to the training data set (70% of the weeks) and validation data set (30% of the weeks). For the RT-agent, this results in 567.360 1-minute training data points and 248.610 validation data points. For the DA-agent, the number of 15-minute data points is a factor 15 smaller.

3.2 Simulation parameters

The main simulation parameters are listed in Table 2. Note that the high maximum continuous rating of the electrolyzer can only be realized by combining multiple electrolyzer stacks (with a MCR of e.g. 5 MW [13, 12]) in parallel. In this work, we make abstraction from the power control of the individual stacks. The RT-agent controls the electrolyzer plant as one electrolyzer unit, with as overall minimum stable operation level the value for one individual alkaline electrolyzer stack.

Table 2: Simulation parameters

Parameter	Value
wind farm rated power (P_R^{WF})	> 250 MW
electrolizer maximum continuous rating (MCR)	20% of P_R^{WF}
electrolizer minimum stable operation level (MSOL)	3% of MCR
electrolizer maximum ramp-up and ramp-down speed	1 MW/s
electrolizer warm up and cool down time (cold start/shut down)	30 min
electrolizer efficiency	20 kg/MWh
hydrogen price ($\lambda_k^{H_2}$)	10 €/kg

3.3 Learning curves and profit

Figure 2 shows, in addition to the learning curves of the DA-agent (denoted as "total profit by the RL-agents") and the RT-agent, the three components of the total profit separately, i.e. the profits through day-ahead trading, hydrogen production and grid balancing. The solid lines are the average values for seven separate training runs, while the colored zones indicate the ranges between the minimum and maximum values from these training runs. It can be seen that the total profit generated jointly by the two agents (green curve) converges to a maximum. Notice also that during training the profit through DA trading is decreased in favor of an increased RT profit, realized both by an augmented hydrogen production and reduction of imbalance costs.

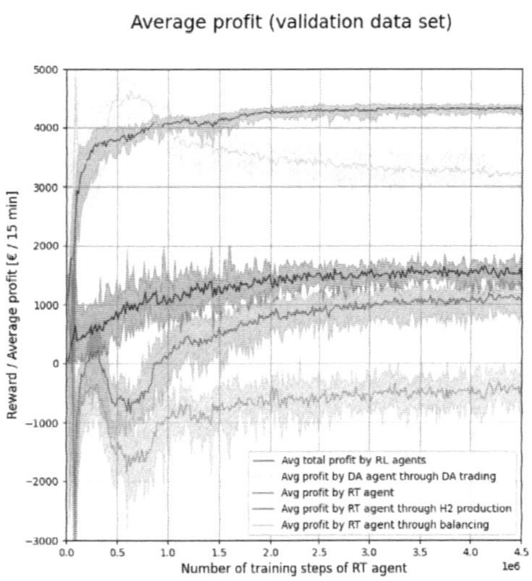

Figure 2: Learning curves of the DA-agent and RT-agent, as well as the profit components

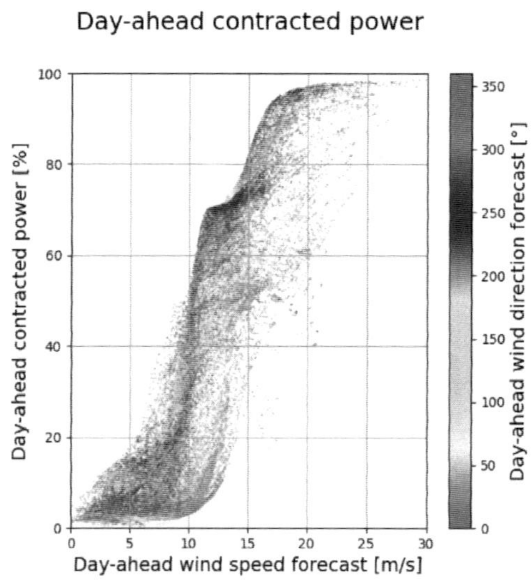

Figure 3: Day-ahead contracted power as function of day-ahead wind speed and wind direction forecasts

In Table 3, the profit components of the RL-agents are compared with those of the three baselines (§2.7). The RL-agents obtain both an increased day-ahead trading profit and real-time profit, compared to baselines 1 and 2. The day-ahead profit is increased only slightly (+2%), whereas the RT profit is increased substantially (resp. +29% and +12%). This results in a significant increase of total profit (resp. +7% and +4%). Note that for a large offshore wind farm, such an increase in operational profit is equivalent to having one to several additional wind turbines. Notice that the RL-agents cannot achieve the profit generated in the hypothetical scenario where the farm owner knows, as only market player, the perfect weather forecast.

Table 3: Comparison of the profit of the RL-agents with three baseline cases: (1) *Plant balancing* (§2.7.1), (2) *No wind farm curtailment* (§2.7.2) and (3) *Perfect weather forecast* (§2.7.3). The euro values are the total profit per 15-minute time block. The percentages are the profits scaled with the corresponding values for baseline 1.

	Baseline 1		**Baseline 2**		**RL-agents**		**Baseline 3**	
	Plant bal.		**No WF curt.**				**Perfect WFC**	
Total profit	4071€	100%	4204€	103%	4364€	107%	4773€	117%
DA trading profit	3271€	100%	3271€	100%	3335€	102%	3552€	109%
RT profit	800€	100%	934€	117%	1028€	129%	1221€	153%
H_2 profit	1255€	100%	1221€	97%	1443€	115%	1221€	97%
Imbalance cost	455€	100%	287€	63%	415€	91%	0€	0%

3.4 *Day-ahead power trading*
Figure 3 gives some insight in the trading behavior of the DA-agent. It shows the contracted power as function of the day-ahead wind speed and direction forecasts. One can recognize the typical S-curve resulting from to the S-shaped power curves of the wind turbines. In fact, one can recognize two S-curves: a first S-curve is going up to 100% of the farm rated power, and a second S-curve is mounting up only to about 80% of the farm rated power. The 20% reduction coincides with the relative power rating of the electrolyzer compared to the wind farm (ref. Table 2). Additionally, the DA-agent has learned the influence of the wind direction on the power production of the wind farm. For the wind directions between 200° and 270° (blue data points) there are no upstream wind farms, whereas for wind directions between 330° and 30° (red data points), upstream wind farms cause important wake power losses.

3.5 *Real-time power control*
Figures 4 to 5 give some insight in the control behavior of the RT-agent. Figure 4 demonstrates that farm power curtailment occurs basically only for positive farm power margins (i.e. when the wind farm can produce more power than has been contracted day-ahead) and that curtailments are more pronounced in case of large positive grid imbalances. This is a good control behavior, as unnecessary wind farm curtailment represents an opportunity cost.

In Figure 5 one can see that in case of large farm power margins there is more hydrogen production, whereas for negative farm power margins basically no hydrogen is produced. In addition, the 3% MSOL, below which the start-up and shut-down of the electrolyzer can be done only slowly, is clearly visible. The figure shows also that even in the case of large positive farm imbalance, the hydrogen production is reduced to compensate for negative grid imbalances.

3.6 *Time series*
Figure 6 presents time series for a complete day, from 0h00 until 24h00. The upper plot shows power signals, i.a. the day-ahead contracted power by the DA-agent, the real-time power production of the wind farm and hydrogen production. The plot shows also the grid imbalance estimated by Elia at the preceding minute, which is an input feature for the RT-agent. Furthermore, the average grid imbalance during the 15-minute time block is shown. However, this value is only known after completion of the 15-minute time block and can, consequently, not be used as input for the RT-agent. Note that the 1-minute signal of the grid imbalance is fluctuating and spiking heavily. It is thus not trivial to do accurate look-ahead forecasting of the 15-minute grid imbalance based on this 1-minute signal.

The middle plot shows the evolution of the day-ahead electricity price during the day, as well as the hydrogen price, which is constant during the day. Notice that during the night, morning and evening the day-ahead electricity prices are very high. In contrast, at noon the electricity price is slightly negative.

The lower plot shows the total profit of the RL-agents, as well as the three profit components related to respectively day-ahead trading, hydrogen production and grid balancing.

In the upper plot, one can see that the DA-agent trades, in general, less energy than the day-ahead power forecast, especially during the hours with low electricity prices. During the hour with negative electricity price (12h00-13h00), the DA-agent does still sell some electricity, resulting in a negative day-ahead revenue. However, as in real-time there happens to be a surplus of energy on the grid, the

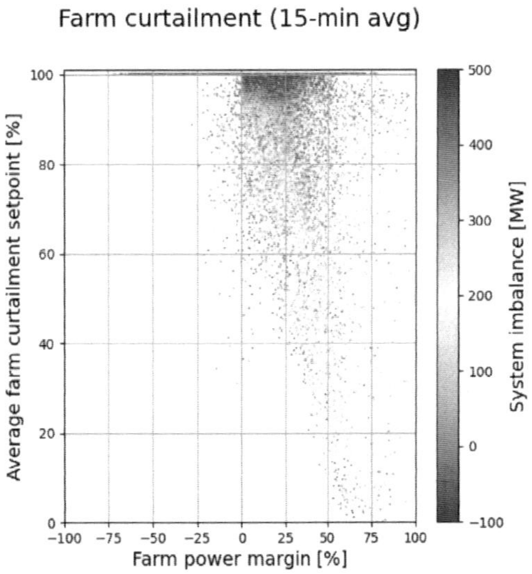

Figure 4: Wind farm curtailment as function of power margin of the wind farm and system imbalance of the Belgian grid

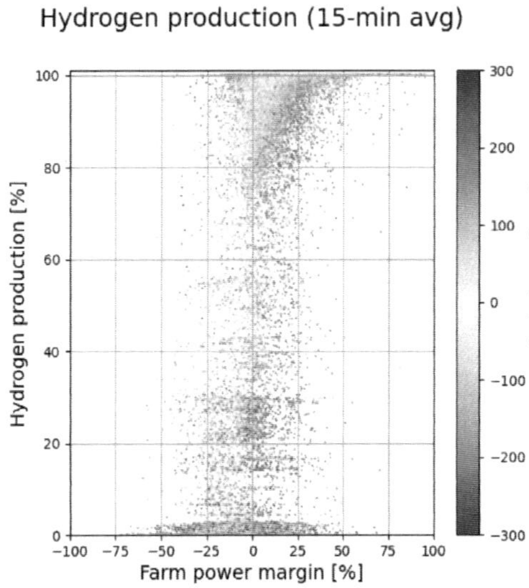

Figure 5: Hydrogen production as function of the farm power margin and system imbalance of the Belgian grid

RT-agent can curtail the farm power below the contracted amount, resulting in a grid balancing profit that outweighs the day-ahead loss. One can see, that the RT-controller is also curtailing substantially between 10h00 and 11h00, during which there is a large energy surplus on the grid. In contrast, during periods with important shortages of energy on the grid, the RT-agent supplies a surplus of electrical energy by stopping, or not starting up, hydrogen production.

4 Conclusions

This work proposes a novel methodology for optimal day-ahead trading and power control for a hybrid wind-hydrogen plant. The objective is to maximize the total profit of the hybrid plant by leveraging three revenue streams: day-ahead electricity trading, hydrogen production and grid balancing.

The methodology employs two reinforcement learning agents that collaboratively maximize the total profit. The first RL agent manages electricity trading on the day-ahead market, while the second RL agent dynamically adjusts the power setpoints of the wind farm and the electrolyzer in quasi-real-time.

The methodology has been applied to a real-world offshore wind farm. The RL agents achieve both an increased day-ahead trading profit and real-time profit, resulting in a significantly higher total profit over conventional strategies. As the share of renewables in the energy mix continues to grow, revenue from balancing services is expected to become even more important in the future.

The proposed RL-based approach offers a foundation for future development. Future work will focus on refining the electrolyzer model, implementing health-aware control, and incorporating subsidy schemes.

Acknowledgements

This research was supported by funding from the Flemish Government under the Flanders AI Research Program, the Belgian Federal Government under ETF project POSEIDON, and Horizon Europe under the SBEP EU project INSPIRE co-funded by FWO (Fonds Wetenschappelijk Onderzoek).

Author contributions

SA: conceptualization, data curation, formal analysis, investigation, methodology, software, validation, visualization and writing; TV: supervision and paper reviewing; AN: Funding acquisition; JH: Funding acquisition and supervision.

Journal of Physics: Conference Series **3025** (2025) 012022 doi:10.1088/1742-6596/3025/1/012022

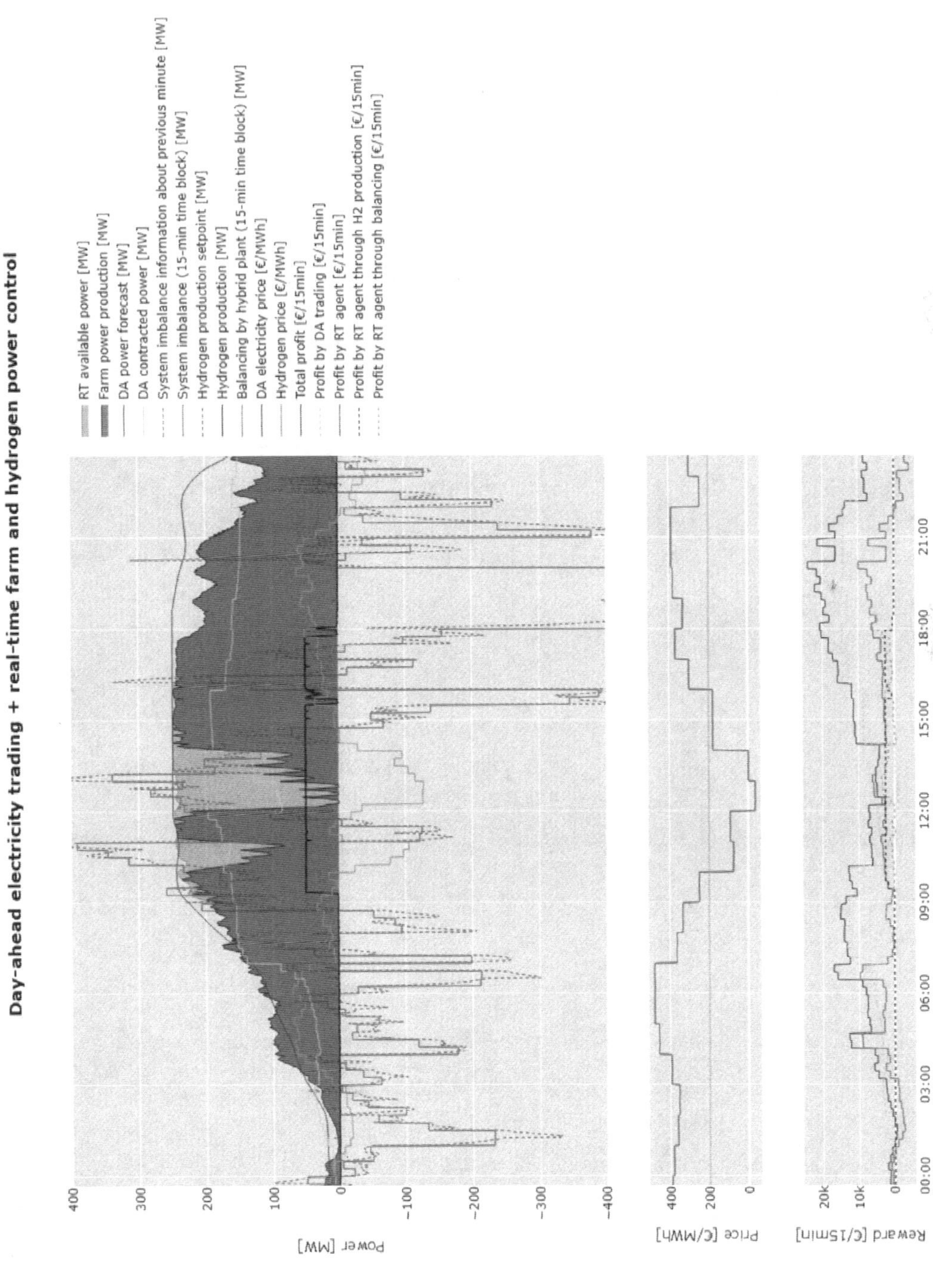

Figure 6: One-day time-series of day-ahead trading and real-time power control of a wind farm and hydrogen production unit

References

[1] Calado G and Castro R 2021 *Applied Sciences* **11** 5561

[2] Brunik K, Thomas J, Clark C, Duffy P, Kotarbinski M, Kee J, Grant E, Starke G, Riccobono N, Koleva M *et al.* 2024 Potential for large-scale deployment of offshore wind-to-hydrogen systems in the united states *Journal of Physics: Conference Series* vol 2767 (IOP Publishing) p 062017

[3] Pinson P, Chevallier C and Kariniotakis G N 2007 *IEEE transactions on Power Systems* **22** 1148–1156

[4] Dai T and Qiao W 2015 *IEEE transactions on sustainable energy* **6** 707–719

[5] Hosseini S A, Toubeau J F, De Greve Z and Vallée F 2020 *Applied energy* **280** 115973

[6] Skajaa A, Edlund K and Morales J M 2015 *IEEE Transactions on power systems* **30** 3181–3189

[7] Kayedpour N, De Kooning J D, Samani A E, Kayedpour F, Vandevelde L and Crevecoeur G 2023 *IEEE Transactions on Sustainable Energy* **15** 276–289

[8] Helgren E, Kazempour J and Mitridati L 2024 *Electric Power Systems Research* **234** 110787

[9] Ally S, Verstraeten T, Daems P J, Nowé A and Helsen J 2024 *Wind Energy Science Discussions* **2024** 1–33

[10] Sutton R S and Barto A G 2018 *Reinforcement learning: An introduction* (MIT press)

[11] Haarnoja T, Zhou A, Abbeel P and Levine S 2018 Soft actor-critic: Off-policy maximum entropy deep reinforcement learning with a stochastic actor *International conference on machine learning* (PMLR) pp 1861–1870

[12] John Cockerill 2023 DQ1000 Alkaline electrolyser `https://hydrogen.johncockerill.com/wp-content/uploads/sites/3/2023/04/dq-1000-def-2-hd-en.pdf` Online; accessed 09-12-2022

[13] Cummins Inc Cummins electrolyzer technologies `https://www.cummins.com/sites/default/files/2023-10/electrolyzer-brochure.pdf` Online; accessed 10-11-2023

[14] Elia 2022 Tariffs for maintaining and restoring the residual balance of individual access responsible parties - period 2020-2023

[15] Raffin A, Hill A, Gleave A, Kanervisto A, Ernestus M and Dormann N 2021 *Journal of Machine Learning Research* **22** 1–8

Journal of Physics: Conference Series **3025** (2025) 012023 doi:10.1088/1742-6596/3025/1/012023

Reliable operation of wind-storage systems for baseload power production

Jenna Iori[1], Michiel Zaaijer[1], Jasper Kreeft[2], Dominic von Terzi[1], Simon Watson[1]

[1]Faculty of Aerospace Engineering,TU Delft, Delft, The Netherlands.
[2]Shell Global Solutions International BV, The Hague, The Netherlands

E-mail: j.iori@tudelft.nl

Abstract. As the penetration of renewable energy increases in the generation mix, the problem of power dispatchability becomes more critical. The co-location of storage systems with wind energy is a promising solution to shift power delivery from periods of high wind resource availability to periods of high electricity demand. Producing baseload power from wind farms all or most of the time is an example of such dispatchability. In this work, we present an optimization-based dispatch strategy to produce baseload power. At every time step, an optimization problem is solved to decide the storage operation, maximize revenues on the electricity market and reach a given reliability target. In order to reduce the impact of forecast uncertainties on the reliability of the power delivery, a robust formulation of the dispatch optimization is used, based on a pessimistic version of the forecast. The proposed method is evaluated for 18 offshore sites with a 100 MW wind farm and storage system, for one year of operation. By using a robust dispatch strategy, the reliability increases by up to 0.9 points, with a minor impact on revenues (+2% on average), compared to the reference dispatch strategy using a regular forecast. Our study demonstrates the feasibility of providing a reliable baseload power from wind energy in the presence of forecast uncertainty.

1 Introduction

The lack of dispatchability of wind energy is a significant challenge for grid stability and reliability of supply, in scenarios of high penetration of renewable energy. The co-location of wind farms with energy storage provides a possible solution. This type of system is effective in mitigating the variability of wind power [1, 2] and compensating forecast error when aiming to produce scheduled or reference power [3, 4, 5]. A relevant use-case for increasing the dispatchability of wind power is the production of baseload power where minimum power is produced most or all of the time [6, 7]. A baseload hybrid power plant pilot, called the Baseload Power Hub, is currently under construction at the Hollandse Kust Noord wind farm as part of the CrossWind innovation program [8]. This system is designed to provide at least 20% of the average power production of a wind turbine 99% of the time and will demonstrate the technical feasibility of this type of power plant.

The relevance of a baseload wind-storage power plant lies in its reliability, i.e. its ability to provide a minimum load consistently regardless of variations of the wind resource. Forecast uncertainties increase the risk of not achieving a steady and sufficient power supply. As such, the dispatch schedule of the storage system is critical. Dispatch methods based on *if-then* rules can be used to mitigate the variability of the wind power [4, 9]. More advanced strategies based on optimization have been developed to take

Content from this work may be used under the terms of the Creative Commons Attribution 4.0 licence. Any further distribution of this work must maintain attribution to the author(s) and the title of the work, journal citation and DOI.
Published under licence by IOP Publishing Ltd

full advantage of forecast information for bidding in the electricity market [10, 11, 5]. However, these approaches do not address the question of reliability considering forecast uncertainties.

In this work, we explore the use of a *robust* approach for the dispatch optimization strategy of the storage system, inspired from the field of robust optimization. A pessimistic version of the power forecast is used as an input for the dispatch optimization, underestimating the expected power production. This approach encourages a conservative operation of the storage system and as such is promising for increasing the reliability of the power delivery. The present paper presents the robust dispatch optimization, and addresses the following research question: To what extent does a robust approach improve the reliability of baseload production for wind-storage systems?

The rest of the paper is divided as follows: The next section reports the models used to represent the storage system, wind power production and forecast. The rest of the section presents the dispatch optimization and its robust version. The numerical experiments designed to answer the research question are reported in Section 3, and the corresponding results shown in Section 4. Finally, Section 5 and 6 provides suggestions for future work and conclusions for the study.

2 Methodology

2.1 Storage system model

The storage system is described with its macro-parameters: power capacity $\bar{P}$, energy capacity $\bar{E}$ and round-trip efficiency η. The operation of the storage system is described with the power P_i and energy E_i at each time step i. We use the convention that the power is positive in discharge and negative in charge. The following model is used to link the storage power in charge or discharge to its energy level, over a given number of time steps n and with a time step Δt:

$$E_{i+1} - E_i = \begin{cases} -\Delta t P_i & \text{if } P_i \geq 0, \\ -\dfrac{\Delta t}{\eta} P_i & \text{else,} \end{cases} \qquad i = 0, ..., n-1, \tag{1}$$

$$-\bar{P} \leq P_i \leq \bar{P} \qquad i = 0, .., n-1,, \tag{2}$$

$$0 \leq E_i \leq \bar{E} \qquad i = 0, .., n. \tag{3}$$

2.2 Wind farm model

A simple wind farm model is used to describe the dependency of the power production on the wind speed u and wind direction θ, represented by the functions

$$f_{\mathrm{WF}}(u, \theta) = n_{\mathrm{WT}} \cdot \min(P_{\mathrm{R}}, (1 - f_{\mathrm{WL}}(\theta)) f_{\mathrm{WT}}(u)), \tag{4}$$

$$f_{\mathrm{WT}}(u) = \begin{cases} \dfrac{1}{2}\rho C_P A u^3 & \text{if } u_{\mathrm{in}} \leq u \leq u_{\mathrm{out}}, \\ 0 & \text{else.} \end{cases} \tag{5}$$

where n_{WT} is the number of wind turbines, P_{R} the turbine rated power, ρ the air density, C_P the power coefficient, u_{in} and u_{out} the cut-in and cut-out wind speeds, A the rotor area, and $f_{\mathrm{WL}}(\theta)$ a wake loss term with a periodicity of 60° and a maximum of 0.25. The model is illustrated in Figure 1. It was developed to approximate the IEA Wind 740-10-MW Reference Offshore Wind Plant with a regular layout [12].

2.3 Forecast

The power forecast is generated based on numerical weather prediction forecast data from ECMWF's Integrated Forecast System. A wind speed forecast is available every 12 hours. For every forecast issue, the forecast is retrieved with a varying resolution, between 1 hour (for lead-times up to 12 h), 2 hours (for lead-time between 12 and 24 h) and 3 hours (for lead-time above 24 h). A cubic interpolation is used as necessary to provide a one-hour resolution for the entire range of lead times. Observation data are obtained from ERA5 [13]. From the data of the wind speed and wind direction (forecast and observation), the corresponding power production from the wind farm is obtained using Equation 4. Let $\hat{u}^{\mathrm{WF}}_{i+j|j}$ and $\hat{\theta}^{\mathrm{WF}}_{i+j|j}$ be the wind speed and direction forecast issued at time j for a lead time i. The corresponding power forecast $\hat{P}^{\mathrm{WF}}_{i+j|j}$ is calculated as

$$\hat{P}^{\mathrm{WF}}_{j+i|j} = f_{\mathrm{WF}}(\hat{u}_{j+i|j}, \hat{\theta}_{j+i|j}). \tag{6}$$

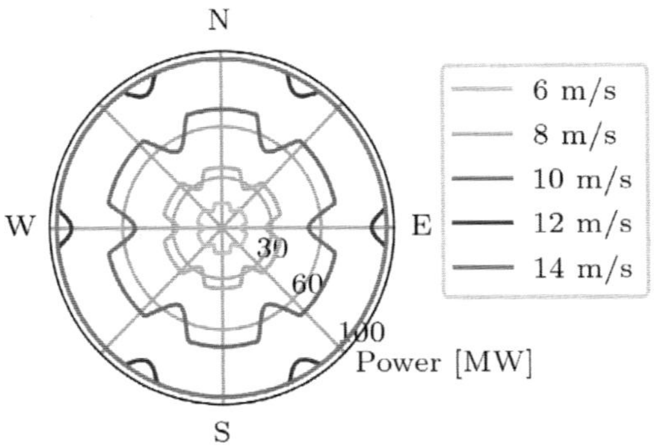

Figure 1: Wind farm power production as a function of wind speed and wind direction.

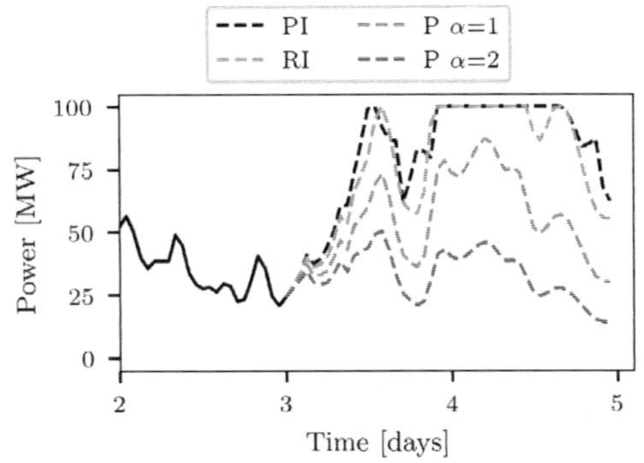

Figure 2: Example of perfect information (PI), real information (RI), and pessimistic (P) forecasts.

2.4 Dispatch optimization

The operation of the storage system is decided using an online optimization, where the results of an optimization problem solved at each time step give the power when charging or discharging the storage system. The objective of the optimization is to maximize revenues over a given time window corresponding to the maximum power forecast lead time. A simple revenue model is used, where a varying electricty price, noted $\boldsymbol{\lambda}$, is associated with the power delivered to the grid. Market bidding mechanisms are not modelled in this work. Furthermore, the dispatch optimization aims at satisfying a minimum baseload power production P_{BL} with a given target reliability $\tilde{r}$.

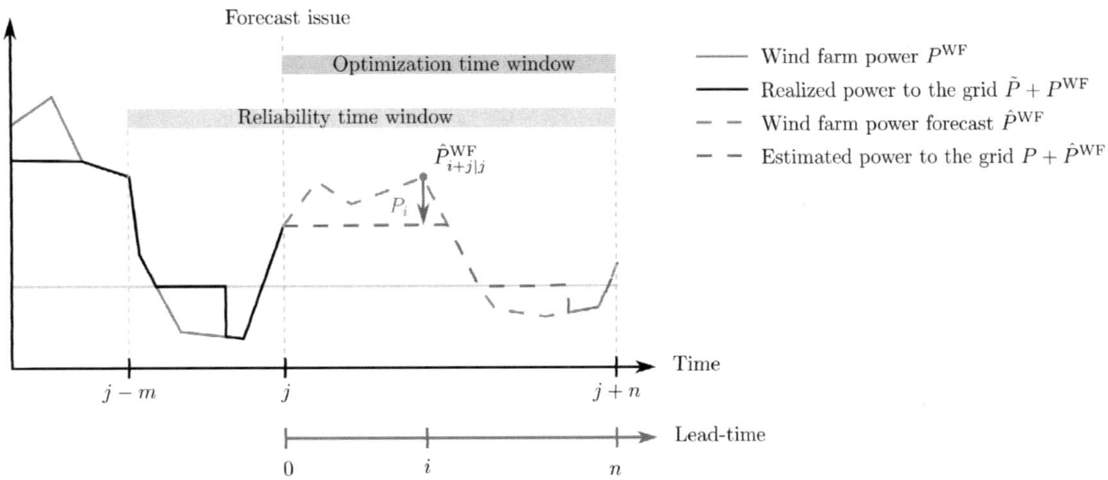

Figure 3: Schematic representation of the dispatch optimization. The time scale related to the forecast and the optimization problem is shown in blue and indexed by i. Real-time is indexed by j.

Figure 3 illustrates the online optimization for a given time step j. A forecast is issued at this time step and gives an estimation of the power produced by the wind farm over the next n time steps. The solution of the optimization problem gives the expected storage operation schedule $P_i,\ i = 0,..n-1$. Finally, the storage system is operated for one time step following the optimum. The realized storage power and energy level are denoted by $\tilde{P}_j$ and $\tilde{E}_j$. Here, two different indexing are used. The indexing related to real-time is denoted by j, whereas the index i is used to refer to the forecast lead-time and the optimization time window.

The reliability of the baseload power production is the proportion of time where the power delivered to the grid is equal or above the baseload level P_{BL}. The reliability should reach its target value when measured over months or years. However, this long time window is difficult to capture within the online optimization, due to the short time scale of the forecast. As a solution, a short-term reliability measure is used within the dispatch optimization, calculated over n time steps ahead of the current time and m time steps before. This time window is illustrated in Figure 3. By considering previous time steps, this metric can better approximate the reliability calculated over months or years. In order to describe the metric in mathematical terms, binary variables denoted by $\boldsymbol{z}$ are used to represent if the baseload power is delivered or not at each time step. Considering h the number of times over the past m time steps where the baseload power is not achieved, the short-term reliability r_j for a given time step j is calculated as

$$r_j(\boldsymbol{z}, h) = \frac{1}{m+n}(h + \sum_{i=0}^{n-1} z_i), \quad \text{where } z_i = \begin{cases} 1 & \text{if } P_i + \hat{P}^{\text{WF}}_{j+i|j} \geq P_{\text{BL}}, \\ 0 & \text{otherwise,} \end{cases} \tag{7}$$

The optimization problem is formulated as a mixed-integer linear program. The corresponding mathematical formulation is reported in Equations (8)-(14), for a given time step j. The design variables are the storage energy level and power over the forecasting time window $\boldsymbol{E}, \boldsymbol{P}$, as well as binary variables $\boldsymbol{z}$ and a penalty term $p \in \mathrm{R}^+$ associated with the reliability constraint. Equation (9) enforces upper and lower bounds on the power delivered to the grid. The binary variable z_i dictates if the lower bound is zero or the baseload power P_{BL} for lead-time i. Equation (10) constrains the reliability r_j to be above the target reliability $\tilde{r}$. Here, the penalty term p serves as a slack variable to ensure that the feasible set for the problem is not empty. Equation (11) ensures that the initial energy level within the optimization problem corresponds to the realized energy level at the considered time step j. Equations (12)-(13) represent the storage system model, in a relaxed form. This choice of formulation reduces the number of binary design variables and in turn the computational burden of solving the problem. Finally, the bounds on $\boldsymbol{P}$ and $\boldsymbol{E}$ are enforced through the last constraint.

$$\underset{p\in\mathrm{R}^+, \boldsymbol{z}\in\{0,1\}^n, \boldsymbol{E}\in\mathrm{R}^{n+1}, \boldsymbol{P}\in\mathrm{R}^n}{\text{maximize}} \quad \sum_{i=0}^{n} \lambda_{i+j} P_i - \mu p + \beta E_n \tag{8}$$

$$\text{subject to} \quad P_{\text{BL}} z_i \leq P_i + \hat{P}^{\text{WF}}_{j+i|j} \leq P_{\max} \qquad i = 0, .., n-1 \tag{9}$$

$$r_j(\boldsymbol{z}, h) \geq (\tilde{r} - p) \tag{10}$$

$$E_0 = \tilde{E}_j \tag{11}$$

$$E_{i+1} - E_i \leq -\Delta t P_i \qquad i = 0, .., n-1 \tag{12}$$

$$E_{i+1} - E_i \leq -\frac{\Delta t}{\eta} P_i \qquad i = 0, .., n-1 \tag{13}$$

$$\text{Eq. (2-3)} \tag{14}$$

The objective function is made up of three terms. The first term represents storage revenues, which include loss of revenues when charging, and added revenues when discharging. The second term is associated with the penalty on the reliability and drives the algorithm to choose a solution where the target reliability is reached and $p = 0$. The last term is added due to the relaxation of the storage model, to ensure the storage system equation holds even if $\hat{P}^{\text{WF}}_{j+i|j} = P_{\max}$, $i = 0, ..., n-1$. The relaxation also requires that the electricity price is positive for all time steps, $\lambda_{i+j} > 0$, $\forall i$. The factors μ and β are tuning parameters. Algorithm 1 reports how the storage dispatch is decided based on the solution of the optimization problem.

2.5 *Robust formulation using pessimistic forecast*

The optimization problem represented by Equations (8)-(14) is written using a deterministic approach with a point forecast $\hat{P}^{\text{WF}}_{j+i|j}$, $i = 0, ..., n-1$. However, due to forecast uncertainties, the actual power can be lower than the forecasted value. In this situation, the constraint related to the baseload power represented by Equation (9) does not consider such uncertainties, and may lead to a lower reliability.

In order to take into account such uncertainties in the dispatch optimization problem, it is reformulated as a robust optimization using a chance-constrained approach [14], with probability $1 - \epsilon$. Considering all possible power scenarios $\boldsymbol{P}^\xi$, $\xi \in \Xi$, the chance-constrained version of Equation (9) is

$$\mathbb{P}_{\xi\in\Xi}\big(P_{\text{BL}} z_i \leq P_i + P^\xi_{j+i} \leq P_{\max}\big) \geq 1 - \epsilon, \ i = 0, .., n-1. \tag{15}$$

Algorithm 1 Dispatch strategy based on an online optimization

Require: $\boldsymbol{\lambda}$, $\tilde{r}$, P_{BL}, $\bar{P}$, $\bar{E}$, n, m, N
  Initialize $\tilde{E}_0$
  for $0 \leq j \leq N$ **do**
    Generate forecast $\hat{P}^{\mathrm{WF}}_{j+i|j}$ for $i = 0, ..., n-1$
    $h \leftarrow 0$
    if $j \leq m$ **then**
      $h \leftarrow \sum_{k=0}^{j-1} \tilde{z}_k$
    else if $j \geq m$ **then**
      $h \leftarrow \sum_{k=j-m}^{j-1} \tilde{z}_k$
    end if
    Solve optimization problem represented by Eqs. (8-14) and obtain the solution $\boldsymbol{E}, \boldsymbol{P}, \boldsymbol{z}$
    Operate the storage for one time step: $\tilde{P}_j \leftarrow P_0$, $\tilde{E}_{j+1} \leftarrow E_1$, $\tilde{z}_j \leftarrow z_0$
  end for

Instead of generating a representative ensemble of the set Ξ for the evaluation of the constraint, a pessimistic version of the forecast, noted $\hat{P}^{\mathrm{WF-}}$, is generated under the assumption that it constitute a lower bound for the power scenarios with a probability $1-\epsilon$, i.e

$$\mathbb{P}_{\xi \in \Xi}\big(\hat{P}^{\mathrm{WF-}}_{j+i|j} \leq P^{\xi}_{j+i}\big) \geq 1-\epsilon, \ i = 0, .., n-1. \tag{16}$$

Using this pessimistic version of the forecast, Equation (17) is rewritten as

$$P_{\mathrm{BL}} z_i \leq P_i + \hat{P}^{\mathrm{WF-}}_{j+i|j} \leq P_{\max}, 1 = 0, .., n-1. \tag{17}$$

We note that the two equations are not equivalent when it comes to the upper bound. However, we assume that this aspect has a limited impact on the performance of the system. The robust dispatch optimization problem has the same formulation as the regular dispatch optimization problem presented before, but using a pessimistic version of the power forecast instead. In other words, solving the robust dispatch optimization problem is equivalent to solving the regular dispatch optimization problem with a pessimistic power forecast signal.

The pessimistic version of the power forecast is generated from a downgraded version of the wind speed forecast. This signal, noted $\hat{u}^{-}$, is obtained by reducing the original forecast value by a factor proportional to the standard deviation of the wind speed forecast error, σ_i, as

$$\hat{u}^{-}_{j+i|j} = (1 - \alpha \frac{\sigma_i}{u_{\mathrm{mean}}}) \hat{u}_{j+i|j}, \ \forall i, j, \tag{18}$$

where u_{mean} is the average wind speed over one year of observation data. The pessimistic power forecast is then calculated as

$$\hat{P}^{\mathrm{WF-}}_{j+i|j} = f_{\mathrm{WF}}(\hat{u}^{-}_{j+i|j}, \hat{\theta}_{j+i|j}). \tag{19}$$

The term α allows to adjust the level of pessimism of the forecast, and in turns the level of robustness of the optimization problem, represented by ϵ. A higher value of α means that the baseload power constraint is satisfied with a higher probability. It is however outside the scope of this work to determine the exact relationship between α and ϵ. Figure 2 illustrates two pessimistic forecasts with varying values of α.

3 Test cases

The regular dispatch optimization strategy is compared to the proposed robust approach. In practice, the dispatch optimization problem is solved using four different forecast signals: a perfect information forecast with zero forecast error, a real information forecast, and two pessimistic versions of the real information forecast with different pessimism levels ($\alpha = 1$ and $\alpha = 2$), illustrated in Figure 2. The results obtained with a perfect information forecast serve as a reference. Furthermore, the methodology is compared to a more simple approach, using a rule-based strategy illustrated in Figure 4. In this strategy the decision to charge or discharge the storage is based on the instantaneous wind power P^{WF}_j and price λ_j, only.

It is tuned by choosing the two threshold parameters for power and price, P_{th} and λ_{th} respectively, to maximize the storage revenues and satisfy the target reliability. In all cases, price forecast is not modeled and the information on price is assumed to be perfect.

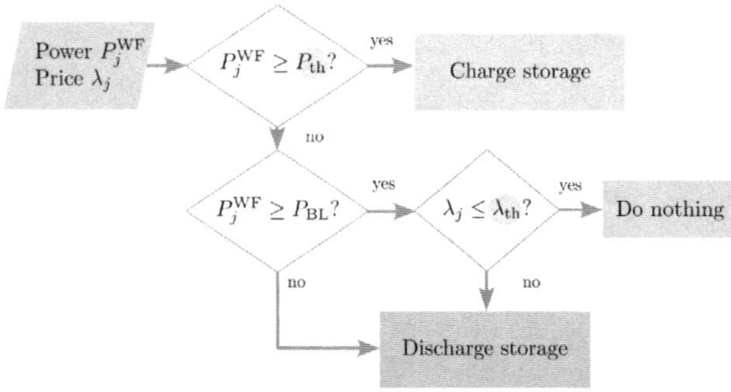

Figure 4: Logic of the rule-based operation strategy.

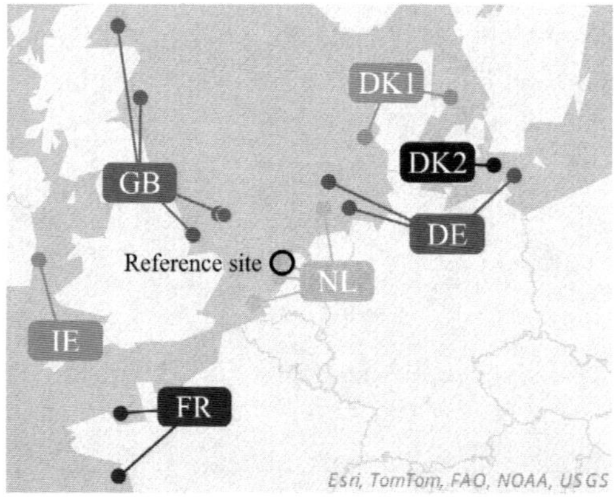

Figure 5: Site locations and associated electricity markets

Two performance metrics are used to compare the different cases: storage revenues and reliability of the baseload power production. Considering the price signal $\boldsymbol{\lambda}$ and the realized storage power $\tilde{\boldsymbol{P}}$, the storage revenues are calculated over n_T time steps as $\sum_{j=0}^{n_T-1} \lambda_j \tilde{P}_j$.

The reliability r of the baseload power production is calculated in a similar fashion as in Equation (20). The difference is that the calculation is applied to the realized storage power $\tilde{\boldsymbol{P}}$, realized wind farm power $\boldsymbol{P}^{\mathrm{WF}}$ and corresponding binary variables $\tilde{\boldsymbol{z}}$:

$$r = \frac{1}{n_T} \sum_{j=0}^{n_T-1} \tilde{z}_j, \quad \text{where } \tilde{z}_j = \begin{cases} 1 & \text{if } \tilde{P}_j + P_j^{\mathrm{WF}} \geq P_{\mathrm{BL}}, \\ 0 & \text{otherwise,} \end{cases} \tag{20}$$

To assess the performance of the proposed robust approach, we consider a 100 MW offshore wind farm in Europe, required to satisfy a baseload level $P_{\mathrm{BL}} = 10$ MW, with a reliability target of 99%. We use $n_{\mathrm{WT}} = 10$, $C_P = 0.45$ and a rotor radius of 90 m. Figure 5 shows the different locations considered, and the reference site. The price signal $\boldsymbol{\lambda}$ corresponds to the day-ahead electricity price from ENTSO-E [15]. All data are extracted for the year 2019, and the storage operation is simulated for approximately one year, with a time-step of one hour and $n = 8600$. A 48-hour forecast window is considered for the dispatch optimization strategy. Since the dimension of the storage required to satisfy the baseload constraint may vary between sites, different storage sizes are used, decided using the sizing method described in [6]. We assume a round-trip efficiency of 85%, corresponding to the efficiency of battery systems. Finally, the reliability window includes one week of operation before the current time step with $m = 168$. The optimization problem is solved using MOSEK [16], with $\mu = (n+m) P_{\mathrm{BL}} \max_i(\lambda_{j+i})$ and $\beta = 10^{-6}$.

4 Results

4.1 Reference site

A comparison is first run for the reference site, represented in Figure 5. For this site, the storage has a power capacity of 10 MW and an energy capacity of 538 MWh. Figure 6 and 7 reports the performance of the rule-based strategy and the regular and robust dispatch optimization strategy. Figure 6 shows that the state-of-charge for the rule-based strategy and robust dispatch optimization strategy are higher than that of the dispatch optimization strategy using a real or perfect information forecast, showing a more conservative or risk-averse behaviour. When the pessimism level of the robust approach, represented by α, increases, the operation obtained with the dispatch optimization tends towards that of the rule-based strategy. Notably, results show that during day 144 the baseload power production is satisfied only when using the rule-based strategy and the robust dispatch optimization strategy with $\alpha = 2$. This results confirms that the use of a pessimistic forecast encourages a conservative storage operation.

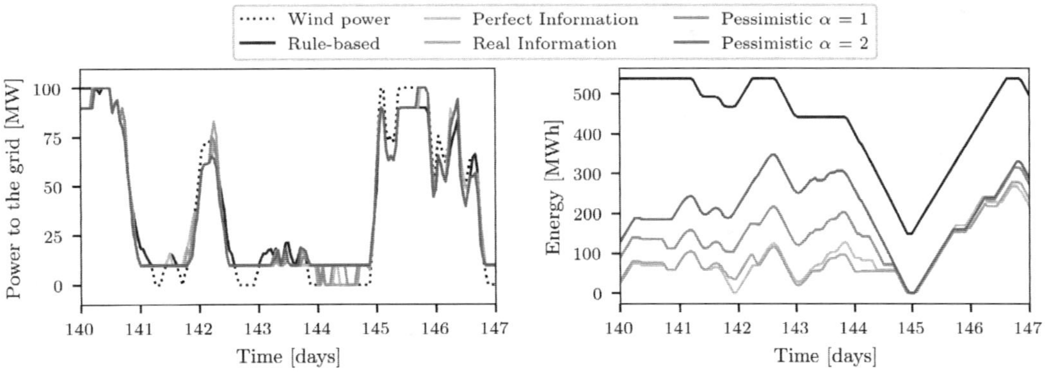

Figure 6: Power sent to the grid and energy level of the storage system for the rule-based strategy and dispatch optimization strategy using different forecast signals for the reference site.

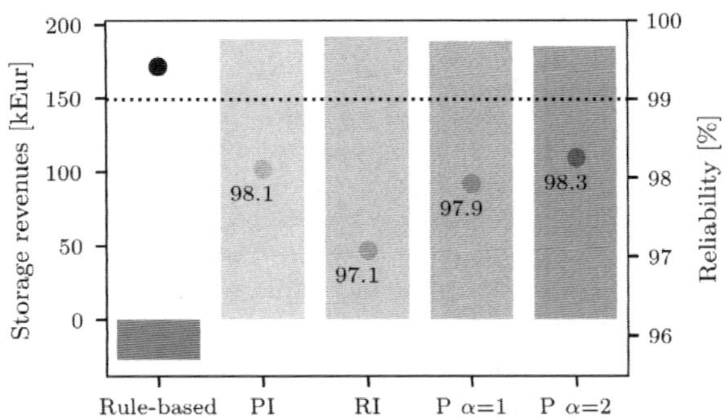

Figure 7: Reliability (dots) and storage revenues (bars) over one year for the reference site for the rule-based and optimization strategies using perfect (PI), real information (RI), and pessimistic (P) forecasts.

The storage revenues and reliability of the baseload power production calculated over the year are shown in Figure 7. The rule-based strategy shows the lowest revenues, due to its tendency to keep the storage full without taking advantage of the forecast information. Instead, the optimization-based strategies all present positive revenues. Interestingly, the storage revenues are not impacted significantly by the type of forecast. This suggest that the baseload constraint is driving the optimal operation of the storage.

In addition, the reliability decrease between the perfect and real information cases due to the impact of errors in the forecast. However, the use of a robust approach with a pessimistic forecast results in an increase in reliability, with a slight change in revenues, showing the effectiveness of the proposed method. We note that the reliability is lower than the target for the cases based on dispatch optimization. This is because the reliability constraint is enforced on a limited time window, and not on the entire year. Furthermore, the reliability is below target even for the perfect information case, showing the negative impact of a limited forecast window.

4.2 All sites

Figure 8 generalizes these results for 18 offshore sites in Europe. For each site, the relative difference in storage revenues and difference in reliability compared to the perfect information case are plotted. Results show that the forecast uncertainties associated with the real information forecast lower the reliability for all sites, between -0.4 and -1.5 points. Similarly to Figure 7, we see an increase in reliability when using a pessimistic version of the forecast. In the more conservative case ($\alpha = 2$) the reliability is higher than for the perfect information forecast for half of the sites. On average, the reliability is increased by 0.7 points for $\alpha = 1$ and by 0.9 points for $\alpha = 2$ compared to the real information case. Instead, the

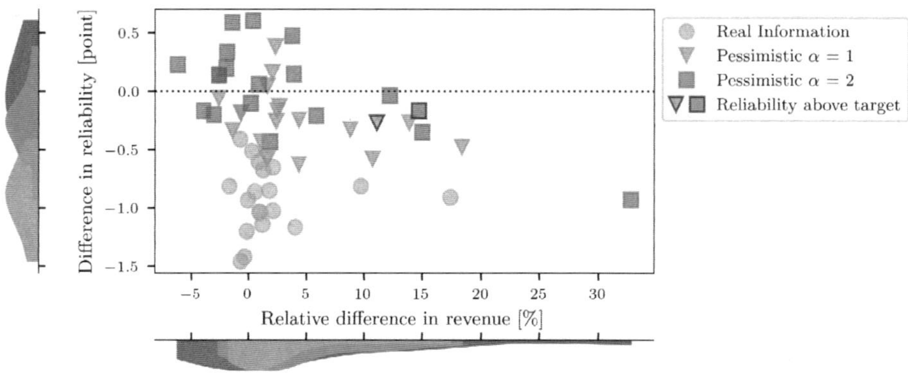

Figure 8: Reliability and revenues of the dispatch optimization based on real information and pessimistic forecasts for 18 offshore sites in Europe, relative to the dispatch optimization using perfect information forecast

impact on revenues does not show a specific trend. On average, the revenues increase by around 2% for the pessimistic cases, but with a large spread in the data. Notably, there is an outlier where the storage revenues with the pessimistic forecast for $\alpha = 2$ are 33% higher than with the perfect information forecast. Overall, these results show that a robust dispatch strategy using a pessimistic forecast can effectively mitigate the impact of forecast error on the reliability of the baseload power production.

The results for the rule-based strategy are not reported in the figure: it is the worst approach for all sites, with a decrease in revenues between -100% and -250% compared to the perfect information case. However, this strategy achieves the target reliability for all sites.

As can be seen on Figure 8, the target reliability of 99% is reached for only one site using the pessimistic forecasts. These results suggest that the storage system is not able to both deliver baseload power with a given reliability and generate positive revenues. In the dispatch optimization problem, the balance between reliability and revenues can be calibrated with the parameter μ. For the results presented here, $\mu = (n+m)P_{\mathrm{BL}}\max_i(\lambda_{j+i})$. This value means that if the reliability is below target, this penalty leads to an apparent loss of revenue equal to selling power to the highest price in the period. As such, our results suggest that satisfying the reliability target would lead to a much larger revenue loss and possibly negative revenues as with the rule-based strategy.

5 Discussion

Future work is needed to understand why the revenues increase for some sites and decrease for others. Furthermore, further development on the formulation of a reliability constraint for dispatch optimization would be beneficial to ensure that the target reliability is reached. Finally, this work uses a simple model for wind farm power production. Using a higher-fidelity model would make our analysis more precise and realistic.

6 Conclusion

This study demonstrates the feasibility of providing a reliable baseload power from wind energy when information about future wind power is limited in time and uncertain. Our results show that the reliability corresponding to a baseload production can be increased when using a dispatch optimization method using a pessimistic version of the power forecast, following a robust formulation of the optimization problem. In particular, the impact of forecast uncertainties can be efficiently mitigated by the proposed method. We further show that the impact on revenues is limited on average.

Finally, our results demonstrates that the proposed robust dispatch method allows the storage dispatch to shift its priority from satisfaction of baseload power in time of need and revenue maximization through arbitrage any other time. This ability to automatically adapt the goal during operation is a clear strength of dispatch optimization strategies over rule-based strategies.

Code and data availability

The wind speed and wind direction forecast data used to generate the results presented in this work are available open-source in the data repository associated with [17]. The research results can be reproduced with the open-source code SHIPP available at https://github.com/jennaiori/shipp [18].

Acknowledgements

The results presented in this work are based on data from the European Centre for Medium-Range Weather Forecasts (ECMWF). This work is part of the Hollandse Kust Noord wind farm innovation program. Funding was provided by CrossWind C.V., and content support was provided by Shell.

References

[1] Abdullah, M. A., Muttaqi, K. M., Sutanto, D., and Agalgaonkar, A. P. "An Effective Power Dispatch Control Strategy to Improve Generation Schedulability and Supply Reliability of a Wind Farm Using a Battery Energy Storage System". en. In: *IEEE Transactions on Sustainable Energy* 6.3 (2015), pp. 1093–1102. DOI: 10.1109/TSTE.2014.2350980.

[2] Jiang, Q. and Wang, H. "Two-Time-Scale Coordination Control for a Battery Energy Storage System to Mitigate Wind Power Fluctuations". en. In: *IEEE Transactions on Energy Conversion* 28.1 (2013), pp. 52–61. DOI: 10.1109/TEC.2012.2226463.

[3] Ding, H., Pinson, P., Hu, Z., and Song, Y. "Integrated Bidding and Operating Strategies for Wind-Storage Systems". en. In: *IEEE Transactions on Sustainable Energy* 7.1 (2016), pp. 163–172. DOI: 10.1109/TSTE.2015.2472576.

[4] Luo, F., Meng, K., Dong, Z. Y., Zheng, Y., Chen, Y., and Wong, K. P. "Coordinated Operational Planning for Wind Farm With Battery Energy Storage System". en. In: *IEEE Transactions on Sustainable Energy* 6.1 (2015), pp. 253–262. DOI: 10.1109/TSTE.2014.2367550.

[5] Das, K., Philippe Grapperon, A. L. T., Sørensen, P. E., and Hansen, A. D. "Optimal battery operation for revenue maximization of wind-storage hybrid power plant". In: *Electric Power Systems Research* 189 (2020). DOI: 10.1016/j.epsr.2020.106631.

[6] Iori, J., Zaaijer, M., Von Terzi, D., and Watson, S. "Design drivers for the storage system of baseload hybrid power plants". English. In: *IET Conference Proceedings*. Publisher: Institution of Engineering and Technology. Azores, Portugal, 2024, 245–250(5). DOI: 10.1049/icp.2024.1844.

[7] Chen, C., Lu, Y., and Xing, L. "Levelling renewable power output using hydrogen-based storage systems: A techno-economic analysis". en. In: *Journal of Energy Storage* 37 (2021), p. 102413. DOI: 10.1016/j.est.2021.102413.

[8] Dutch Minister of Economic Affairs and Climate Policy. *Ministerial Order for the permitting of Offshore Wind Energy Hollandse Kust (noord) Wind Farm Site V, (WJZ/19201387)*. 2019.

[9] Nguyen, C.-L., Lee, H.-H., and Chun, T.-W. "Cost-Optimized Battery Capacity and Short-Term Power Dispatch Control for Wind Farm". en. In: *IEEE Transactions on Industry Applications* 51.1 (2015), pp. 595–606. DOI: 10.1109/TIA.2014.2330073.

[10] Pinson, P., Chevallier, C., and Kariniotakis, G. N. "Trading Wind Generation From Short-Term Probabilistic Forecasts of Wind Power". en. In: *IEEE Transactions on Power Systems* 22.3 (2007), pp. 1148–1156. DOI: 10.1109/TPWRS.2007.901117.

[11] Crespo-Vazquez, J. L., Carrillo, C., Diaz-Dorado, E., Martinez-Lorenzo, J. A., and Noor-E-Alam, M. "A machine learning based stochastic optimization framework for a wind and storage power plant participating in energy pool market". en. In: *Applied Energy* 232 (2018), pp. 341–357. DOI: 10.1016/j.apenergy.2018.09.195.

[12] Kainz, S., Quick, J., Souza de Alencar, M., Moreno, S., Dykes, K., Bay, C., Zaaijer, M., and Bortolotti, P. *IEA Wind TCP Task 55: The IEA Wind 740-10-MW Reference Offshore Wind Plants*. English. IEA, 2024. DOI: 10.2172/2333634.

[13] Hersbach, H., Bell, B., Berrisford, P., Hirahara, S., Horányi, A., Muñoz-Sabater, J., Nicolas, J., Peubey, C., Radu, R., Schepers, D., et al. "Complete ERA5 from 1940: Fifth generation of ECMWF atmospheric reanalyses of the global climate". In: *Copernicus Climate Change Service (C3S) Data Store (CDS)* 10 (2017).

[14] Roald, L. A., Pozo, D., Papavasiliou, A., Molzahn, D. K., Kazempour, J., and Conejo, A. "Power systems optimization under uncertainty: A review of methods and applications". en. In: *Electric Power Systems Research* 214 (2023), p. 108725. DOI: 10.1016/j.epsr.2022.108725.

[15] European Network of Transmission System Operators for Electricity. *ENTSO-E Transparency Platform.* https://transparency.entsoe.eu/dashboard/show. [Online].

[16] MOSEK ApS. *The MOSEK Optimizer API for Python 10.1.2.* 2024.

[17] Iori, J. *Data associated with the publication "Reliable operation of wind-storage systems for baseload power production".* 2025. DOI: 10.4121/52056a89-9324-486f-a16b-93c0a1ceae2a.

[18] Iori, J. *SHIPP - A design optimization software for hybrid power plants.* 2025. DOI: 10.4121/2EE36148-369F-4E1F-B770-C86752D7DCA4. URL: https://github.com/jennaiori/shipp.

First experience in operation and data acquisition at the Hydrogen Lab Bremerhaven

Jannes Vervoort[1*], Anna Heuschmann[1], Kevin Schalk[1], Marieke Stein[1], Karsten Korsch[1] and Nora Denecke[1]

[1] Hydrogen Labs and Field Tests, Fraunhofer Institute for Wind Energy Systems, Bremerhaven, Germany

*jannes.vervoort@iwes.fraunhofer.de

Abstract. The Hydrogen Lab Bremerhaven (HLB), operated by Fraunhofer IWES, supports the energy transition by enabling the large-scale integration of green hydrogen. As the hydrogen market evolves, reliable operational data is essential. HLB provides a versatile infrastructure, including two electrolysers, a fuel cell, a hydrogen CHP plant, and storage systems. Connected to wind turbines via a Power Purchase Agreement (PPA), it ensures green hydrogen production and supports offshore electrolysis and desalination research. During its initial phase, HLB gathered key data on system performance, grid fluctuations, and hydrogen integration. Insights include electrolyser dynamics, load ranges, and power forecasting to minimize intraday trading. Offshore conditions were simulated in desalination tests using waste heat from electrolysis. Ongoing research at HLB is crucial for advancing hydrogen technologies, focusing on grid integration, offshore applications, and large-scale system validation to establish performance benchmarks and regulatory standards.

1. Introduction

Hydrogen is increasingly recognized as a critical energy carrier, as reflected in numerous national adoption goals and the EU-wide targets [1]. Ensuring the reliable and efficient production of hydrogen while addressing grid integration challenges within the future energy system is essential to achieving these ambitious objectives. Certification of hydrogen production systems plays a key role in ensuring safety, efficiency, and environmental sustainability. This requires a comprehensive assessment of the strategies applied in the design, construction, operation, and maintenance of hydrogen production facilities.

The hydrogen industry is currently increasing, and numerous smaller and larger projects are being planned. Large-scale projects in particular are experiencing a high market demand as most of the announced projects by 2030 are planned with a capacity > 100 MW [5]. Consequently, the responsible grid operators, who must ensure a smooth integration of the technology, but also regulation authorities and technical committees question the future role and tasks of hydrogen systems in the energy system. With increasing number of hydrogen plants and projects they recognise the need to further develop the applicable standards [5, 6, 7, 8]. The HLB focuses on grid connection standards and regulations, as well as the ongoing development of JIP protocols.

Regulatory requirements for hydrogen systems, such as electrolysers and hydrogen parks, are becoming increasingly specific, setting clear criteria that individual systems must meet for

Content from this work may be used under the terms of the Creative Commons Attribution 4.0 licence. Any further distribution of this work must maintain attribution to the author(s) and the title of the work, journal citation and DOI.
Published under licence by IOP Publishing Ltd

efficient operation. Two key factors in this context are grid connection and the required hydrogen purity. For the operation of the electrolyser systems at the HLB, compliance with local distribution grid regulations—TAR 4110 (Technical Requirements for the Connection and Operation of Customer Installations in the Medium-Voltage Network) and TAR 4105 (Technical Requirements for the Connection and Parallel Operation of Generators in Low-Voltage Distribution Networks)—is crucial for effective integration into the medium-voltage grid and for evaluating system performance at internal grid connection points [2, 3]. For the commissioning and operation of larger electrolyser plants, the technical requirements set by the four German transmission system operators (TSOs) play a critical role [4].

This paper provides a brief overview of the current standardization status of the existing JIP test protocols to highlight the need for new testing and validation methods for electrolysers. Furthermore, it examines the implications of these requirements for the design and operation of the Hydrogen Lab Bremerhaven (HLB) and outlines the key research priorities in this context. It provides an overview of the operational experiences at the HLB, highlighting key challenges, results, and future objectives. The insights gained by Fraunhofer IWES in recent years regarding the operation of electrolysers are summarized, with a particular focus on the evaluation of JIP protocols. Executions of the test protocols are shown for individual example days for the alkaline electrolyser. Furthermore, the article outlines the necessary next steps to enhance hydrogen production in the coming years.

2. Methodology: Standardizing and testing of electrolysers

The planning of the Hydrogen Lab Bremerhaven (HLB) began in 2018, with construction starting in 2021. The first commissioning took place in 2022. Since then, Fraunhofer IWES has operated the HLB and continuously expanded its capabilities. The lab focuses primarily on testing megawatt-scale electrolysers, with a strong emphasis on existing regulations and their practical implementation. In addition to electrolysers, the HLB enables the testing of other critical systems required for hydrogen production in both onshore and offshore environments, such as desalination plants. During the commissioning phase and the initial two years of operation, Fraunhofer IWES gathered valuable insights into the operation of electrolysers, hydrogen storage, power reconversion, and water treatment. The HLB was specifically designed to thoroughly analyze grid requirements for electrolysers and to test their operation under real-world conditions. Initial tests based on the mentioned protocols have already been conducted at the HLB using the available electrolysers. This led to the first data sets for the electrolysers which are collected and analyzed. Multiple protocols were tested at the HLB and systematically analysed to assess the dynamic operability of electrolysers. The SCADA data from the systems was used for the initial evaluations; no additional measurement technology was installed. For this paper the dynamic operation of the alkaline electrolyser was investigated as an example. All shown values and times may vary depending on the circumstances of the electrolyser operation around the tests shown. Only the electrolysers are considered here; to produce green hydrogen, other systems are of course also required, such as storage systems and, if necessary, compressors.

3. EU harmonized test protocols for electrolyzers

The JIP test protocols provide an effective approach for evaluating the performance of electrolysers. A key aspect of the harmonized EU protocols for testing low-temperature water electrolysers is the assessment of the time required for the system to transition between minimum and maximum power levels and vice versa [6].

In the future, the developed protocols are expected to provide valuable input for regulations and standards governing the operation of electrolysers primarily powered by renewable energy sources. The following points outline key criteria defined in the JRC technical report [6]:

- Facilitating the identification and application of relevant technical standards by providing a comprehensive framework that defines the scope and extent of activities for certifying the safety, structural integrity, quality, and reliability of systems and components.
- Providing guidelines for the procedure to assess electrolyser performance.
- Serving as a guide for system developers and electrolyser manufacturers in accordance with the latest technological advancements.

Figure 1 presents an example from the JIP protocols, which aim to ensure that future electrolyser systems operate reliably, safely, and cost-efficiently alongside renewable energy sources. These protocols represent a significant milestone in the development of standardized testing for electrolysers. The shown test aims to impose high current ramp rates and simulates applications where the system is expected to be subject to frequent load fluctuations. The steps can be extended over the entire available electrolyser output to test the system at minimum and maximum load. Each step shall be held for a minimum of 15 minutes or until the power stabilizes, whichever occurs first. The performance can be determined in each step, as well as the ramp rates in between. A schematic of the test steps to be carried out can be found in the following figure:

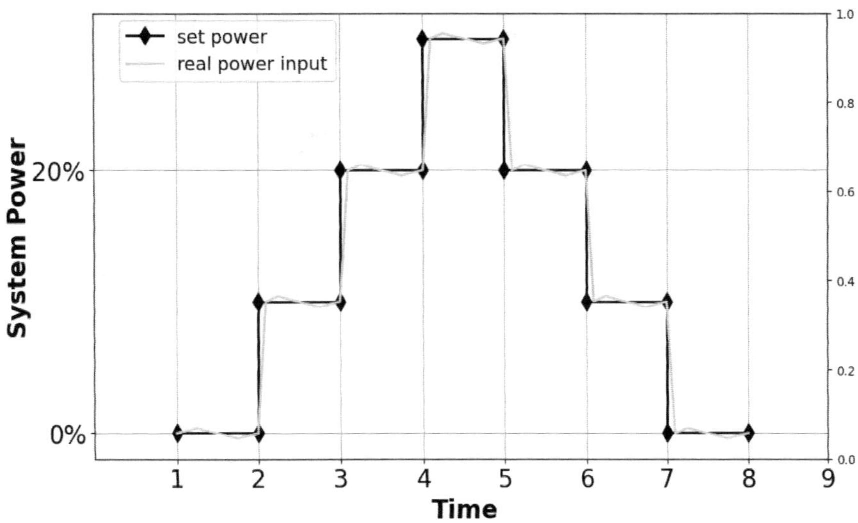

Figure 1. Example of a test protocol which is defined in the JIP document. Step-up and step-down test profile [6, 9].

4. Structure of the Hydrogen Lab Bremerhaven

Fraunhofer IWES operates three hydrogen labs across Germany, located in Bremerhaven, Leuna, and Görlitz. This paper specifically examines the test infrastructure at the Bremerhaven facility. At the Hydrogen Lab Bremerhaven (HLB), the focus is on the interaction between wind turbines and electrolytic hydrogen production. The increasing decentralization of power generation, driven by the integration of renewable energy sources, places significant demands on power grids, which were originally designed for the parallel operation of large, centralized generators. The HLB will also focus on integrating hydrogen production, storage, and utilization to support the development of a local hydrogen economy. In addition to on-site H_2 gas tanks and fuel cells, local project partners will explore hydrogen applications in the food industry, as well as its role in producing synthetic natural gas (SNG) and advancing methanization processes.

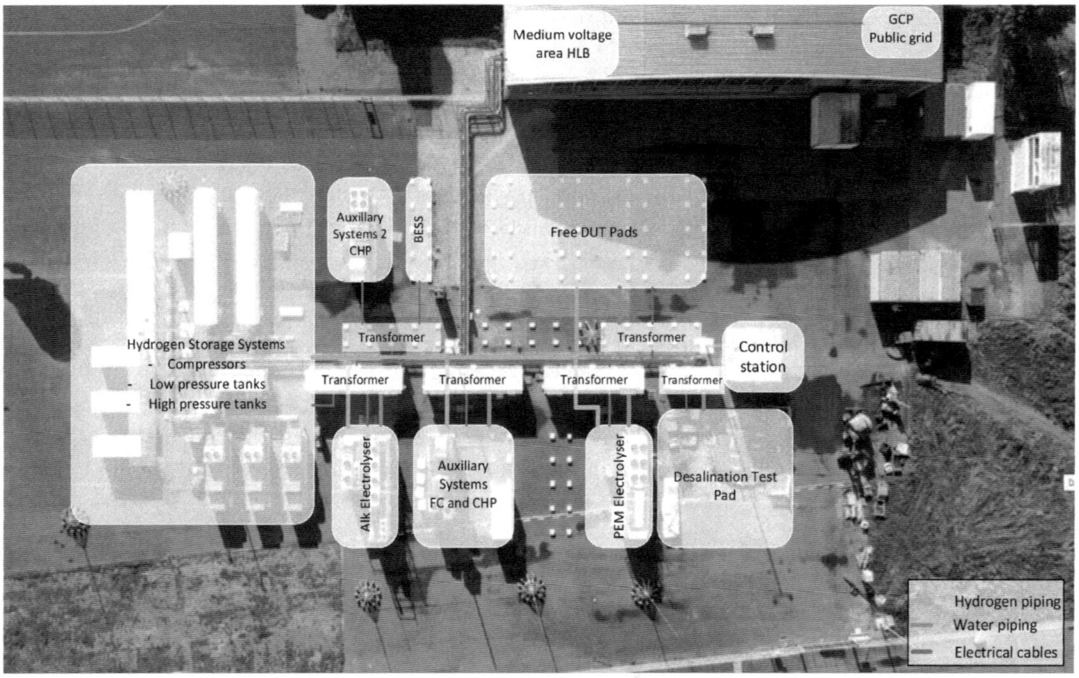

Figure 2. Overview structure of the Hydrogen Lab Bremerhaven (HLB).

The table below provides an overview of all currently installed components at the HLB, along with their respective research focuses. The components at the HLB have been carefully selected to address a wide range of future challenges related to hydrogen storage and utilization across various sectors. To fully address the entire hydrogen value chain, storage plays a crucial role and should not be overlooked. The HLB provides both high- and low-pressure storage systems to support comprehensive research and testing.

Table 1. List of units at the HLB.

Unit	Nominal power	Description
PEM Electrolyser	1 MW	PEM electrolyser with nine stacks and nominal H2 production rate of 210 Nm3/h. Heat extraction is provided for seawater desalination.
Alkaline Electrolyser	1.3 MW	Alkaline electrolyser with three stacks and nominal H2 production rate of 270 Nm3/h.
2x CHP unit	170 kW and 350 kW	The combined heat and power unit (CHP) contains a hydrogen powered generator to supply electrical energy and heat
Fuel cell	200 kW	The fuel cell converts hydrogen back into electrical energy, which can be fed into the grid.
BESS	500 kWh / 50 kW_{el}	The BESS can be used for stabilising the grid.
DUT-Pads	Up to 5.0 MW	The device under test (DUT) pads can flexibly be used to connect all kinds of PGUs and PCUs.

5. Results of the dynamic operation of the electrolysers on the HLB

Both alkaline and PEM electrolysers are in principle capable of meeting the grid service requirements needed to operate in the markets studied in terms of dynamic operating capabilities, although various factors such as capacity, flexibility, dynamic performance, etc. would limit their grid performance. For location-dependent/capacity-oriented services like congestion management and capacity management, capacity requirement might be a barrier if these services are requested by the TSO. This capacity barrier is normally much lower in the distribution grid; therefore, would not limit the potential of electrolysers providing such services. The potential of offering voltage control and power quality to the grid operators is much dependent on the ability of grid inverters of a production or consumption plant, which has not been much considered by the current design of inverters for electrolysers. In special applications such as microgrids with high penetration of intermittent renewables, offering this service to local grid operator should be highly feasible if the design of grid inverters is improved [9].

To better evaluate the electrolyser performance, step-up and step-down tests were conducted with one of the stacks of the alkaline electrolyser at the HLB in accordance with the QualyGrids and JRC protocols [6, 9]. The step-up test took place from 12:45 to 14:00 local time, followed by the step-down test from 14:00 to 15:30. The used production setpoints were 20, 40, 60, 80 and 100 % of the maximum possible production volume of one stack. The maximum hydrogen production of one alkaline electrolyser stack is: 92 Nm^3/h. All setpoints were held for at least 20 min. Figures 3 and 4 present the results of the test.

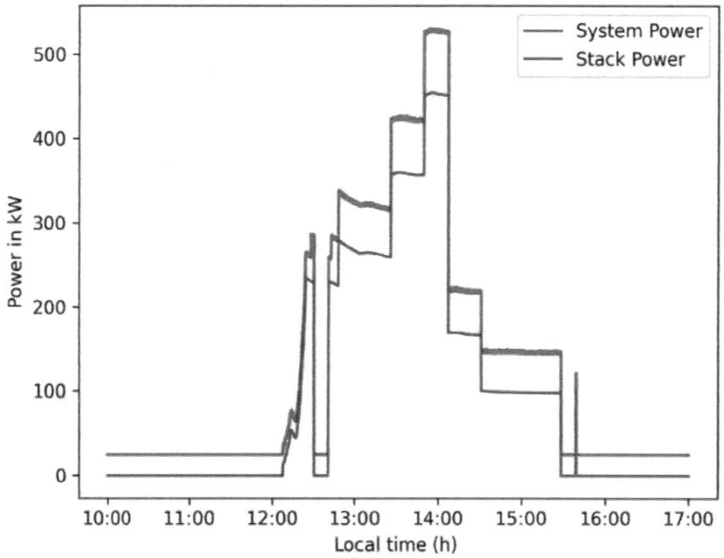

Figure 3. Actual System Power and Stack Power of the alkaline electrolyser during the tests on January 7, 2025. The red solid line shows the system power in kW and the green solid line the stack power in kW.

The system power, and hydrogen production shown in Figures 3 and 4 are measured values, recorded at a sampling rate of 10 Hz and shown as an average value over 1 s. The Stack Power is calculated according to equation (2), the efficiencies according to equations (1a) and (1b) and both values are based on the 1 s mean values shown in the figures. The integrated electrolyser system sensors were used. The measurement acquisition system of the electrolyser is connected to the control station and a corresponding server of the HLB.

Figure 3 shows a clearly visible offset for the power consumption of the periphery which is not negligible. For further investigation regarding electrolyser performance and efficiency it is important to know which power is currently being considered.

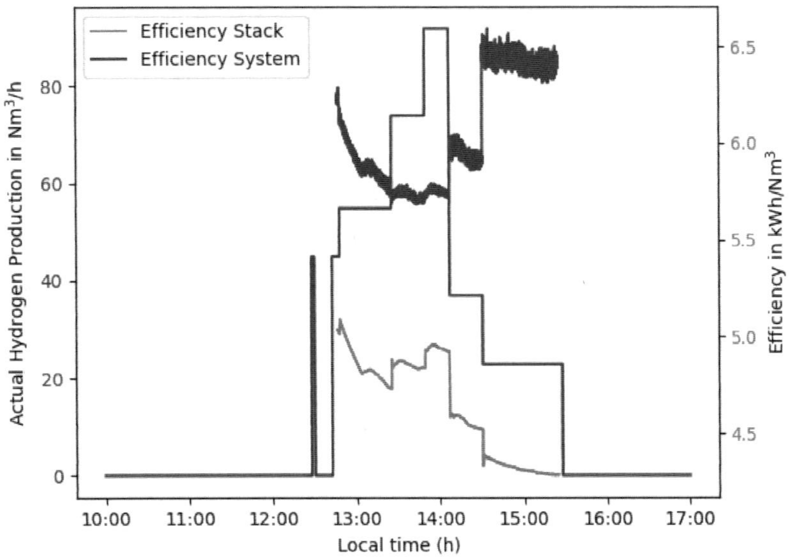

Figure 4. Actual Hydrogen Production and Efficiency of the alkaline electrolyser during the tests on January 7, 2025. The blue solid line shows the hydrogen production in Nm^3/h, the dark red solid line the system efficiency in kWh/Nm^3 and the red solid line the stack efficiency in kWh/Nm^3.

The first peak in the curve in Figure 4 shows a first bit of hydrogen production before the system falls back into standby because of internal automation and control issues. Therefore, the efficiency is calculated from the second start of the hydrogen production on.

The efficiency of the system is calculated using either the stack power or the system power, shown in Figure 3, and the hydrogen production from Figure 4 according to the following equations:

$$Efficiency\ Stack\ (kWh/Nm^3) = \frac{Stack\ Power\ (kW)}{Hydrogen\ Production\ (Nm^3/h)} \quad (1a)$$

$$Efficiency\ System\ (kWh/Nm^3) = \frac{System\ Power\ (kW)}{Hydrogen\ Production\ (Nm^3/h)} \quad (1b)$$

Stack power is calculated from internally measured Stack Voltage and Stack Current according to the following equation:

$$Stack\ Power\ (W) = Stack\ Current(A) * Stack\ Voltage\ (V) \quad (2)$$

In accordance with the QualyGrids and JRC protocols [6, 9], multiple tests were conducted also with one of the stacks of the alkaline electrolyser at the HLB. As a second example from the test campaign, a step-down test was selected for the paper. Figures 5 and 6 present the results of the test. The goal is to reduce the power and the hydrogen production step by step and analyse the stability and efficiency of the system and the stack.

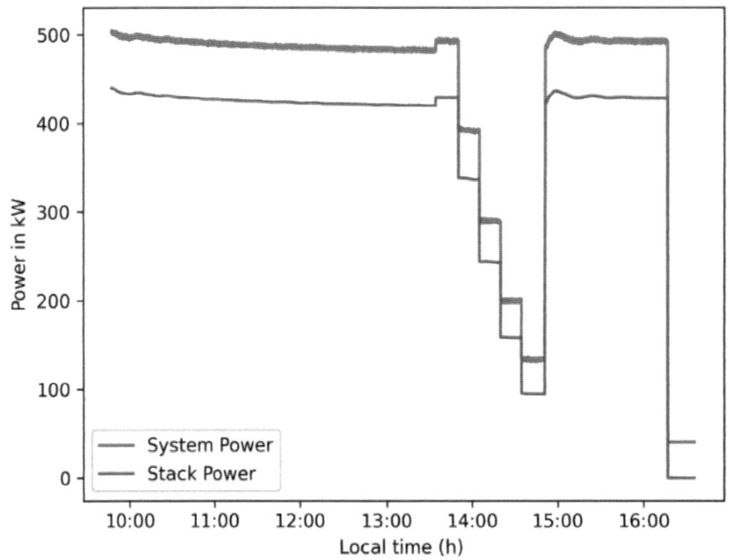

Figure 5. Actual System Power and Stack Power of one stack of the alkaline electrolyser during the tests on November 13, 2024. Stack Power is calculated from internally measured Stack Voltage and Stack Current. The red solid line shows the system power in kW and the green solid line the stack power in kW.

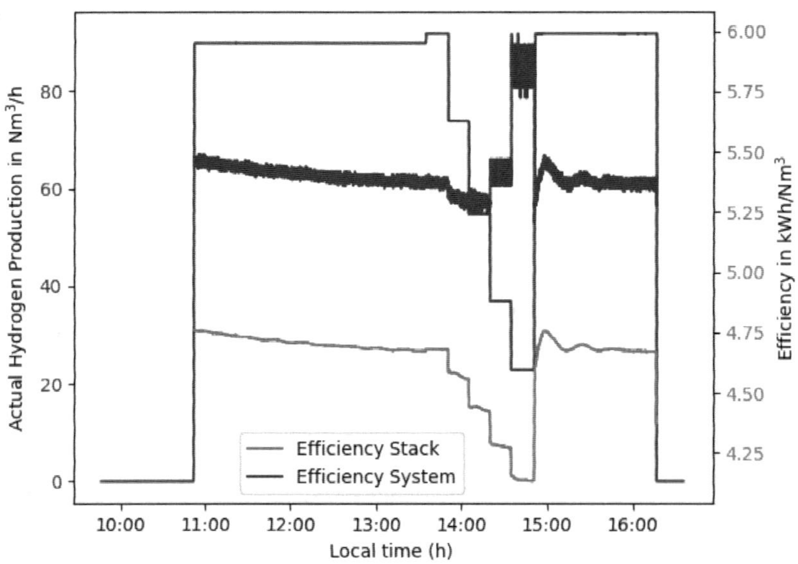

Figure 6. Actual Hydrogen Production and Efficiency of one stack of the alkaline electrolyser on November 13, 2024. Efficiency is calculated with the stack power from Figure 5. The blue solid line shows the hydrogen production in Nm^3/h, the dark red solid line the system efficiency in kWh/Nm^3 and the red solid line the stack efficiency in kWh/Nm^3.

Both Figures 4 and 6 show an increase in stack efficiency during the ramp-down tests, while the system efficiency increases for lower production setpoints due to the additional load of the periphery. The more detailed correlation of both efficiencies requires further investigation through repeated tests following the same protocol.

6. Conclusion

The initial findings from dynamic operation at the HLB demonstrate that the test protocols defined by JRC and QualyGrids provide a reliable framework for evaluating electrolysers and assessing their suitability for specific applications. These protocols are designed to support the production of green hydrogen, which inherently involves fluctuating power inputs. As a result, step-up and step-down tests, along with the corresponding efficiency measurements, represent critical criteria for assessing electrolyser performance. The improvement of the efficiency based on the stack power needs to be confirmed with further tests and behaves differently if the total energy consumption is considered because of the additional periphery load as it can be seen in the presented figures. Correlations of the measured data and more precise statements can be made with more testing and data acquisition in the future and test protocols need to be repeated multiply times as all measured can vary depending on several factors such as the operation or standstill times before the performed test protocol.

For the test campaign at the HLB, the system's built-in sensors were used to assess performance. However, to obtain more accurate insights, significantly higher sampling rates are required. If the existing sensors do not support these rates, additional sensors must be installed to enable precise power quality measurements. Power quality measurements play a critical role in grid connection, particularly in relation to converter efficiency. While this aspect was not the primary focus of the measurement campaign and was therefore not analysed in detail, it remains essential for the operation and integration of electrolysers into medium- and high-voltage grids. Current regulations now provide clear guidelines, requiring electrolysis systems to contribute to grid stability.

7. Acknowledgements

This work is supported by the German Federal Ministry for Economic Affairs and Climate Action, the European Regional Development Fund, and the Federal State of Bremen

References

[1] Europäische Kommission: 'A hydrogen strategy for a climate-neutral Europe', Brüssel, Juli 2020.
[2] VDE: 'VDE-AR-N 4105 - *Generators connected to the low-voltage distribution network - Technical requirements for the connection and parallel operation with low-voltage distribution networks*', 2018
[3] VDE: 'VDE-AR-N 4110 - *Technical requirements for the connection and operation of customer installations to the medium voltage network (TAR medium voltage)*', 2023
[4] 50Hertz, amprion, TenneT, TransnetBW: 'Technical requirements for the connection of electrolyser facilities', February 2024
[5] IEA: 'Global Hydrogen Review 2023', December 2023
[6] European Commission: 'JRC Technical Report – *EU harmonised protocols for testing of low temperature water electrolysers*', 2021
[7] The European Clean Hydrogen Alliance: 'REPORTS OF THE ALLIANCE ROUNDTABLES ON BARRIERS AND MITIGATION MEASURES', October 2021
[8] 'H2Giga', https://www.wasserstoff-leitprojekte.de/projects/h2giga, accessed 15 April 2024
[9] QualyGrids., 'Qualifying tests of electrolysers for grid services,' (QualiGrids, 2023), p. 114

AUTHOR INDEX

Abrahamsen, Asger B. 105
Ally, Stijn 220
Andrade, Nathianne 64
Angelis, Georgo 127
Bak, Christian 85
Barber, Sarah 136
Baumgärtner, David 156, 184
Bechmann, A. 1
Beckerlee, Jimmie 85
Berezyak, M. 147
Besseling, Jannik 20
Blickwedel, Lucas 20
Callies, Doron 40
Carvalho, Gustavo 64
Courtney, Michael 53
Cousin, Alexis 165
Crevecoeur, Guillaume 74
Daems, Pieter-Jan 11
De Baets, Patrick 31
De Geest, Roeland 31
Denecke, Nora 241
Dimitrov, Nikolay 105
Doyle, Tom 210
Dupoiron, Marine 165
Faria, Bruno R. 105
Fauconnier, Dieter 31
Figueiredo, R. 147
Filho, Fernando S. 64
Fischer, Andreas 85
Frings, Dustin 20
Gharbia, Ibtihel B. 165
Gloria, Nathalie 40
Golden, Danny 210
Guillemin, Fabrice 165
Haag, Michael D. 85
Halber, Joscha 20
Hammer, Florian 136
Hannesdóttir, Ásta 53
Hasan, Syed 210
Heidari, Jalal 74
Helsen, Jan 11, 220
Heuschmann, Anna 241
Hildebrand, Tabea 40
Hoffmann, Jil 20
Hollas, Christian 94
Ildvedsen, Sigurd 85
Iori, Jenna 231
Jacobs, Georg 20, 94
Jäger, Florian 40
Kaczenski, Jonas 156
Kayedpour, Nezmin 74
Khvoenkova, Nina 165
Kolios, Athanasios 105
Korsch, Karsten 241
Kraasch, Corinna 176
Krasnov, Alexander 40
Kreeft, Jasper 231
Krudde, Laurens 127
Langebæk, Jonas R. 85
Lima, Samuel 64
Lorenz, Torge 184
Ludwig, Paula 176
Lylloff, Oliver 85
Mackeprang, Jürgen 156
Manner, Ernst 94
Martins, J. 147
Mechler, Sebastian 176
Michaelis, Luis 40
Nejad, Amir R. 116
Nowé, Ann 220
O'Rourke, Fergal 198
Oliveira, C. 147
Olsen, Anders S. 85
Pacheco, J. 147
Pal, Sumit K. 116
Pauscher, Lukas 40
Pereira, C. 147
Perez, Victor 105
Poletto, Jean C. 31
Pracejus, Thomas 156
Queiroz, Rodrigo 64
Quenedey, Emilie 165
Quick, J. 1
Reichartz, Thorsten 20
Renger, Fabian 94
Riechert, Jens 40
Rinker, Jennifer M. 53
Röder, Julian 94
Ryan, Peter 198
Sagredo, Esperanza S. 53
Santos, Guido 64
Schalk, Kevin 241
Schelenz, Ralf 20
Schläger, Christian 184
Schmitz, Lukas 94
Sharma, Ashutosh 176

Stein, Marieke 241
Sterckx, Jonathan 11
Szczepaniec, Krzysztof 198
Teixeira, N. 147
Tona, Paolino 165
Van Binsbergen, Diederik 11
Van Maele, Djordy 31
Vandevelde, Lieven 74
Verstraeten, Timothy 11, 220
Vervoort, Jannes 241
Von Terzi, Dominic 231
Watson, Simon 116, 231
Weber, Stefan 85
Wegner, Anna 176
Zaaijer, Michiel 231
Zappalá, Donatella 116
Zweiffel, Maximilian 94